Smart Innovation, Systems and Technologies

Volume 38

About this Series

The Smart Innovation, Systems and Technologies book series encompasses the topics of knowledge, intelligence, innovation and sustainability. The aim of the series is to make available a platform for the publication of books on all aspects of single and multi-disciplinary research on these themes in order to make the latest results available in a readily-accessible form. Volumes on interdisciplinary research combining two or more of these areas is particularly sought.

The series covers systems and paradigms that employ knowledge and intelligence in a broad sense. Its scope is systems having embedded knowledge and intelligence, which may be applied to the solution of world problems in industry, the environment and the community. It also focusses on the knowledge-transfer methodologies and innovation strategies employed to make this happen effectively. The combination of intelligent systems tools and a broad range of applications introduces a need for a synergy of disciplines from science, technology, business and the humanities. The series will include conference proceedings, edited collections, monographs, handbooks, reference books, and other relevant types of book in areas of science and technology where smart systems and technologies can offer innovative solutions.

High quality content is an essential feature for all book proposals accepted for the series. It is expected that editors of all accepted volumes will ensure that contributions are subjected to an appropriate level of reviewing process and adhere to KES quality principles.

More information about this series at http://www.springer.com/series/8767

Gordan Jezic · Robert J. Howlett
Lakhmi C. Jain
Editors

Agent and Multi-Agent Systems: Technologies and Applications

9th KES International Conference, KES-AMSTA 2015 Sorrento, Italy, June 2015, Proceedings

Springer

Editors
Gordan Jezic
Faculty of Electrical Engineering and Computing, Department of Telecommunications
University of Zagreb
Zagreb
Croatia

Lakhmi C. Jain
Faculty of Education, Science, Technology and Mathematics
University of Canberra
Canberra
Australia

Robert J. Howlett
KES International
Shoreham-by-Sea
UK

ISSN 2190-3018 ISSN 2190-3026 (electronic)
Smart Innovation, Systems and Technologies
ISBN 978-3-319-19727-2 ISBN 978-3-319-19728-9 (eBook)
DOI 10.1007/978-3-319-19728-9

Library of Congress Control Number: 2015940746

Springer Cham Heidelberg New York Dordrecht London

Printed on acid-free paper

Springer International Publishing AG Switzerland is part of Springer Science+Business Media (www.springer.com)

Preface

This volume contains the proceedings of the 9th KES Conference on Agent and Multi-Agent Systems—Technologies and Applications (KES-AMSTA 2015) held in Sorrento, Italy, between June 17 and 19, 2015. The conference was organized by KES International, its focus group on agent and multi-agent systems and University of Zagreb, Faculty of Electrical Engineering and Computing. The KES-AMSTA conference is a subseries of the KES conference series.

Following the successes of previous KES Conferences on Agent and Multi-Agent Systems—Technologies and Applications, held in Chania, Greece, (KES-AMSTA 2014), Hue, Vietnam (KES-AMSTA 2013), Dubrovnik, Croatia (KES-AMSTA 2012), Manchester, UK (KES-AMSTA 2011), Gdynia, Poland (KES-AMSTA 2010), Uppsala, Sweden (KES-AMSTA 2009) Incheon, Korea (KES-AMSTA 2008), and Wroclaw, Poland (KES-AMSTA 2007), the conference featured the usual keynote talks, oral presentations, and invited sessions closely aligned to the established themes of the conference.

The aim of the conference was to provide an internationally respected forum for scientific research in the technologies and applications of agent and multi-agent systems. This field is concerned with the development and evaluation of sophisticated, AI-based problem-solving and control architectures for both single-agent and multi-agent systems. Current topics of research in the field include (amongst others) agent-oriented software engineering, BDI (beliefs, desires, and intentions) agents, agent cooperation, coordination, negotiation, organization and communication, distributed problem solving, specification of agent communication languages, agent privacy, safety and security, formalization of ontologies, and conversational agents. Special attention is paid to the feature topics: learning paradigms, agent-based modeling and simulation, business model innovation and disruptive technologies, anthropic-oriented computing, serious games and business intelligence, design and implementation of intelligent agents and multi-agent systems, digital economy, and advances in networked virtual enterprises.

The conference attracted a substantial number of researchers and practitioners from all over the world who submitted their papers for two main tracks covering the methodology and applications of agent and multi-agent systems, and six invited sessions on specific topics within the field. Submissions came from 22 countries. Each paper was peer reviewed by at least two members of the International Programme Committee and International Reviewer Board. 38 papers were selected for oral presentation and publication in the volume of the KES-AMSTA 2015 proceedings.

The Program Committee defined the following main tracks: Mobile Agent Security and Multi-Agent Systems, and Agent-based Algorithms, Simulations and Organizations. In addition to the main tracks of the conference there were the following invited sessions: Agent-based Modeling and Simulation (ABMS), Learning Paradigms and Applications: Agent-based Approach (LP:ABA), Anthropic-Oriented Computing (AOC), Business Model Innovation and Disruptive Technologies, Serious Game and Business Intelligence through Agent-based Modeling, and Design and Implementation of Intelligent Agents and Multi-Agent Systems.

Accepted and presented papers highlight new trends and challenges in agent and multi-agent research. We hope that these results will be of value to the research community working in the fields of artificial intelligence, collective computational intelligence, robotics, dialogue systems and, in particular, agent and multi-agent systems, technologies, tools, and applications.

We would like to express our thanks to the keynote speaker, Prof. Marcin Paprzycki from Systems Research Institute of the Polish Academy of Sciences, Warsaw, Poland, for his interesting and informative talk of a world-class standard.

The Chairs' special thanks go to special session organizers, Dr. Roman Šperka, Silesian University in Opava, Czech Republic, Prof. Mirjana Ivanović, University of Novi Sad, Serbia, Prof. Costin Badica, University of Craiova, Romania, Prof. Zoran Budimac, University of Novi Sad, Serbia, Prof. Manuel Mazzara, Innopolis University, Russia, Max Talanov, Kazan Federal University and Innopolis University, Russia, Radhakrishnan Delhibabu, Kazan Federal University, Russia, Salvatore Distefano, Politecnico di Milano, Italy, Jordi Vallverdú, Universitat Autònoma de Barcelona, Spain, Dr. Yun-Heh (Jessica) Chen-Burger, The Heriot-Watt University, Edinburgh, UK, Assoc. Prof. Setsuya Kurahashi, University of Tsukuba, Japan, Prof. Takao Terano, Tokyo Institute of Technology, Japan, Prof. Hiroshi Takahashi, Keio University, Japan, Dr. Arnulfo Alanis Garza, Instituto Tecnológico de Tijuana, México, and Dr. Lenin G. Lemus-Zúñiga, Universitat Politècnica de València, España for their excellent work.

Thanks are due to the Program Co-chairs, all Program and Reviewer Committee members, and all the additional reviewers for their valuable efforts in the review process, which helped us to guarantee the highest quality of selected papers for the conference.

We cordially thank all authors for their valuable contributions and all of the other participants in this conference. The conference would not be possible without their support.

April 2015

Gordan Jezic
Robert J. Howlett
Lakhmi C. Jain

Organization

KES-AMSTA 2015 was organized by KES International—Innovation in Knowledge-Based and Intelligent Engineering Systems.

Honorary Chairs

I. Lovrek, University of Zagreb, Croatia
L.C. Jain, University of South Australia

Conference Chair

G. Jezic, University of Zagreb, Croatia

Executive Chair

R.J. Howlett, University of Bournemouth, UK

Program Chair

M. Kusek, University of Zagreb, Croatia
R. Sperka, Silesian University in Opava, Czech Rupublic

Publicity Chair

D. Katusic, University of Zagreb, Croatia

Keynote Speakers

Prof Marcin Paprzycki
Systems Research Institute of the Polish Academy of Sciences, Warsaw, Poland
Software Agents as Resource Brokers in Grid

International Program Committee

Dr. Richard Adams, University of Surrey, UK
Prof. Arnulfo Alanis Garza, Departamento de Sistemas y Computación—Instituto Tecnológico de Tijuana, Mexico

Mr. Kyle Alves, University of Surrey, UK
Dr. Walid Adly Atteya, University of Bradford, UK
Dr. Ahmad Taher Azar, Faculty of Computers and Information, Benha University, Egypt
Costin Badica, University of Craiova, Romania
Dr. Marina Bagić Babac. University of Zagreb, Faculty of Electrical Engineering and Computing, Croatia
Dr. Dariusz Barbucha, Gdynia Maritime University, Poland
Dr. Iva Bojic, Massachusetts Institute of Technology, USA
Dr. Grażyna Brzykcy, Poznan University of Technology, Institute of Control and Information Engineering, Poland
Prof. Zoran Budimac, Univeristy of Novi Sad, Serbia
Assoc. Prof. Frantisek Capkovic, Institute of Informatics, Slovak Academy of Sciences, Bratislava, Slovakia
Dr. Jessica Chen-Burger, The Heriot-Watt University, Scotland, UK
Prof. Ireneusz Czarnowski, Gdynia Maritime University, Poland
Prof. Paul Davidsson, Malmö University, Sweden
Prof. Radhakrishnan Delhibabu, Kazan Federal University, Russia
Prof. Salvatore Distefano, Politecnico di Milano, Italy
Dr. Nicola Dragoni, Technical University of Denmark, Denmark, and Örebro University, Sweden
Dr. Trong Hai Duong, International University—Vietnam National University HCMC
Dr. Konrad Fuks, Poznan University of Economics, Poland
Prof. Chihab Hanachi, University of Toulouse, IRIT Laboratory, France
Prof. Ronald L. Hartung, Franklin University, USA
Dr. Quang Hoang, Hue University, Vietnam
Prof. Tzung-Pei Hong, National University of Kaohsiung, Taiwan
Prof. Mirjana Ivanovic, University of Novi Sad, Serbia
Prof. Dr. sc. Dragan Jevtic, University of Zagreb, Croatia
Prof. Gordan Jezic, University of Zagreb, Croatia
Prof. Joanna Józefowska, Poznan University of Technology, Poland
Dr. Arkadiusz Kawa, Poznan University of Economics, Poland
Prof. Petros Kefalas, The University of Sheffield International Faculty, City College, Greece
Dr. Adrianna Kozierkiewicz-Hetmańska, Wrocław University of Technology, Poland
Dr. Dariusz Krol, Wroclaw University of Technology, Poland
Dr. Konrad Kułakowski, AGH University of Science and Technology, Poland
Prof. Setsuya Kurahashi, University of Tsukuba, Tokyo
Prof. Mario Kusek, University of Zagreb, Croatia
Prof. Kazuhiro Kuwabara, Ritsumeikan University, Japan
Dr. Fang-Pang Lin, National Center for High-Performance Computing, Taiwan
Dr. David Lopez-Berzosa, University of Surrey, UK
Dr. Marin Lujak, University Rey Juan Carlos, Spain

Prof. Manuel Mazzara, Innopolis University, Russia
Dr. Daniel Moldt, University of Hamburg, Department of Informatics, Germany
Assist. Prof. Vedran Podobnik, University of Zagreb, Faculty of Electrical Engineering and Computing, Croatia
Dr. Bhanu Prasad, Florida A&M University, Tallahassee, USA
Prof. Radu-Emil Precup, Politehnica University of Timisoara, Romania
Dr. Ewa Ratajczak-Ropel, Gdynia Maritime Academy, Poland
Dr. Roman Sperka, Silesian University in Opava, Czech Republic
Prof. Ryszard Tadeusiewicz, AGH University of Science and Technology, Krakow, Poland
Prof. Hiroshi Takahashi, Keio University, Japan
Prof. Yasufumi Takama, Tokyo Metropolitan University, Japan
Prof. Max Talanov, Kazan Federal University and Innopolis University, Russia
Prof. Takao Terano, Tokyo Institute of Technology
Dr. Krunoslav Trzec, Ericsson Nikola Tesla, Croatia
Prof. Taketoshi Ushiama, Kyushu University, Japan
Prof. Jordi Vallverdú, Universitat Autònoma de Barcelona, Spain
Prof. Toyohide Watanabe, Nagoya Industrial Science Research Institute, Japan
Dr. Ching-Long Yeh, Tatung University, Taipei, Taiwan
Dr. Mahdi Zargayouna, IFSTTAR, France

Workshop and Invited Session Chairs

Agent-Based Modelling and Simulation
Dr. Roman Šperka, Silesian University in Opava, Czech Republic

Learning Paradigms and Applications: Agent-based Approach
Prof. Mirjana Ivanovic, University of Novi Sad, Serbia
Prof. Zoran Budimac, University of Novi Sad, Serbia
Prof. Costin Badica, University of Craiova, Romania
Prof. Lakhmi Jain, University of South Australia, Australia

Anthropic-Oriented Computing
Prof. Manuel Mazzara, Innopolis University, Russia
Max Talanov, Kazan Federal University and Innopolis University, Russia
Radhakrishnan Delhibabu, Kazan Federal University, Russia
Salvatore Distefano, Politecnico di Milano, Italy
Jordi Vallverdú, Universitat Autònoma de Barcelona, Spain

Business Model Innovation and Disruptive Technologies
Dr. Yun-Heh (Jessica) Chen-Burger, The Heriot-Watt University, Edinburgh, UK
Serious Game and Business Intelligence through Agent-based Modelling
Assoc. Prof. Setsuya Kurahashi, University of Tsukuba, Japan
Prof. Takao Terano, Tokyo Institute of Technology
Prof. Hiroshi Takahashi, Keio University, Japan

The design and implementation of Intelligent Agents and Multi-Agent Systems
Dr. Arnulfo Alanis Garza, Instituto Tecnológico de Tijuana. México
Dr. Lenin G. Lemus-Zúñiga, Universitat Politècnica de València. España

Software Agents as Resource Brokers in Grid

Marcin Paprzycki
Systems Research Institute of the Polish Academy of Sciences, Warsaw, Poland

Abstract Years of middleware development have resulted in creation of multiple platforms for Grid computing (e.g. gLite, Globus, EGEE, Unicore, etc.). Furthermore, an attempt is currently under way to com- bine them within a single meta-system (the European Middleware Ini- tiative; EMI). However, practical experiences show that neither of these approaches fully succeeds with the users.

The aim of the presentation is to outline our attempt at combining soft- ware agents with semantic technologies for development of a different meta-level computational Grid middleware. In our approach, software agents use their "semantic brain" to manage the Grid "brawn" (to refer- ence a seminal paper: I. Foster, C. Kesselman, N. Jennings, Brain Meets Brawn: Why Grid and Agents Need Each Other; 2004).

Contents

Part VI IS: The Design and Implementation of Intelligent Agents and Multi-Agent Systems

Part VII IS: Serious Game and Business Intelligence Through Agent-Based Modelling

Part I
Mobile Agent Security and Multi-Agent Systems

MASIR: A Multi-agent System for Real-Time Information Retrieval from Microblogs During Unexpected Events

Imen Bizid, Patrice Boursier, Jacques Morcos and Sami Faiz

Abstract Microblogs have proved their potential to attract people from all over the world to express voluntarily what is happening around them during unexpected events. However, retrieving relevant information from the huge amount of data shared in real time in these microblogs remain complex. This paper proposes a new system named MASIR for real-time information retrieval from microblogs during unexpected events. MASIR is based on a decentralized and collaborative multi-agent approach analyzing the profiles of users interested in a given event in order to detect the most prominent ones that have to be tracked in real time. Real time monitoring of these users enables a direct access to valuable fresh information. Our experiments shows that MASIR simplifies the real-time detection and tracking of the most prominent users by exploring both the old and fresh information shared during the event and outperforms the standard centrality measures by using a time-sensitive ranking model.

Keywords Multi-agent systems · Microblogs · Real-time · Information retrieval · Unexpected events

---

I. Bizid (✉) · S. Faiz
LTSIRS Laboratory, Tunis, Tunisia
e-mail: imen.bizid@univ-lr.fr

S. Faiz
e-mail: sami.faiz@insat.rnu.tn

I. Bizid · P. Boursier · J. Morcos
L3i Laboratory, University of La Rochelle, La Rochelle, France

P. Boursier
e-mail: patrice.boursier@univ-lr.fr; patrice@iumw.edu.my

J. Morcos
e-mail: jacques.morcos@univ-lr.fr

P. Boursier
IUMW, Kuala Lumpur, Malaysia

G. Jezic et al. (eds.), *Agent and Multi-Agent Systems: Technologies and Applications*, Smart Innovation, Systems and Technologies 38,
DOI 10.1007/978-3-319-19728-9_1

1 Introduction

Since the emergence of Twitter, microblogs have experienced a new social aspect allowing users to express in real time their opinions, perceptions and expectations. The usefulness of such networks has been demonstrated during the Arab Spring revolutions and the last memorable natural disasters. Microblogs were the main source of information used by official organizations and news channels in the different event phases (i.e. event detection, situation awareness, alert dissemination) [3]. Since, these online social platforms have become indispensable to acquire and monitor information in real time during unforeseen events.

Unexpected events, in the context of this paper, refer to the events which occur without any warning or preparation such as natural and human disasters. Information retrieval during this kind of events is more complex than during simple ones as it is not possible to gain a priori information before its occurrence. Most of the proposed IR approaches in this context are based on both machine and human intelligence [6]. Such techniques are time consuming and require the presence of an important number of volunteers for the collection and the annotation of relevant information tasks.

Indeed, to enhance situation awareness of such events, it is important to find new ways coping with the limited data access using microblogs APIs on the one hand and to be able to retrieve valuable information shared in real time on the other hand [8]. Identifying and tracking prominent users who are behind the relevant and exclusive information shared during the targeted event can offer a rapid access to the required information. The prominence of users in these events relies on the relevance and freshness of their information independently of their social influence. Hence, using traditional techniques would fail to track and identify such users in real time due to the strict microblogs conditions for tracking users and the complexity of this type of events. The original contributions of this paper are twofold. First, we propose a multi-agent system to insure a real-time identification and an extensive tracking of prominent users in the context of managing unexpected events. Second, we propose a novel approach to detect prominent users based on both their geographic and social positions over time during an unexpected event.

This paper is organized as follows: Sect. 2 reviews existing approaches for information retrieval from social networks. Section 3 describes our proposed system and its functional model. Section 4 presents our experiments and evaluates our results. Section 5 summarizes our conclusions and future work.

2 Related Work

Tracking of social media users for information retrieval has been studied using mainly three ways: social networks' interface crawling through public and false profiles and phishing techniques [4], social networks applications such as Netvizz [10], as well as social networks Application Programming Interfaces (APIs) [1]. These techniques have provided a direct access to an important number of users. However, they have no selection strategy to evaluate the prominence of tracked users.

To the best of our knowledge, the issue of prominent users identification has never been explored in the context of unexpected events. However, there have been several attempts proposing new measures identifying influential users and domain experts in microblogs and specially Twitter [7, 12, 13]. Most of the proposed approaches identifying social media influencers are based on standard centrality measures such as eigenvector centrality and its variants HITS [2] and PageRank [7]. These adapted measures to microblogs specificities (e.g. number of tweets, mentions, retweets ...) have yielded promising results for the identification of influential users [11, 14]. However, they are computationally expensive and sensitive to well-connected users (e.g. celebrities, communication channels...) [5]. On the other hand, low research works have proposed domain experts identification based on efficient ranking models using a set of features describing the activities of users in particular topics [9, 15]. These approaches are efficient to identify in real time topical authorities sharing relevant information based on their previously shared information.

While somewhat similar to [9, 15], prominent users identification in the context of unexpected events differs in different points. Firstly, there is no a priori information describing the user's prominence that may be extracted before the occurrence of the event. In our context, information related to interested users in an event has to be explored in real time. Secondly, many features describing the prominence of a user in specific events have never been explored in the prior approaches such as the geolocation position, the time of the first user's activity regarding the event and the social position of the user in both the network and the event. Thirdly, our approach is tested in a real case from the identification of all users interested in the event to the detection and the tracking of the most prominent ones.

The problem of prominent users identification has been studied only theoretically using test databases. Microblogs APIs constraints related to extensive crawling have never been taken into account. Moreover, existing approaches for microblogs users tracking have focused only on the quantity of users while neglecting their qualities. In this paper, we propose a multi-agent system for relevant and exclusive information retrieval in Twitter by identifying and tracking the most prominent users in real time cases.

3 Decentralized Collaborative System for Information Retrieval

This section describes the operation details of MASIR for information retrieval from microblogs during unexpected events. MASIR is based on a multi-agent approach compliant with Twitter APIs specifications. The idea behind modeling a multi-agent system for information retrieval relies on its ability to retrieve information using parallel processing which makes the analysis process computationally feasible in real time. Moreover, the multiple used agents boost the number of monitored users in parallel and ease the detection of the most prominent users during the event.

Figure 1 describes the decentralized structure of our system. MASIR is composed of 6 different kinds of agents designed to execute a well-defined related tasks. The process starts when the keywords and/or hashtags representing the targeted event were specified to the Stream Retrieval Agent (SRA). Using these parameters, SRA searches for the list of new users sharing real-time information about the event and sends it continuously to the Historic Listener Agents Manager (HLAM). HLAM assigns a Historic Listener Agent (HLA) to each identified user by SRA in order to extract and store his historic in the Historic and Social Information Base (HSIB). The collected data is analyzed by the Prominent Users Detector (PUD) in order to detect the most prominent users that have to be tracked in real time by the Stream Listener Agents (SLAs). These agents are described in detail in the following sub-sections.

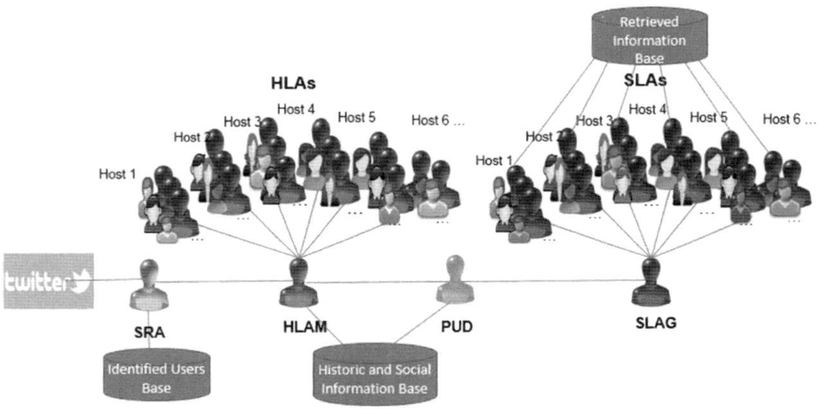

Fig. 1 A decentralized multi-agent system for real-time information retrieval from Twitter

3.1 The Stream Retrieval Agent (SRA)

SRA retrieves the tweets published in real time about an event and extracts the identities of users who are sharing it by following these monitoring operations:

1. *Streaming search:* SRA remains connected to Twitter during the event in order to search in real time for new tweets using the assigned hashtags or keywords identifying the targeted event.
2. *Users' identification:* SRA extracts the identity of users sharing on-topic tweets.
3. *Users' filter:* SRA applies dynamically new filters with reference to the Identified User Base (IUB) in order to force the streaming search to retrieve only tweets shared by new users.

4. *Users' storage:* SRA stores in IUB the identifier of any new detected user posting information related to the event.
5. *List of users sending:* SRA has to send the list of new detected users interested in the event every 30 s to HLAM.

3.2 The Historic Listener Agents Manager (HLAM)

HLAM manages the extraction process of the social and historic information from the identified users profiles. It controls multiple HLA agents which are in charge of the historic and social information collection from each identified user profile. HLAM can undergo different transitions according to the processed operations:

1. *Users assignment:* When HLAM receives the list of users sent by SRA, it adds the new identified users in a waiting list. Then, HLAM assigns each user to one of the available HLAs by respecting the FIFO (First In First Out) principle.
2. *Information reception:* This operation is processed after the reception of a message from HLA precising that the historic extraction process was accomplished. Then, at this stage HLAM saves the returned information collected by HLA in HSIB.
3. *HLA status change:* Once HLAM has received all the extracted information from a HLA, it sets this HLA status to "free" in order to be able to assign it to a new user.

3.3 The Historic Listener Agents (HLAs)

HLAs have to extract historic information shared by each assigned user. Once a HLA has finished the extraction of the needed information related to a user, it sends a message to HLAM to store the collected information in HSIB. Then, the HLAM will change this HLA status to "free". Each HLA have to be able to process the three following operations:

1. *Receiving a user's identity:* When HLA status is set to "free", HLA could be assigned to a unique user recognized by his unique identifier.
2. *Historic information extraction:* HLA extracts all the historic information shared by the assigned user.
3. *Social information extraction:* HLA extracts the list of the followers and followees of the assigned users.
4. *Extracted information Sending:* HLA sends all the information collected to HLAM in order to store it in HSIB and to change its status.

3.4 The Prominent Users Detector (PUD)

PUD acts as the intermediary between the historic extraction process and the streaming process. This agent detects the most prominent users using the data collected during the historic extraction process. The identification of these prominent users insures the attribution of the limited number of parallel SLAs to the most central users during the event. PUD detects the most prominent users by calculating and updating periodically the Prominence Score (PS) of the already watched users. This final score is estimated according to the geo-location and social positions of the user and the recency of his first interaction regarding the event. PS is computed using the following ranking model:

$$PS(u) = w_1 * RS(u) + w_2 * GPS(u) + SPS(u) \tag{1}$$

where 0.38 and 0.02 are the weights reflecting the importance of RS and GPS. All the weights' values (from w_1 to w_6) used during the calculation of PS and SPS were estimated a priori through a user study evaluating the active Twitter users in the South Korea ferry disaster. This study was conducted by a group of volunteers who have evaluated the Twitter users according to the relevance and recency of their information about the disaster. These volunteers have noted these users from 1 to 10 according to their prominence. These notes were used for fitting a linear regression model composed of the different predictor scores proposed in this paper to evaluate the prominence of each user. The weights evaluating each predictor were normalized to form the sum 1 for all the weights.

The Recency Score (RS) indicates the recency of the first on-topic information shared by the user regarding the time of occurrence of the event (t_{event}). The difference in time between the first shared information (t) and (t_{event}) is measured in minutes.

$$RS(u) = \frac{1}{t - t_{event} + 1} \tag{2}$$

The Geo-location Position Score (GPS) indicates the inclusion rate of the geo-location(i.e. longitude, latitude) specified by the user in the territory concerned by the event. The event area is represented by a polygon or a set of polygons (*Pe*) that may include many distant zones. For each user *u*, we extract from his different historic tweets collected by HLAs the set of his geo-locations (*Cu*). For example, if all the geolocations specified by the user are included in the event area, his GPS will be set to 1.

$$GP(u) = \frac{Cu \cap Pe}{Cu \cup Pe} \tag{3}$$

The Social Position Score (SPS) indicates how much the user's followers (F) and followees (Fe) are interested in the analyzed event. The more a user has many on-topic followers (OnF) and followees ($OnFe$) having a high RS, the more his final SPS is high. As well-connected users such as CNN and BBC may have a large number of OnF and $OnFe$ even they are not sufficiently prominent regarding the analyzed event, these numbers are adjusted by F and Fe which makes our SPS insensitive to well connected users. SPS is computed as follows using the social information already extracted by HLA and stored in HSIB:

$$SP(u) = w_3 * \frac{\sum_{i=1}^{OnF} RS(i)}{log(OnF + 1)} + w_4 * \frac{OnF}{log(F)} + w_5 * \frac{\sum_{i=1}^{OnFe} RS(i)}{log(OnFe + 1)} + w_6 * \frac{OnFe}{log(Fe)} \tag{4}$$

where $w_3 = 0.21$, $w_4 = 0.1$, $w_5 = 0.23$ and $w_6 = 0.04$ are the weights reflecting the importance of the different predictors contributing in the final SPS score of each user.

3.5 The Stream Listeners' Agents Generator (SLAG)

SLAG manages the tracking process of the most prominent users during the event. It starts the generation and management process when it receives the list of prominent users by PUD. SLAG generates a SLA for each user in the list. These SLAs are generated in different hosts in order to avoid the risk of IP banning by Twitter. Hence, SLAG processes the following operations :

1. *Receiving detected users:* SLAG receives periodically an updated list of the most prominent users that have to be tracked in real time.
2. *Killing existing SLAs:* After receiving the updated list, LAG kills SLAs which are tracking users who do not exist in the new list. By killing these SLAs, SLAG will release the place in some hosts in order to be able to track the new prominent users.
3. *Generating a new SLA:* After liberating the place for the new prominent users, SLAG generates SLAs for these users.

3.6 Streaming Listener Agents (SLAs)

While HLA extracts the historic and social information of one assigned user and once it has finished this task, it listens to another user, SLAs differ in various points. First, SLAs keep listening to a user profile in order to detect any new update. Second, SLAs are dynamically generated by the SLAG. Each SLA is in charge of tracking the assigned user profile in real time. SLAs store in real time any new detected information shared by its assigned user in the Retrieved Information Base (RIB). RIB contains the tweets extracted in real time from the most prominent users' profiles.

4 Experiments and Results

The architecture of MASIR was implemented using Java Agent DEvelopment framework (JADE). Using this framework, each agent was created in a running instance named container. MASIR agents were executed in various containers distributed in 5 hosts connected via a Virtual Private Network. Our system was implemented using two Twitter APIs; the Search API for the historic information extraction process and the Streaming API for the real-time tracking of prominent users.

As the Streaming and Search APIs limit the number of the crawled profiles simultaneously to around 5, MASIR used to encounter this limit by distributing SLAs and HLAs in various hosts and by using different Twitter accounts. This distribution aims not only to avoid IP banning when the authorized crawling limit rate is reached but also to boost the number of listened profiles. The 5 used hosts incorporate all a main container in order to enable SLAG to monitor all the created agents and automate the agents generation process according to the number of available hosts. Using this strategy, HLAM managed up to 75 HLAs (15/host) and SLAG generated up to 175 SLAs (35/host).

MASIR was run during the Herault floods which have occurred in the Herault country in France. These floods have lasted two days from 29 to 30 September 2014. They have caused important damages estimated between 500 and 600 million Euros. MASIR was launched after a while from the official announcement of the event using the hashtag “Herault” which was used by Twitter users to refer to the event. Our system has collected 41.064 historic tweets and 22.136 fresh tweets shared respectively by 3.143 users listened by HLAs and 604 users tracked by SLAs. Hence, MASIR has coped with the limits imposed by the Twitter APIs by tracking an important number of users in real time.

In order to evaluate the quality of the users tracked in real time by MASIR. We have collected all the tweets shared by the identified users by SRA from the announcement of the event to its end using the Search API after two days of the disaster. Then, these users were evaluated by a group of volunteers to define our ground-truth. This group were asked to note these users from 1 to 10 according to the relevance and freshness of their tweets. The top 175 users were selected in order to evaluate the users tracked by MASEA periodically.

Table 1 presents the total number of prominent and non prominent users identified by MASIR during the two days of the disaster and the number of the true prominent users tracked at each period of time with reference to the ground truth results.

According to these results, an important number of the ground-truth prominent users were identified by SRA and tracked by SLAs from the first day of the disaster. We also note that the precision of our detection process was improved during the end of the second day by tracking 46 % of the ground-truth prominent users continuously. We compared the used model by MASIR to detect prominent users with three baseline algorithms: the eigenvector centrality, PageRank and HITS algorithms. These measures were chosen as they were used in the literature for the detection of prominent users in various contexts.

Table 1 The simulation results of the identified and the true detected users by MASIR

		Identified users by SRA	Ground-truth prominent users	True prominent users listened by SLAs
1st day	12 am–00 pm	1254	157	67
	00 pm–12 am	1264	157	67
2nd day	12 am–00 pm	2433	173	57
	00 pm–12 am	3143	175	81

To measure the quality of results returned by each baseline in each period of time, we calculated the precision of the returned prominent users by each algorithm. The obtained results are shown in Fig. 2.

Compared with the time consuming centrality measures, our model gains a significant increase in performance at the different stages of the event. We also note that the performance of the baseline measures decreases over time as they are sensitive to well-connected users. Moreover, we note that the centrality measures are not suitable for the detection of prominent users in the context of unexpected events.

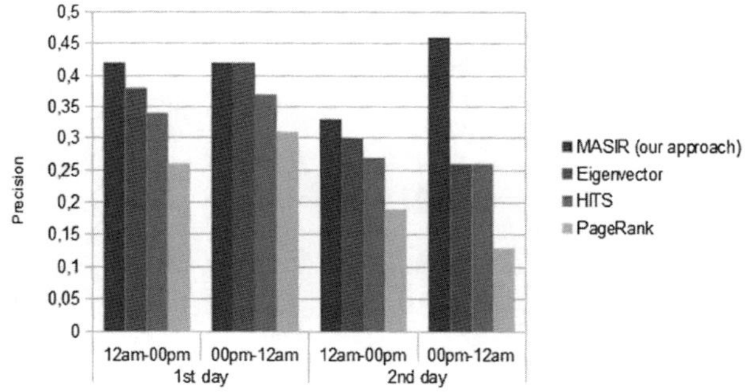

Fig. 2 The precision of *MASIR* vs *Eigenvector, Hits* and *PageRank* during the disaster

According to the results of our experiments, MASIR outperforms the centrality measures in terms of performance and time as the most prominent users were detected at an early stage of the event. Moreover, the used distributed architecture and our time sensitive ranking model have made the detection and the tracking of these users feasible in real time. Even MASIR has not detected all the top 175 users, these results are promising. Similar to recommendation and information retrieval engines, we can argue that if from the 9 recommended users, 8 are bad and 1 is extremely good and is sharing the required information, there is a high chance that the user will be satisfied by the retrieved information in real time.

5 Conclusion and Future Work

This paper highlights the power of multi-agent systems based architecture for real-time information retrieval from microblogs during unexpected events. MASIR uses various collaborative agents enabling a real-time detection of the most prominent users who tend to share valuable information. This first research effort to deal with the detection and tracking of prominent users in real unexpected events cases has provided promising results. The straightforward time sensitive measures used by our ranking model have outperformed the standard centrality measures. Moreover, the employed multi-agent architecture has coped with the Twitter APIs limits to be able to track in real time 175 users using only 5 hosts.

For future work, we aim to propose new features reflecting the user behavior during the event in order to improve the detection process of prominent users. In addition, we would like to optimize the distribution of our agents and minimize the number of exchanged messages between the different agents.

References

1. Abdulrahman, R., Neagu, D., Holton, D., Ridley, M., Lan, Y.: Data extraction from online social networks using application programming interface in a multi agent system approach. In: Nguyen, N. (ed.) Transactions on Computational Collective Intelligence XI. Lecture Notes in Computer Science, vol. 8065, pp. 88–118. Springer, Berlin (2013)
2. Agichtein, E., Castillo, C., Donato, D., Gionis, A., Mishne, G.: Finding high-quality content in social media. In: Proceedings of the 2008 International Conference on Web Search and Data Mining, pp. 183–194. WSDM '08, ACM, New York, NY, USA (2008)
3. Bizid, I., Faiz, S., Boursier, P., Yusuf, J.: Integration of heterogeneous spatial databases for disaster management. In: Parsons, J., Chiu, D. (eds.) Advances in Conceptual Modeling. Lecture Notes in Computer Science, vol. 8697, pp. 77–86. Springer (2014)
4. Canali, C., Colajanni, M., Lancellotti, R.: Dataacquisition in social networks: Issues and proposals. In: Proceedings of the International Workshop on Services and Open Sources, SOS'11 (2011)
5. Cappelletti, R., Sastry, N.: Iarank: ranking users on twitter in near real-time, based on their information amplification potential. In: Proceedings of the 2012 International Conference on Social Informatics, pp. 70–77. SOCIALINFORMATICS '12, IEEE Computer Society, Washington, DC, USA (2012)
6. Imran, M., Castillo, C., Lucas, J., Meier, P., Rogstadius, J.: Coordinating Human and Machine Intelligence to Classify Microblog Communications in Crises, pp. 712–721. The Pennsylvania State University (2014)
7. Kwak, H., Lee, C., Park, H., Moon, S.: What is twitter, a social network or a news media? In: Proceedings of the 19th International Conference on World Wide Web, pp. 591–600. WWW '10, ACM, New York, NY, USA (2010)
8. Nakamura, H.: Effects of social participation and the emergence of voluntary social interactions on household power-saving practices in post-disaster Kanagawa, Japan. Energy Policy 54(0), 397–403 (2013) (decades of Diesel)
9. Pal, A., Counts, S.: Identifying topical authorities in microblogs. In: Proceedings of the Fourth ACM International Conference on Web Search and Data Mining, pp. 45–54. WSDM '11, ACM, New York, NY, USA (2011)

10. Rieder, B.: Studying facebook via data extraction: the netvizz application. In: Proceedings of the 5th Annual ACM Web Science Conference, pp. 346–355. WebSci '13, ACM, New York, NY, USA (2013)
11. Silva, A., Guimarães, S., Meira, Jr., W., Zaki, M.: Profilerank: finding relevant content and influential users based on information diffusion. In: Proceedings of the 7th Workshop on Social Network Mining and Analysis, pp. 2:1–2:9. SNAKDD '13, ACM, New York, NY, USA (2013)
12. Smailovic, V., Striga, D., Mamic, D.P., Podobnik, V.: Calculating user's social influence through the smartsocial platform. In: 22nd International Conference on Software, Telecommunications and Computer Networks (SoftCOM), pp. 383–387 (Sept 2014)
13. Smailovic, V., Striga, D., Podobnik, V.: Advanced user profiles for the smartsocial platform: reasoning upon multi-source user data. In: Web Proceedings of the 6th ICT Innovations Conference 2014, pp. 258–268 (2014)
14. Weng, J., Lim, E.P., Jiang, J., He, Q.: Twitterrank: finding topic-sensitive influential twitterers. In: Proceedings of the Third ACM International Conference on Web Search and Data Mining, pp. 261–270. WSDM '10, ACM, New York, NY, USA (2010)
15. Xianlei, S., Chunhong, Z., Yang, J.: Finding domain experts in microblogs. In: Proceedings of the Tenth International Conference on Web Information Systems and Technologies, WEBIST'14 (2014)

Towards a Distributed Multiagent Travel Simulation

Matthieu Mastio, Mahdi Zargayouna and Omer Rana

Abstract With the generalization of real-time traveler information, the behavior of modern transport networks becomes harder to analyze and to predict. It is now critical to develop simulation tools for mobility policies makers, taking into account this new information environment. Information is now individualized, and the interaction of a huge population of individually guided travelers have to be taken into account in the simulations. However, existing mobility multiagent and micro-simulations can only consider a sample of the real volumes of travelers, especially for big regions. With distributed simulations, it would be easier to analyze and predict the status of nowadays and future networks, with informed and connected travelers. In this paper, we propose a comparison between two methods for distributing multiagent travelers mobility simulations, allowing for the consideration of realistic travelers flows and wide geographical regions.

1 Introduction

Transport systems are more and more complex and they have to evolve to integrate more connected entities (mobile devices, connected vehicles, etc.). Indeed, we can now provide optimal routes for the travelers but we are also able to update these routes in real time based on new network status (congestions, accidents, bus down, canceled carpooling, etc.). Giving information to the traffic network users is generally good and allows the improvement of the global network traffic flows. However,

M. Mastio (✉) · M. Zargayouna
Université Paris-Est, IFSTTAR, GRETTIA, Boulevard Newton, Champs Sur Marne, 77447 Marne la Vallée Cedex 2, France
e-mail: matthieu.mastio@ifsttar.fr

M. Zargayouna
e-mail: hamza-mahdi.zargayouna@ifsttar.fr

O. Rana
School of Computer Science & Informatics, Cardiff University, Cardiff, UK
e-mail: ranaof@cardiff.ac.uk

G. Jezic et al. (eds.), *Agent and Multi-Agent Systems: Technologies and Applications*, Smart Innovation, Systems and Technologies 38,
DOI 10.1007/978-3-319-19728-9_2

without control, the massive spread of information via billboards, radio announcements and individual guidance may have perverse effects and create new traffic jams. Indeed, with this generalization of real-time traveler information, the behavior of modern transport networks becomes harder to analyze and to predict. It is then important to model and simulate a realistic number of travelers to correctly observe these effects.

The ability to run a traffic simulator with real volumes of travelers at a city, a region or a country scale, would allow to observe the consequences of different information strategies on the status of multimodal traffic before implementing them in the real world. The management of millions of travelers in real time requires considerable computational power and current mobility simulations do not scale up in a way that would make it possible to predict the effects of regulation and information actions on the network. Our main objective in this paper is to test the scalability of this kind of simulators and develop scenarios at actual city scale. For this purpose, we aim to split the simulation between several servers on a grid and balance the load optimally between the servers while minimizing inter-server communications.

To have a generic distribution pattern, we propose a reference mobility simulator based on the multiagent paradigm. The multiagent paradigm is relevant for the modeling and simulation of transport systems [1]. This is why the multiagent approach is often chosen to model, solve and simulate transport problems. This approach is particularly relevant for the simulation of individual travels since the objective is to take into account human behaviors that interact in a complex, dynamic and open environment.

The remainder of this paper is structured as follows. In Sect. 2, we present the previous proposals for travelers mobility simulation and the existing distributed multiagent platforms. Section 3 presents a formal definition of the multiagent environment. In Sect. 4, we describe two methods for distributing simulations over several hosts. Section 5 explains our experimental setup and provides a comparison of the two proposed methods, before concluding and describing some further work we are pursuing.

2 Related Work

There exists several multiagent simulators for travelers mobility. For instance, MATSim [2] is a widely known platform for mobility micro-simulation. However, the mobile entities in MATSim are passive and their state is modified by central modules, which limits its flexibility and its ability to integrate new types of (proactive) agents. Transims [3] simulates multimodal movements and evaluates impacts of policy changes in traffic or demographic characteristics while Miro [4] reproduces the urban dynamics of a French city and proposes a prototype of multiagent simulation that is able to test planning scenarios and to specify individuals' behaviors. AgentPolis [5] and SM4T [6] are also multiagent platforms for multimodal transportation.

Finally, SUMO [7] is a widely used microscopic simulator mainly focused on traffic. However, none of these proposals considers the distribution problem.

Some general-purpose multiagent platforms have been specifically developed for large scale simulation in the last years. RepastHPC [8], a distributed version of Repast Simphony, uses the same concepts of projections and contexts and adapts them for distributed environments. Pandora [9] is close to RepastHPC and automatically generates the code required for inter-server communications. GridABM [10] is based on Repast Simphony but takes another approach and proposes to the programmer general templates to be adapted to the communication topology of his simulation.

However, these distributed platforms do not offer fine controls on how the communications between hosts are performed. Indeed, the communication layer is transparent for the programmer, which makes it easier for him to develop distributed simulations, but prevents him from optimizing the distribution. The best way to manage the communications depends of the application and using such general platforms for a travel simulator would not produce optimal results. More theoretical works study general methods to address this problem. In [11] and [12] the authors propose to relax some synchronization constraints to achieve a better scalability by reducing the time the hosts wait for each other. In [13] and [14] the authors discuss the issues related to multiagent simulation in a distributed virtual environment. In the present paper, we propose specific approaches to distribute traffic-based simulations, which could be of great benefit to the work on travelers mobility simulations.

3 The Multiagent Environment

3.1 The Model

The multiagent environment of a travel simulation is made of the transportation network in which the traveler agents evolve. We model the transport network with a graph $G(V, E)$ where $E = \{e_1, ..., e_n\}$ is a set of edges representing the roads and $V = \{v_1, ..., v_n\}$ is a set of vertices representing the intersections. A set of agents A is traveling in this network from origins to destinations trying to minimize their travel time. The travel time of an edge at time t depends on the number of agents using it. To calculate this time, we use a triangular fundamental diagram of traffic flow that gives a relation between the flow q (vehicles/hour) and the density k (vehicles/km). The fundamental diagram suggests that if we exceed a critical density of vehicles k_c, the more vehicles are on a road, the slower their velocity will be. Here is the equation we use to model this phenomena (Fig. 1):

$$q = \begin{cases} \alpha k & \text{if } k \leq k_c \\ -\beta(k - k_c) + \alpha k_c & \text{if } k > k_c \end{cases} \quad (1)$$

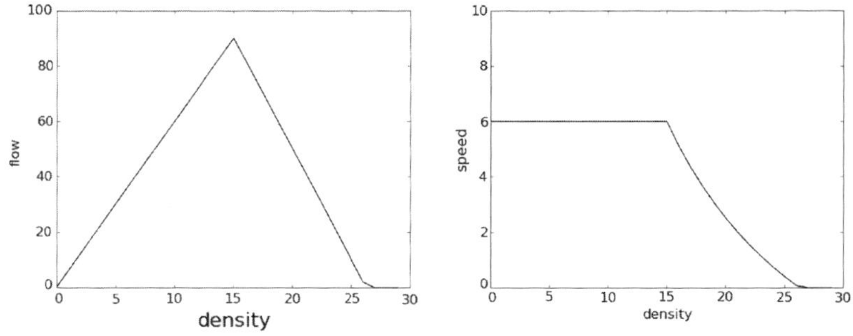

Fig. 1 Fundamental diagram with $\alpha = 6$, $\beta = 8$ and $k_c = 15$ (left). Speed in function of density (right)

This equation is parametrized with α the free flow speed on this road, β the congestion wave speed and k_c the critical density. As $v = \frac{q}{k}$:

$$v = \begin{cases} \alpha & \text{if } k \leq k_c \\ \frac{-\beta(k-k_c)+\alpha k_c}{k} & \text{if } k > k_c \end{cases} \tag{2}$$

Thus we can define a cost function that returns a travel time per distance units $(1/v)$ in function of the number of agents $|A_e|$ on this edge:

$$cost(|A_e|) = \begin{cases} \frac{1}{\alpha} & \text{if } |A_e| \leq k_c \\ \frac{|A_e|}{-\beta(|A_e|-k_c)+\alpha k_c} & \text{if } |A_e| > k_c \end{cases} \tag{3}$$

Both edges and vertices are weighted with positive values evolving dynamically with the number of agents present on them. Given A_v, the set of agents on a vertex v, and A_e, the set of agents on a edge e, we have $|v| = |A_v|$ representing the weight of a vertex and $|e| = |A_e|$ the weight of an edge. $|V| = \sum_{v \in V} |v|$ is the weight of a subset of vertices. In the same way, if $|e|$ is the weight of a vertex, $|E| = \sum_{e \in E} |e|$ is the weight of a subset of edges.

3.2 The Simulator

We have developed a reference simulator where each agent represents a traveler evolving in a multiagent environment as described in the previous paragraph. Agents appear nondeterministically with an origin and a destination vertex. They compute the shortest path based on the current status of the network before to start traveling. They ask for a new shortest path each time they reach a vertex in their path, to check wether a new shortest path becomes possible, following the dynamics of the network. At each time step of the simulation, if the agents are currently on an edge, they go forward as far as the road state (current mean speed) allows them to go. The simulation ends when all the travelers have reached their destinations or when a time step threshold is reached.

4 Proposed Approach

To launch our simulation at a city scale, we need a large memory and computing power. This is why we aim to deploy it on several machines. A simulation running on a cluster is typically SPMD (Single Program Multiple Data), i.e. each processor runs the same program but owns only a part of the program data in its private memory, and all the processors are connected by a network. Communications are explicitly declared by the programmer. The advantage of this approach is its high scalability; it can be implemented on most parallel architectures and we can deploy the same simulation on larger systems if we need more power.

In mobility simulations, traveler agents are moving on a transport network. To distribute such simulations, we have to split the workload between the available servers efficiently. In our model, the main workload is generated by the calculation of the shortest path. An agent's path has to be recalculated each time this agent reaches a new intersection (because the travel times evolve dynamically), thus an agent can be considered as a unit of workload. Therefore, in order to distribute this model we need to split the agents between the servers. To do so, we can either distribute the environment, or the agents.

4.1 Agent Distribution

One approach would be to cut the set of agents in k equal parts (with k the number of available servers), distribute each subset on a server and run the simulation. As the travel times depend on the number of agents on each arc, all the agents need to know at every step how many agents there are on each arc in order to compute their shortest path. In our implementation, at each time step, if an agent managed by a server quits or arrives on an edge, this server communicates the information to all other servers. Thus at any time step, all the servers know the state of the entire network. At the moment, it is the only communication needed for the simulation, so the total communication cost is $k|E|I$, with I representing the size of an integer.[1]

[1]Coding the number of agents arriving or leaving an edge.

4.2 Environment Distribution

The second distribution pattern tries to keep agents that are close in the graph on the same server. Instead of distributing the agents, we distribute the vertices and the outgoing edges (and thus the agents located on these vertices and edges) so the agents that are in the same place of the graph are on the same server.

Each server is only aware of what is happening on the part of the graph that it is managing. So at each time step, before the agents act, all the servers need to synchronize with the servers running in other processes. There are now two types of communications: the servers have to communicate the weight of their edges (the number of agents on it) and when an agent moves to a vertex that is not on his current server he has to move to that vertex's server. Let C be the cost of the edge communications and M the agents migration cost. At each step, the total communication cost is given by $T = C + M$. The cost of the edges weight communications is: $C = |E| \times I$. An agent could be coded with three integers (ID, current location and destination). Let n be the number of migration for one step. Thus the agents migration cost is: $M = 3In$. There are on average $|A|/|E|$ agents per vertex. So with E_c the set of edges between different servers we have on average $n = |A|/|E| \times |E_c|$. Thus $T = I(|E| + 3|A|/|E| \times |E_c|)$. As we could expect, the less edges there are between two servers and the less the communication cost there is (Fig. 2).

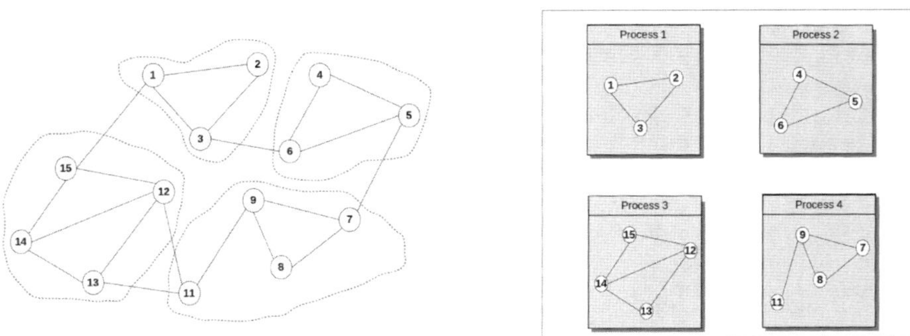

Fig. 2 The graph is parted and each part are distributed between the available processes

So for the environment distribution method to be effective, we need to split the vertices into k disjoint sets such that each set has approximately the same vertex weight and such that the cut-weight, the total weight of edges cut by the partition, is minimized. This problem is known as the $(k, 1 + \epsilon)$-balanced partitioning problem, that is the problem of finding a collection of disjoint subsets $V_1, \ldots, V_k$ that cover V, i.e., $V = V_1 \cup \ldots \cup V_k$ such as each part contains at most $(1 + \epsilon)\frac{|A|}{k}$ and $|E_c|$ is minimized.

The problem of partitioning a network has been widely studied in the scientific literature. As demonstrated in [15] this is a NP-hard problem so trying to find an optimal solution with, for example, integer programming is not an option for large graphs. This is why some heuristics have been proposed to solve this problem in reasonable time. The multilevel partitioning method has been recognized as a very powerful method that offers a more global vision on graphs than traditional techniques. As the complexity of the partitioning problem is dependent on the size of the partitioned graph, the simple idea of multilevel partitioning is to regroup the vertices and to work with the groups instead of the independent vertices. The multilevel partitioning has been formalized in a generic framework by Walshaw in [16]. To distribute the environment, we use a slightly modified version of the Differential Greedy algorithm [17]. We modified this algorithm to use it with weighted vertices and to produce more connected partitions.

5 Experiments and Results

5.1 Implementation

To test the effectiveness of our approach, we have implemented our model and deployed it on an actual cluster. We choose Python to develop the model since this language is efficient for quick prototyping. Python is a mature portable language with a lot of well tested scientific libraries and is along with C and Fortran one of the most used languages for high performance computing [18].

For the inter-process communications we use MPI, that is the *de facto* standard language for parallel computing with a huge community of users. MPI offers a simple communication model between the different processes in a program and has many efficient implementations that run on a variety of machines. Moreover MPI4PY is an efficient interface that allows to use MPI with Python.

5.2 Results

We have launched the distributed simulations on the Cardiff University cluster. For our tests, we used eight hosts under CentOS Linux (kernel version 2.6.32-220) on a processor Intel Xeon CPU E5-2620 (12 cores at 2Ghz) with 32GB of memory. We ran the simulation on three configurations: the first is a sequential version of the program on a single host (conf1), the second is a distributed version on the eight hosts (conf2), and the last is run on the eight hosts using the 12 cores of each one (conf3). The simulation is performed for 100 time steps on a 200 nodes power-law graph generated with the Barabasi-Albert model [19]. We compared the two methods of distribution (agent-based and environment-based distributions) on the different configurations with an increasing number of agents (from 1,000 to 40,000) (Table 1).

Table 1 Computational times (in seconds) for a 100 time steps simulation on a 200 nodes scale free graph

number of agents	1000	5000	10000	20000	30000	40000
conf1 (1 core)	10	27	43	67	104	140
conf2 agent distribution (12 cores)	3	6	8	12	17	23
conf3 agent distribution (96 cores)	2	3	3	4	4	5
conf2 environment distribution (12 cores)	6	13	17	22	31	41
conf3 environment distribution (96 cores)	5	10	11	11	15	17

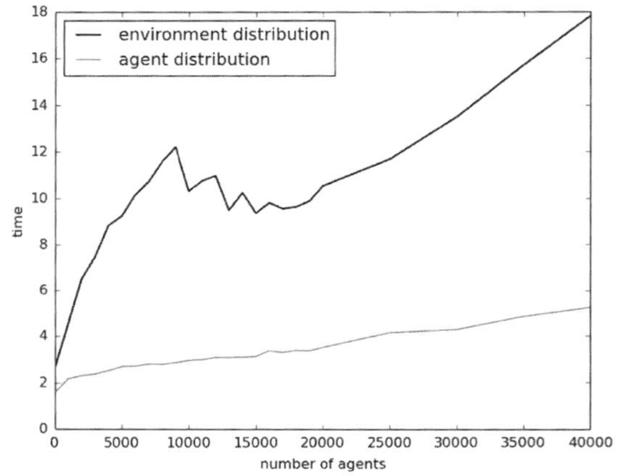

Fig. 3 Comparison of computational times between the different distribution methods with conf3

As we could expect, the agent-based distribution is more effective than the environment-based distribution with the proposed simulation model.[2] Indeed, at the moment we have not defined any local interactions in our model. As a consequence, we are in an perfect case for the agent base distribution since the amount of inter-server communications will be limited. But with the further implementation of local interactions (for example pursuit model or vehicle to vehicle communications) the environment-based approach will be able to take advantage of the collocation in the same server of physically close agents. Furthermore, there is at the moment no dynamic load balancing mechanism for the environment distribution. Indeed, if an important number of agents are concentrated in the same part of the network, they will be nevertheless in the same server. It will hence take more time for this server to calculate all the shortest paths and, all the other servers will have to wait for it. Thus, if a server is overloaded it can slow down all the simulation (Fig. 3).

[2]The computational times are not strictly growing with the number of agents for the environment-based method. This is more likely due to the random origins and destinations of the agents. Therefore the simulation could sometimes be more complex with fewer agents.

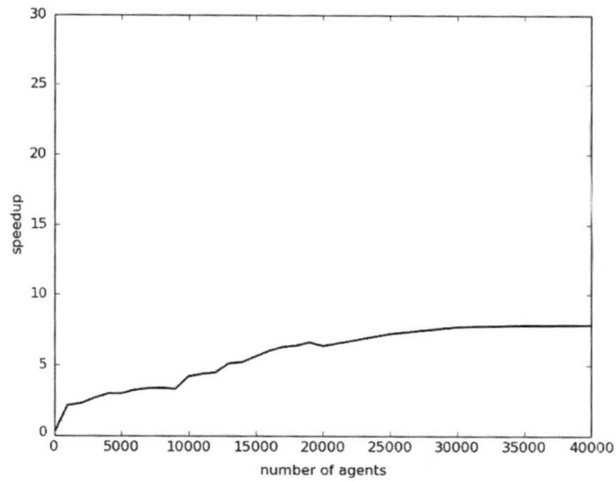

Fig. 4 Speedup between conf1 and conf3 with environment distribution

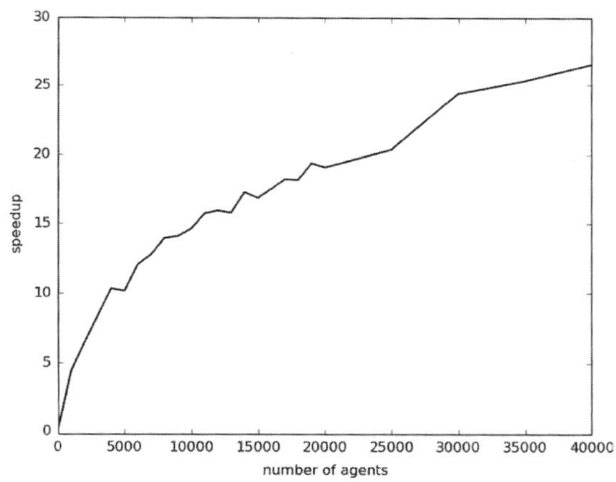

Fig. 5 Speedup between conf1 and conf3 with agent distribution

The speedups between the sequential run and the two distribution methods performed on conf3 are shown on Figs. 4 and 5. The speedup is a measures of how much faster the simulation is on conf3 (96 cores) than on conf1. As we can see on these figures, with 40000 agents for example, the simulation is 8 times faster with the environment based distribution and 28 times faster with the agent based distribution. Both of these methods improve largely the execution time of our multiagent traffic simulator. As explained above, the agent-based distribution method shows better results due to the relative simplicity of the current model but after adding inter-agents communications the difference in performance between the two methods should shrink.

6 Conclusions and Perspectives

In this paper, we have presented two distribution methods of a multiagent travel simulation over several hosts. The two methods are efficient to scale the simulation up with the number of agents. With our current simulation model, the agent-based method is more efficient than the environment-based distribution method. Our simulation model is very general, our proposals and findings are applicable to all the state-of-the-art travel simulators.

Our future works will deal with two aspects. The first concerns the simulation model. We will tackle more specific travel simulation models, where all types of communications are present (local, global and community-based communications). We expect the environment-based distribution method to show better performance than with the current model. The second aspect concerns the environment-based distribution method. The distribution is currently done statically at the beginning of the simulation and we believe that the speedup could be largely improved by adding dynamic load balancing mechanisms.

References

1. Badeig, F., Balbo, F., Scemama, G., Zargayouna, M.: Agent-based coordination model for designing transportation applications. In: 11th International IEEE Conference on Intelligent Transportation Systems, 2008 (ITSC 2008), pp. 402–407, IEEE, 2008
2. Maciejewski, M., Nagel, K.: Towards multi-agent simulation of the dynamic vehicle routing problem in matsim. In: Proceedings of the 9th International Conference on Parallel Processing and Applied Mathematics—vol. Part II, PPAM'11, pp. 551–560. Springer, Berlin, Heidelberg (2012)
3. Nagel, K., Rickert, M.: Parallel implementation of the transims micro-simulation. Parallel Comput. **27**(12), 1611–1639 (2001)
4. Chipeaux, S., Bouquet, F., Lang, C., Marilleau, N.: Modelling of complex systems with AML as realized in MIRO project. In: LAFLang 2011, Workshop of the International Conference WI/IAT (Web Intelligence and Intelligent Agent Technology), pp. 159–162. IEEE Computer Society, Lyon, France (2011)
5. Jakob, M., Moler, Z., Komenda, A., Yin, Z., Jiang, A.X., Johnson, M.P., Pechoucek, M., Tambe, M.: Agentpolis: towards a platform for fully agent-based modeling of multi-modal transportation (demonstration). In: International Conference on Autonomous Agents and Multiagent Systems, AAMAS 2012 (3 volumes), pp. 1501–1502, Valencia, Spain, 4–8 June 2012
6. Zargayouna, M., Zeddini, B., Scemama, G., Othman, A.: Simulating the impact of future internet on multimodal mobility. In: The 11th ACS/IEEE International Conference on Computer Systems and Applications AICCSA'2014. IEEE Computer Society (2014)
7. Behrisch, M., Bieker, L., Erdmann, J., Krajzewicz, D.: SUMO - simulation of urban MObility - an overview. In: SIMUL 2011, The Third International Conference on Advances in System Simulation, pp. 55–60 (2011)
8. Collier, N., North, M.: Repast HPC: a platform for large-scale agent-based modeling. In: Dubitzky, W., Kurowski, K., Schott, B. (eds.) Large-Scale Computing, pp. 81–109. Wiley (2011)
9. Angelotti, E.S., Scalabrin, E.E., Avila, B.C.: PANDORA: a multi-agent system using paraconsistent logic. In: Proceedings of Fourth International Conference on Computational Intelligence and Multimedia Applications, 2001 (ICCIMA 2001), pp. 352–356 (2001)

10. Gulyas, L., Szemes, G., Kampis, G., de Back, W.: A modeler-friendly API for ABM partitioning. In: Proceedings of ASME 2009 International Design Engineering Technical Conferences and Computers and Information in Engineering Conference, pp. 219–226 (2009)
11. Mengistu, D., Lowis, MV.: An algorithm for optimistic distributed simulations. In: Modelling, Simulation, and Identification/658: Power and Energy Systems/660, 661, 662. ACTA Press (2011)
12. Scheutz, M., Schermerhorn, P.: Adaptive algorithms for the dynamic distribution and parallel execution of agent-based models. J. Parallel Distrib. Comput. **66**(8), 1037–1051 (2006–2008)
13. Ng, B., Si, A., Lau, R.W.H., Li, F.W.B.: A multi-server architecture for distributed virtual walkthrough. In: Proceedings of the ACM Symposium on Virtual Reality Software and Technology, VRST '02, pp. 163–170. ACM (2002)
14. Rihawi, O., Secq, Y., Mathieu, P.: Effective distribution of large scale situated agent-based simulations. In: ICAART 2014 6th International Conference on Agents and Artificial Intelligence, vol. 1, pp. 312–319. SCITEPRESS Digital Library (2014)
15. Bui, T.N., Jones, C.: Finding good approximate vertex and edge partitions is NP-hard. Inf. Process. Lett. **42**(3), 153–159 (1992)
16. Walshaw, C.: Multilevel refinement for combinatorial optimisation: boosting metaheuristic performance. In: Hybrid Metaheuristics, number 114 in Studies in Computational Intelligence, pp. 261–289. Springer, Berlin (2008)
17. Fiduccia, C.M., Mattheyses, R.M.: A linear-time heuristic for improving network partitions. In: 19th Conference on Design Automation, 1982, pp. 175–181 (June 1982)
18. Langtangen, H.P., Cai, X.: On the efficiency of python for high-performance computing. In: Modeling. Simulation and Optimization of Complex Processes, pp. 337–357. Springer, Berlin (2008)
19. Barabási, A.-L., Albert, R.: Emergence of scaling in random networks. Science **286**(5439), 509–512 (1999)

Security of Mobile Agent Platforms Using Access Control and Cryptography

Hind Idrissi, El Mamoun Souidi and Arnaud Revel

Abstract Mobile Agents are autonomous software entities able to move from one host to another. However, this mobility is not all the time safe, as a hosting platform may receive agents with malicious behaviors. In this paper, we attempt to deal with this security problem by proposing a solution based on a strengthened cryptographic authentication and an access control policy. The proposed authentication process is performed through a resistant MITM Diffie-Hellman key exchange protocol, while the resources access control policy is elaborated basing an enhanced DAC model where Shamir-Threshold Scheme is used to manage and share access rights. We have conducted detailed experiments and practical investigations to evaluate the security of our approach and its effectiveness to resist face to some well known attacks.

Keywords Mobile agent platform · Security · Cryptography · Access control · DHKE-DSA · Shamir's threshold scheme

1 Introduction

The field of agents becomes one of the more dynamic research areas in recent years. It is increasingly integrated in several evolving disciplines, such as personal information management, electronic commerce, telecommunications, etc. Mobile agents are specific software agents able to move from one hosting platform to another, with a set of actions to execute (code), resources to deploy (data), and a state of execution.

H. Idrissi (✉) · E.M. Souidi
LabMIA, Faculty of Sciences, University of Mohammed-V, Rabat, Morocco
e-mail: hind.idrissi@univ-lr.fr

E.M. Souidi
e-mail: souidi@fsr.ac.ma

H. Idrissi · A. Revel
L3I, University of La Rochelle, La Rochelle, France
e-mail: arnaud.revel@univ-lr.fr

G. Jezic et al. (eds.), *Agent and Multi-Agent Systems: Technologies and Applications*, Smart Innovation, Systems and Technologies 38,
DOI 10.1007/978-3-319-19728-9_3

They are carrying their sensitive information, which makes them not bound to the system where they begin their execution.

However, mobility of agents introduces new security vulnerabilities, especially for hosting platforms, that could receive malicious agents seeking to exploit security weaknesses and launche a wide variety of attacks. In order to address these security issues, this paper proposes a solution based on two techniques: an authentication process based on a Diffie-Hellman key exchange, which is resistant to the Man-In-The-Middle attack ,and an access control policy consolidated with a cryptographic threshold scheme, to manage the resources access rights. This contribution allows the hosting platform to grant access only to the authenticated mobile agents, and ensures secure communication between the both via the generation of a session key, used to encrypt and decrypt the exchanges. Besides, it monitors the accesses to the platform resources in a flexible and interoperable manner, which guarantees the confidentiality and lessens damages if a malicious agent is received.

Section 2 of this paper provides an overview of the mobile agent security and discusses some related works. Section 3 describes in detail the proposed solution to protect hosting platform from agent attacks. Section 4 presents an evaluation of the proposed solution that was implemented using JADE [1] framework, with a respect to the FIPA standard [2] specified for software interoperability among agents and agent-based applications. The feasibility and security effectiveness of the proposed solution were evaluated by three factors: time performance, authorization performance and the resistance to some well known attacks. Finally, a results interpretation and perspectives are provided in the conclusion.

2 Security Problem

This section briefly explains the security issues that a hosting platform may encounter while receiving a malicious agent, and introduces some well known related works trying to deal with this problem.

Compared to the traditional Client/Server architecture, mobile agent paradigm presents many qualities such as: overcoming network delay, reducing network load and providing innate heterogeneity. However, it raises many new security issues due to the unsafe contact among these entities. According to [3], the security risks in a mobile agent system are classified into three categories: Agent-to-Platform, Platform-to-Agent and Agent-to-Agent. But, this paper tries to address those related to the first category, when a hosting platform needs to be secured against threats from malicious agents. This includes [3]: Masquerading, Denial of Service, Alteration and particularly Unauthorized Access, which exposes the hosting platform or other agents to serious harms by exploring access rights to different resources, including the residual data.

Many efforts have been devoted to investigate the issue of mobile agent platform security. Indeed, several solutions have been proposed in the literature, where authentication and access authorization of mobile agents are very important. Towards this, in [4] many security techniques have been introduced. Safe code interpretation [5] is required when the mobile agent's code is interpreted. During execution of the interpreted code, it cures an unsafe command or simply ignores the command. Proof carrying code [6] makes the author of an agent generates a proof that guarantees the safety of the agent code. Then the host can verify the agent using the proof transmitted with the agent. Path histories [7] is to make agents have a record of the prior hosts that they have previously visited. State appraisal [8] aims to guarantee that an agent's status has not been modified. In Model carrying code [9], the host forms a model that captures the security relevant behavior of code, rather than a proof, by using information accompanying the untrustworthy code. Code consumers are able to know the security needs of untrustworthy code more precisely. Sand-boxing [10] is a method of isolating agents into a limited domain enforced by software. However, these solutions present some limitations in use and they did not reach the security level requested, as they still suffer from many flaws that require their association with other mechanisms, which is burdening the system.

3 Proposed Solution

In this section, we describe in details our proposed solution consisting of two parts. The first one presents our contribution to ensure a strengthened authentication using a fixed and an enhanced version of Diffie-Hellman key exchange. The second one presents our policy to manage the access to the platform resources, through using an extended Discretionary Access Control model (DAC) along with Shamir-threshold sharing scheme. First of all, let us present the architecture of the proposed solution, which is composed of three main parts: the native platform that we would like to secure, the remote platform sending the mobile agent and the trusted third party (TTP). Figure 1 illustrates the structure of the adopted framework via an UML class diagram, showing the links between the different components.

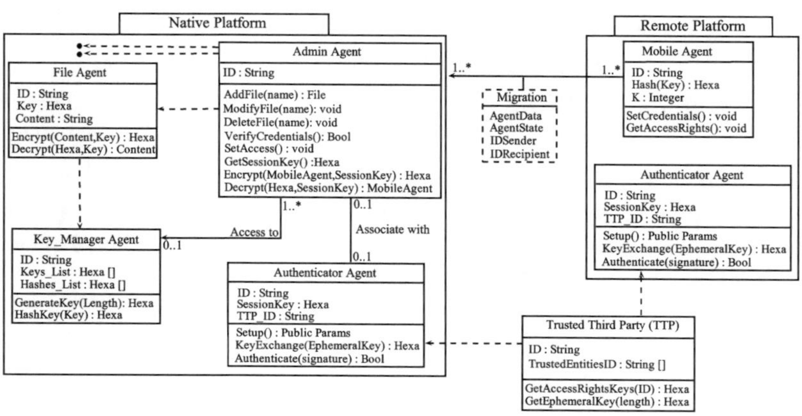

Fig. 1 UML class diagram of the proposed architecture

- *Key_Manager_Agent*: It is an agent that makes use of the pseudo random number generator ISAAC+ [11], to automatically generate a key of 256 bits for each file created in the environment. These keys and their corresponding hash obtained using SHA-3 [12] are assembled in two separate lists.
- *Admin_Agent*: It is responsible of the communication among all intern components and also with remote entities. Its tasks include mathematical operations to authenticate platforms, verify the keys and give access, in addition to managing the data of the environment and controlling their use.
- *Authenticator_Agent*: It is charged with the authentication of any external entity claiming access to the platform. It collaborates with the TTP to perform a Key exchange integrating digital signature to prevent MITM attacks and generate a session key.
- *File_Agents*: These agents contain the sensible files and data of the environment. They are protected and encrypted.

In the remote platform, we consider an "Authenticator_Agent" with the same properties of that in the native platform, and a mobile agent containing two important information:

- *H(Key)*: The mobile agent may contain one or more specific keys for accessing one or more specific files. But this key is not given in clear, even the mobile agent did not know its value. The access keys are hashed using SHA-3, before they are provided along with an integer K to the mobile agent, via the TTP and through a secure channel.
- *K*: It is an integer referring to the number of concatenated access keys contained in the hash. It is useful for the native platform when receiving an authenticated mobile agent, to compute the arrangements of keys, that may construct the same hash.

In this context, it is necessary to notice that each mobile agent follows a preliminary itinerary defined by its own platform. Thus, the agent knows in advance the identities and locations (IP Address) of the platforms it will visit.

3.1 Authentication Between Native Platform and Remote Platform

To prevent vulnerabilities arising due to the unavailability of authentication, an agent charged with performing a mutual authentication is employed. This process inspired from [13] is based on an enhanced Diffie-Hellman key Exchange protocol integrated with digital signature DSA, which makes it well resistant to the Man-In-The-Middle Attacks. Figure 2 describes the authentication process used, which is composed of four phases:

1. *Key Request*: The remote platform communicates with the TTP in order to request the access keys to perform specific tasks. This request is then transferred to the native platform, which analyzes the required tasks and defines the concerned

resources with their corresponding access keys. These latter are first concatenated and hashed using SHA3, then transferred to the TTP along with the integer K. The TTP in turn, relays this response to the requesting remote platform. We should mention that communications in this phase are carried out via a secure channel, through which, the TTP sends to the both platforms an ephemeral key (T) to be involved in the key exchange protocol.

2. *Setup*: Both interacting platforms agree on a set of basic parameters to perform computations. This set includes large primes (p, q and h) to define the rank of the finite field (F_p) to be used, and a generator (g) on this field to achieve exponentiation.
3. *Key Exchange*: In this phase, every platform generates randomly a private key (X_i), which is used to calculate the public key (Y_i) on the basis of setup parameters. Fixing the security of the protocol at this stage requires two main characteristics: the exchange of public key should not be in clear, and the number of passes must be minimized to prevent the intruder from collecting sensitive data. Thus, each platform chooses randomly an ephemeral secret (v and w), which is involved in the computation of the values (M_i and N_i) basing the public key. The overall exchange consists of two passes concluded with the generation of a session key, while its robustness is due to the hardness of the discrete logarithm problem.
4. *Authentication*: At this stage, each platform creates a digital signature through hashing a set of concatenated values. This set includes the parameters and calculated values of the previous phases, in addition to the relevant TTP identity and the appropriate IP address. Both signatures are then exchanged and verified to validate the authentication of communicating platforms.

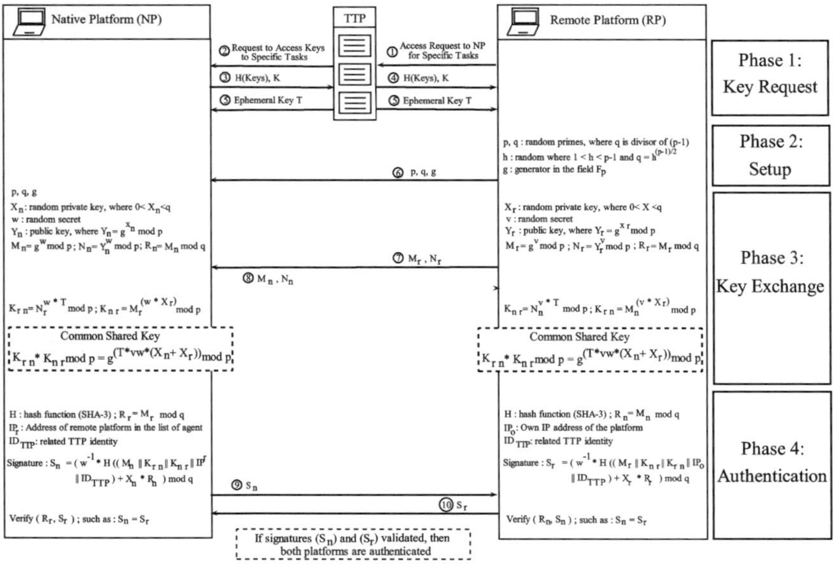

Fig. 2 Authentication process between the native and remote platforms using TTP and key exchange mechanism

3.2 Access Control of the Platform Resources

In Order to define an access control policy for our agent platform, we need to model our architecture as compounded of objects and subjects, such that the objects are the files stored and the subjects are the agents. There exist several access control models and policies to address different types of requirements [14]: Discretionary Access Control (DAC), Mandatory Access Control (MAC), Role-Based Access Control (RBAC) and many others.

In this Paper, we choose to use DAC model for many reasons. First of all, it is a famous model widely used in information and operating system such as Linux. Also, it is flexible and appropriate for the systems where information sharing is important without compromising security. In our work, data are shared with the mobile agents and we need a flexible and light-weighted security policy that will not slow the exchanges. The use of RBAC model here is not adequate, as it is based on roles directly associated with permissions. Thus, the users having the same role possesses necessarily the same privileges which reduce the flexibility. To apply this model, we need absolutel to classify the agents into roles, where those having the same role must have the same access rights on the same files, which is not achievable because each mobile agent has its specific access rights on specific files. In addition, there are no common rules among the agents (such as giving to the AgentA the right to access the files of AgentB, because any other agent having access to agentA will get automatically the right to access the AgentB). This Access Control policy consolidates the security, as each agent is independent with its own access rights, and there is no common relations that might be a source of security attacks.

DAC model is designed such as the authorizations are controlled at the discretion of users, who are the controller or owner of some resources and assign permissions to other users. The typical access rights defined here are: Read, Write, Execute and Own. The meaning of the first three of these is self evident. Ownership is concerned with controlling who can change the access permissions for the files or keys, then the access right Own includes in reality the three others. This model is based on Access Matrix showed in Table 1.

Table 1 Access matrix for the adopted architecture

	File A	File B	..	Keys list	SessionKey
File_Agent A	Own				
File_Agent B		Own			
:			Own		
Admin_Agent	Write,read	Write,read	Write,read	Read	Read
Key_Manager_Agent				Own	
Authenticator_Agent					Own

In order to specify each access right related to each file, we make use of Shamir's (n,t) Threshold Sharing Scheme [15] to decompose the key of each file into sub-keys according to which access rights are granted. This scheme is designed such that each secret S in some field F, is shared among n parties by creating a random polynomial $P \in F(x)$ of a degree t, such that $P(0) = S$. The i-th gets the share $(i, P(i))$. Given any $t + 1$ shares $P(x_0), ..., P(x_t)$, it is possible to recover $P(0)$ (the secret S) using Lagrange Interpolation: $P(0) = \sum_{i=0}^{t} \lambda_i P(x_i)$, where: $\lambda_i = \prod_{j \neq i} \frac{x_j}{(x_j - x_i)}$.

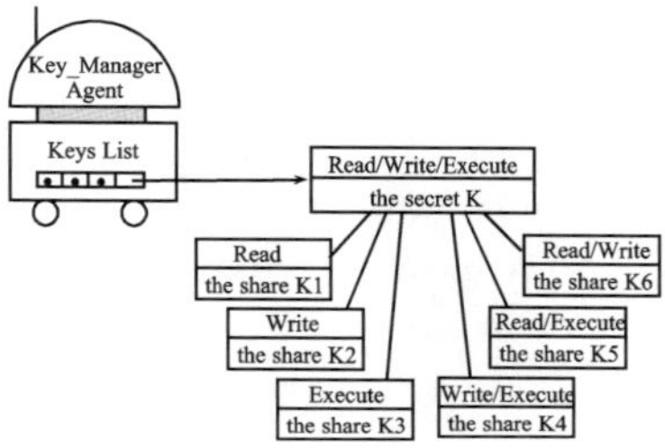

Fig. 3 Sharing the access rights keys using shamir's threshold scheme

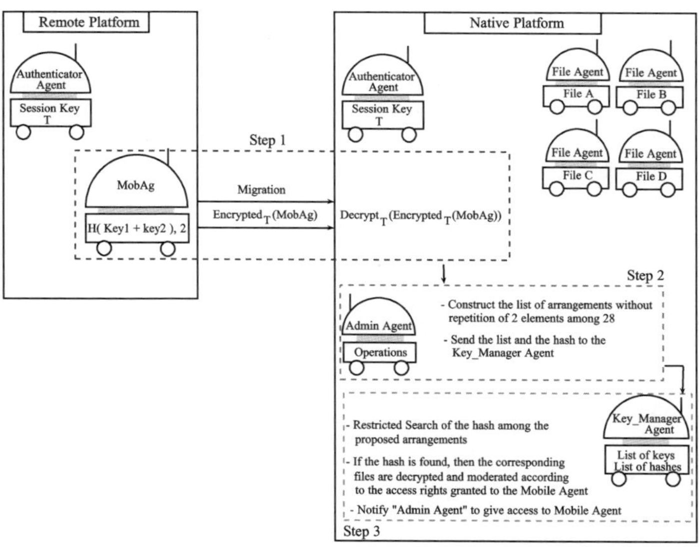

Fig. 4 The process of the proposed access control policy for the native platform

Figure 3 defines $n = 6$ access rights extracted from the main three single ones, with a threshold $t = 4$ to reconstruct the original key, and with $l = 2$ the number of levels (level 1 with one access right and level 2 with two concatenated access rights). This sharing structure may be extended such as: if s is the number of main single access rights, then the number of shares is $n = \sum_{i=1}^{s-1} C_s^i$, while the number of levels is $s - 1$. Let us consider a simulation scenario where a mobile agent is carrying two concatenated keys (H(Key1 + Key2), k = 2) to access two files with specific access rights. We suppose that the native platform contains four files, and the mutual authentication is successfully performed, Fig. 4 describes the processing steps, in normal conditions, of the proposed access control policy.

- *Step 1*: The mobile agent is encrypted using the session key and AES algorithm [16], then it migrates to the native platform. We choose AES-256 due to its very convenient use as it consumes little memory, for its reduced complexity and for its easy implementation. Once arriving to the native platform, the mobile agent is decrypted by the relevant "Admin_Agent" to get the credentials: H(Key1 + Key2) and k = 2. At this moment, there is no risk that the mobile agent will be duplicated because all the system is fault tolerant. When an unexpected failure happened, the mobile agent will resume its execution and recover the last saved state. Besides, the mobility from one platform to other is transactional and each mobile agent can only be executed in one node at a time.
- *Step 2*: In possession of mobile agent credentials, and knowing that for each file in the platform there is 7 keys specifying 6 access rights and a global secret key, the "Admin_Agent" has to construct a list of arrangements without repetition of two elements among the 28 keys (7 keys for 4 files). The order of these arrangements is very important due to the use of hash function, where: $H(Key1 + Key2) \neq H(Key2+Key1)$. It is worth to mention, that the two keys are belonging to different files, which means hat the arrangements of keys for the same file are eliminated. Thus, the final list contains effectively $A_{28}^{2} - 4 \times A_{7}^{2} = 588$ possibilities, and it is sent to the "Key_Manager_Agent" along with the hash of the two concatenated keys.
- *Step 3*: Once receiving the list of key-arrangements, the "Key_Manager_Agent" performs a restricted search for the corresponding hash. If this later is found, then the files subject of this arrangement are recognized with the corresponding access rights. Thereafter, the "Key_Manager_Agent" decrypts the relevant files, adjusts them according to the access rights assigned, and notifies the "Admin_Agent" to allow the mobile agent to execute its tasks on the named files. Else, if the hash is not found, then an acquittal is sent to the "Admin_Agent" in order to terminate the mobile agent.

4 Evaluation

In this section, we present the experimental investigations to prove the feasibility, effectiveness and the security of proposed solution. Our evaluation is based on three factors: Time Performance, Authorization Performance and Security Analysis. Thus, we have designed and implemented the proposed architecture basing real machines of Core i7 at 2.7 GHz, with 4 Go of RAM. All machines are equipped with JADE agent platform in its 4.3.3 version.

4.1 Time Performance

To compute the time cost of our solution, we need to compute the time of each operation performed during the round trip of the mobile agent to access the native platform data. These calculated times include the following operations: access keys request via the TTP (T_{AKR}), key exchange and authentication (T_{KEA}), encryption and decryption of the mobile agent (T_{EDMA}), migration of the agent (T_{Mig}), construction of the arrangement list (T_{CAL}), hash search (T_{HS}), decryption with moderation of the corresponding files (T_{DMF}), mobile agent execution (T_{MAE}) and finally its return in encrypted format to its own platform ($T_{EMA} + T_{Return}$). Let T_{total} be the execution time for the previously mentioned operations:

$$T_{Total} = T_{AKR} + T_{KEA} + T_{EDMA} + T_{Mig} + T_{CAL} + T_{HS} + T_{DMF} + T_{MAE} + T_{EMA} + T_{Return} \quad (1)$$

Knowing that T_{Mig} and T_{Return} are approximately equal, and the time to encrypt an agent is the same as to decrypt it, then:

$$T_{Total} = T_{AKR} + T_{KEA} + \frac{3}{2}T_{EDMA} + 2T_{Mig} + T_{CAL} + T_{HS} + T_{DMF} + T_{MAE} \quad (2)$$

Table 2 presents the measured times of the operations mentioned for the previous scenario, where a remote platform sends a mobile agent with two concatenated keys to get access to two files among four ones, and the total time is calculated according to Eq. 2. These time costs has been measured for the system setup of two real platforms, where a native platform has interacted with a remote one for 100 times. Likewise, the use of JADE as agent framework adopting Java programming language, allows us to take benefit from the services of its security providers and its java cryptography extension (JCE). The process is reiterated with two files of 500 Kbytes and 1000 Kbytes, then the cost time of each operation is calculated as the average of its 100 iteration times.

Table 2 Time of the different operations performed during the process of the proposed solution for one mobile agent with K = 2

	T_{AKR}	T_{KEA}	T_{EDMA}	T_{Mig}	T_{CAL}	T_{HS}	T_{DMF}	T_{MAE}	$T_{total}(K = 2)$
Time (ms)	33.7	8.6	30	46.6	3.4	2.1	62.3	23.8	301.1

We find it interesting to extend our experiments, such that we first launch a baseline test without security operations, and then a test of our proposed solution. This will allow us to know the overhead of the security added to the system, as well as to show its effectiveness and its reliability to detect malicious entities. Both experiments are performed through the configuration of a system architecture composed of six real platforms, where a native platform hosting 20 files of different sizes is interacting with five remote platforms. In each test, every remote platform sends 20 mobile agents to get different access rights to specific files. Among the total 100 mobile agents there are 40 which are malicious. It should be noticed that for both tests, the process is launched 100 times and the average time value is kept.

Figure 5 shows the time costs of the baseline test and the security test, regarding to different and growing values of the parameter (K). Besides, Fig. 6 gives the total time that each test spent to interact with increasing number of mobile agents. The obtained results show the ability of our solution to scale in front of large conditions, namely the expanded number of mobile agents requesting to access platform services, the set up of growing databases and especially the advantage to continue performing in non-connected areas. Moreover, the overhead of security added is about 40 % of the the overall cost, which is credible and highly beneficial for the platform to counter external vulnerabilities (Fig. 7).

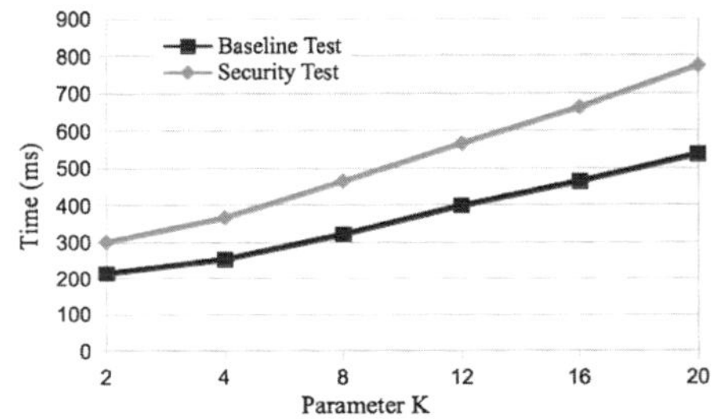

Fig. 5 Time cost of baseline test and security test face to the increase of parameter K

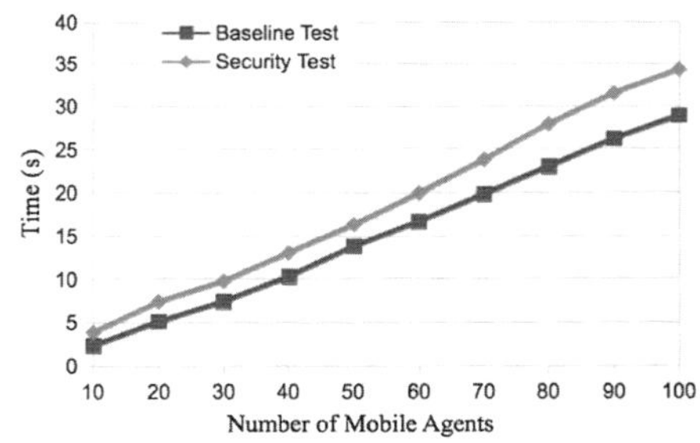

Fig. 6 Time cost of baseline test and security test face to the increase of interacting mobile agents

For comparative purposes, we are referred to the method of LI in [17], where the authentication and access control to the system resources are controlled by the mobile agent platform. This allows transparent and dynamic way to grant access rights among cooperating agents through access control capabilities, while the authentication mechanism is ensured using a Certification Authority (CA). Figure 8 presents the measured time costs for a round-trip of this method compared to the our, for different file sizes: 500 Kbytes, 1000 Kbytes, 1500 Kbytes and 2000 Kbytes respectively. It demonstrates the effectiveness of our solution with a security overhead of 40 % versus 67 % for the other one.

It was worthy at this stage to evaluate the capacity of our solution to detect maliciousness related mainly to unauthorized access attacks. Figure 7 compares the performance of both proposed tests to detect harmful mobile agents. When the Baseline Test was able to detect only 10 malicious agents among 40, which represents 25 % of the total malicious agents number, our scheme was able to detect the forty agents trying to get access without being authorized, which means 100 % of the total malicious agents number.

4.2 *Security Analysis*

In this part of evaluation, the resistance of the proposed solution against some well-known attacks is investigated.

Conspiracy Attack Consider 3 mobile agents belonging to 3 different remote platforms, can these platforms work together to collect correlated parameters of the native platform? If the 3 mobile agents claim access to the same file with different modes (Write, Read, Execute), can they collaborate within the native platform to reach a higher access right than they possess?

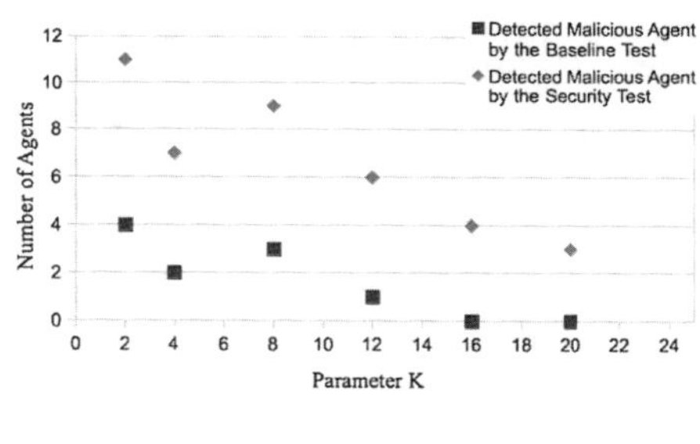

Fig. 7 Comparaison of the malicious agent detection performance of the baseline test and the security test

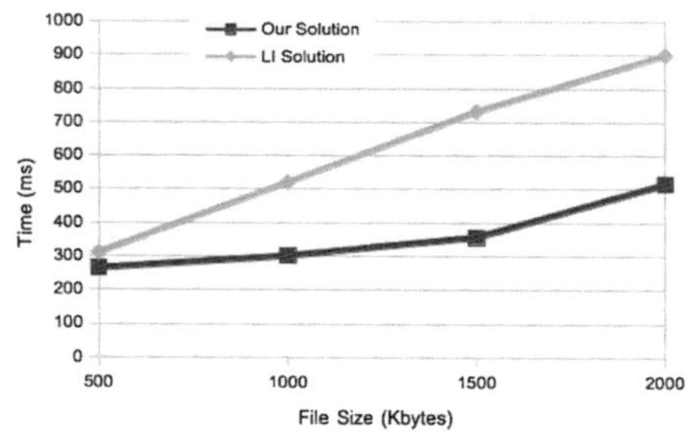

Fig. 8 Comparison of time performance of our solution versus to the solution in [17]

First of all, our contribution benefits from the security and ability of Shamir's Threshold scheme to prevent this attack. Therefore, our security policy is fine-grained according because the Admin_Agent, as the mediator among agents, performs all necessary computations without disclosing the private parameters of the environment. Besides, the received mobile agents are transporting hashes without knowing their real values, which means they need absolutely to resolve the barrier created by exponential operation, large prime number application and one way hash function. Thus, this attack will fail.

MITM Attack The authentication process adopted for our proposed solution proves its resistance against the Man-In-The-Middle attack. This is due to the fixation of the integrated DHKE-DSA using ephemeral keys, such as the computations of the two session keys $K_r n$ and $K_n r$ depend on the ephemeral keys r and n as well as T shared by the TTP via a secure channel. It provides two important properties. The first one is perfect forward secrecy, because in the long-term, if a private key of any party is exposed, previous session keys cannot be computed since the ephemeral keys of that session are unknown. The second one is key freshness, as every session key is a function of ephemeral keys, which means no party can predetermine the session key's value since he would not know the other party's ephemeral key.

5 Conclusion

In this paper, we deal with the security issues that a hosting platform may face when receiving mobile agents. Among the serious concerns in this context are the unauthorized access attacks, that could harm the entire system and be extended to external entities. We have associated Access control and cryptographic mechanisms in a new combination that shows its effectiveness to fill the majority of security needs: Authentication via DHKE-DSA, Confidentiality via AES Encryption by session key, Integrity by SHA3 hush function and access control with availability management via DAC and Shamir's threshold scheme. Our solution proves its scalability and reliability in detection of malicious mobile Agents. As perspective, we will consider other Access control models for highly administrative and hierarchical system, where interoperability and security are strongly requested.

References

1. Giovanni Caire, F.C.: JADE tutorial: jade programming for beginners. TILAB (2007)
2. Foundation for Intelligent Physical Agents, Geneva, Switzerland, FIPA ACL Message Structure Specification (2003)
3. Jansen, W., Karygiannis, T.: Mobile Agent Security, pp. 800–819. NIST Special Publication (1999)
4. Jung, Y., Kim, M., Masoumzadeh, A., Joshi, J.B.: A survey of security issue in multi-agent systems. Artificial Intelligence Review **37**(3), 239–260 (2012)
5. Borselius, N.: Mobile agent security. Electron. Commun. Eng. J. **14**(5), 211–218 (2002)
6. Pirzadeh, H., Dub, D., Hamou-Lhadj, A.: An extended proof-carrying code framework for security enforcement. In: Transactions on Computational Science XI, pp. 249–269. Springer, Berlin (2010)
7. Cao, C., Lu, J.: Path-history-based Access Control for Mobile Agents. International Journal of Parallel, Emergent and Distributed Systems **21**(3), 215–225 (2006)

8. Tsiligiridis, T.A.: Security for mobile agents: privileges and state appraisal mechanism. Neural Parallel Sci. Comput. **12**(2), 153–162 (2004)
9. Tuohimaa, S., Laine, M., Leppnen, V.: Dynamic rights in model-carrying code. In: Proceedings of the International Conference on Computer Systems and Technologies, pp. 1–7 (2006)
10. Arun, V., Shunmuganathan, K.L.: Secure sand-box for mobile computing host with shielded mobile agent. Indian J. Appl. Res. **3**(9), 296–297 (2013)
11. Aumasson, J.: On the pseudo-random generator ISAAC. IACR Cryptology ePrint Archive, 2006:438
12. Jaffar, A., Martinez, C.J.: Detail power analysis of the SHA-3 hashing algorithm candidates on xilinx spartan-3E. Int. J. Comput. Electr. Eng. **5**(4), 410–413 (2013)
13. Phan, R.W.: Fixing the integrated Diffie-Hellman-DSA key exchange protocol. IEEE Commun. Lett. **9**(6), 570–572 (2005)
14. Ennahbaoui, M., Elhajji, S.: Swot analysis of access control models. Int. J. Secur. Appl. **8**(3), 407–424 (2014)
15. Blakley, G.R., Kabatiansky, G.: Shamirs threshold scheme. In: Encyclopedia of Cryptography and Security, pp. 1193–1194. Springer, US (2011)
16. Announcing the Advanced Encryption Standard (AES). Federal Information Processing Standards Publication 197, NIST (2001)
17. Ismail, L.: A secure mobile agents platform. J. Commun. **3**(2), 1–12 (2008)

Data Filtering in Context-Aware Multi-agent System for Machine-to-Machine Communication

Pavle Skocir, Hrvoje Maracic, Mario Kusek and Gordan Jezic

Abstract Energy efficiency is an important aspect of Machine-to-Machine (M2M) systems since large number of devices is not connected to unlimited power supply. To tackle this problem, in our context-aware multi-agent system for M2M communication devices are in low-energy mode whenever they do not have any tasks to perform. Context information is exchanged between devices so that each device knows when other devices will be available for communication. On each device in the system, agents are deployed which exchange context information and adjust their wake-up times. In this paper we focus on decreasing energy consumption of M2M Devices which collect measurement data and forward it to back-end system. A filtering algorithm is developed to find repetitive data and to increase the interval for the next transmission. We assumed that when data values are similar to the previously observed values, it is not necessary to forward them to the gateway and back-end system so often. This approach was implemented on Libelium Waspmote devices and showed significant decrease in energy consumption during the period of 24 h.

Keywords Energy efficiency · Data filtering · Machine-to-machine system

P. Skocir (✉) · H. Maracic · M. Kusek · G. Jezic
Faculty of Electrical Engineering and Computing, University of Zagreb, Unska 3, HR-10000 Zagreb, Croatia
e-mail: pavle.skocir@fer.hr

H. Maracic
e-mail: hrvoje.maracic@fer.hr

M. Kusek
e-mail: mario.kusek@fer.hr

G. Jezic
e-mail: gordan.jezic@fer.hr

G. Jezic et al. (eds.), *Agent and Multi-Agent Systems: Technologies and Applications*, Smart Innovation, Systems and Technologies 38,
DOI 10.1007/978-3-319-19728-9_4

1 Introduction

Machine-to-Machine (M2M) communication is gaining more and more influence since it enables communication without direct human intervention between nowadays ubiquitous intelligent devices which have the ability to sense and interact with their environment [1]. M2M communication is considered as a key component for developing future smart buildings and cities. There are several research challenges in the area of M2M communication: resolving signal interference and packet collisions in both wireless and cellular networks, creating solutions for device mobility, autonomy and security, and achieving interoperability between different devices [2].

Achieving energy efficiency is also one of the challenges of M2M communication [3]. Research activities which deal with energy consumption in M2M systems mainly focus on energy harvesting, deriving energy from external sources, e.g. solar power or wind energy [4], or decreasing energy consumption [5]. The challenge of decreasing energy consumption is in focus since many systems typically use at least some battery-powered devices which would drain their batteries soon if their consumption is not taken into account. Battery-powered devices are widely used also for smart home services, where it is more convenient to implement new sensors with their own power-supply than to connect them to the electrical system [6]. To enable operation of their services, HGI[1] operators expect certain lifetime of various types of devices: 2 years for home electrical appliance remote controller and sensor devices and 10 years for metering devices [7]. To measure up to those standards, enablers of M2M services must manage the energy consumption of those devices and propose techniques to prolong their lifetime.

This work focuses on enhancing energy efficiency in context-aware multi-agent system presented in [8] where Rich Presence Information (RPI) was introduced to enable sharing data between different devices which contain information about when a certain device will be available for communication. By implementing this approach, devices are planning their awake periods based on the tasks they have to perform. Afterwards, they are sending information about these periods to the devices they communicate with. The main benefit of sharing presence information is in saving energy by not sending messages to other devices which are unable to receive them since they are not awake. In this paper we propose a data filtering approach which enhances energy efficiency in the context-aware system with RPI by comparing measurements executed by the devices and, according to current value and historical ones, adjust the operation schedule.

The rest of the paper is organized as follows. Section 2 introduces current achievements regarding data filtering in M2M systems. In Sect. 3 we present the context-aware mobile agent network model, while Sect. 4 describes the process of improving energy efficiency in our system by introducing data filtering. Section 5 presents the results of energy consumption comparison between the RPI system and RPI system with data filtering. Section 6 concludes the paper.

---

[1]HGI - Home Gateway Initiative.

2 Related Work

Data filtering is a wide range of strategies or solutions for refining data sets into what a user needs, without including data that can be repetetive or irrelevant [9]. Since in M2M systems data transmission usually consumes more energy than data processing, filtering on motes has been identified as a promissing solution to enhance energy efficiency [10].

There are various approaches to data filtering in M2M systems. Zoeller et al. [10] propose an on-mote filtering approach which is performed on M2M devices, before sending information to the gateway. The authors introduced a concept for local data filtering based on forecasting, i.e. estimating the future evolution of measured sensor data, and a method for determining the value of information in sensor measurements. The focus is on user perspective for determining the value of information, they try to reduce the number of data transmissions while still trying to provide a sufficient data fidelity for users' needs.

Papageorgiou et al. [11] propose data filtering on gateways before collecting large data sets at back-end M2M systems. They use different filtering methods - general purpose filtering (finding erroneous data and outliers), quality of information assessment and data classification. The choice of the aforementioned methods depends on the information available about application specifications. Step one of the proposed system analyses application requirements in order to determine which filters to use. In step two data is being collected, filtered and sent to the back-end system.

Mendonca Costa et al. [12] support the belief that data filtering is an important aspect in context-aware M2M system. The authors use data filter to reduce the amount of data sent by the end devices in M2M network. They use three types of thresholds to trigger data measurements and transmissions: hard threshold, soft threshold and reporting threshold. Hard threshold is predefined, and when data source value reaches this threshold, the devices switch on the radio module and transmit the sensed data to the server. Soft threshold represents small variations in the data source. Whenever the data source changes by an amount greater or less than the defined threshold from the last measurement, the device transmits that data. Reporting threshold is the maximum interval arrival time of a device. If the data source does not meet neither hard nor soft threshold and the interval since the last transmission exceeds the reporting threshold, the device transmits the last sensed value. Additionally, the system has a time threshold which defines time after which the device demands a new context modification request.

Mathew et al. [13] are trying to co-optimize information quality and energy efficiency. They state that increasing sensor sampling rate improves information quality, but costs energy consumption due to more data being transmitted. The authors introduce a quality/energy efficient metric and present a generic quality and energy adapting system (QEAS) which adapts to network changes and improves energy efficiency. In their model, each sensor is capable to detect noise and change sampling rate, while QEAS is implemented at the base station.

Data filtering can be used in M2M system for purposes other than achieving energy efficiency. Some of these purposes mentioned in [14] are failure detection, processing overhead reduction and predicting values (e.g. energy consumption data, temperature, humidity, pressure and airflow data). The authors proposed an algorithm which deals with the problem of computational overhead required for prediction when using Kelman filtering scheme. Their solution is data filtering algorithm based on statistical data analysis.

In this paper we focus on filtering which takes place on motes and reduces both the time in which the device is operational and the number of data transmissions. Measurements are trigered by a threshold similar to a soft threshold from Mendonca Costa et al. [12] The model on which we base our approach is described in the following sections.

3 Context-Aware Mobile Agent Network Model

Our model of M2M system is based on ETSI architecture and consists of two main domains: network domain, and device and gateway domain [15]. We focus on device and gateway domain which is modeled as a multi-agent system placed in a network of nodes and is shown in Fig. 1. Devices and gateway are represented by nodes that host agents which execute services and communicate between each other [8].

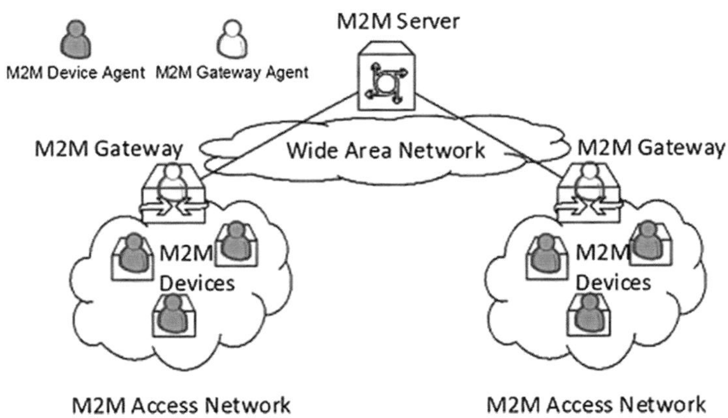

Fig. 1 M2M device and gateway domain

The model is formalized by using a quadruple $\{A, S, N, C\}$ where A represents a set of agents co-operating and communicating in the environment defined by S and N. S denotes a set of processing nodes in which agents perform dedicated services, and N is a network defined as an undirected graph that connects processing nodes and allows agent communication and mobility. C represents a set of context data handled by the agents. This context data is derived from Rich Presence Information (RPI) for machines which contains information about device's location, load, battery level, charging type and availability.

There are two types of agent A: M2M Device Agent which is defined as *presentity* and M2M Gateway Agent which acts as a mediator between end devices and applications is defined as *watcher*. Context information C of each M2M entity is defined as $C = \{ctx_1, ctx_2, \ldots, ctx_j, \ldots, ctx_{nctx}\}$ where ctx_j represents one RPI element. Context information ctx_j is defined as $ctx_j = \{c_{state}, c_{ta}, c_{dc}, c_w, c_p\}$ where c_{state} is the state of the entity that can have four values: *on, off, sleep, hibernate*. c_{ta} is the absolute time when entity will change state, c_{dc} is the time period of an entity in the current state, c_w is a set of *watchers* that the entity will inform about the change in state, while c_p is a set of *presentities* that the *watcher* receives information from.

The functionality of A is defined by a set of elementary services $ES = \{es_1, es_2, \ldots, es_j, \ldots, es_n\}$. Each of these services can be provided by a single as well as by multiple competing or collaborating agents. A set of services that we have defined are: es_1 executing task (e.g. temperature measuring), es_2 sending measurement, es_3 sending context information, es_4 collecting measurement, es_5 collecting context information, es_6 creating measurement report and es_7 sending measurement report to server. Services es_1, es_2 and es_3 are supported in *presentity* nodes, while activities es_4, es_5 and es_6 are supported in *watcher* nodes.

In common operation mode M2M Devices are in low-energy state (hibernate, sleep or deep sleep) or in on state when they carry out their tasks. In one operating cycle, the nodes wake up from hibernate state, carry out their measurement, exchange RPI with gateway and send their measurement and go back to low-energy state. In our model, we used hibernate for low-energy state since it consumes smallest amounts of energy. In previously conducted simulations the M2M Devices had the same operating cycle duration (time between two wake-ups), but different times at which they started their operations. By exchanging RPI and using synchronization mechanism, the M2M Devices adjusted ther operating cycles and grouped together so that M2M Gateway could receive their data and go to hibernate mode instead of always being on and waiting for messages. The simulation showed that reduction of energy consumption by using this approach during 24 h was of about 40 % [8].

In this work we focus on enhancing energy efficiency on M2M Devices. Previously, measurement data was not taken into account and operating cycle was always the same. In the next section it is described how energy efficiency of our system was enhanced by monitoring measurement data and adjusting operating cycle duration.

4 Data Filtering

As mentioned in Sect. 2, data filtering can be used in M2M systems to reduce the number of transmissions of repetitive data and to remove faulty data. We focus on finding repetitive data in sensor measurements on M2M Devices. M2M Device Agent is deployed on each M2M Device and monitors measured values. If it is

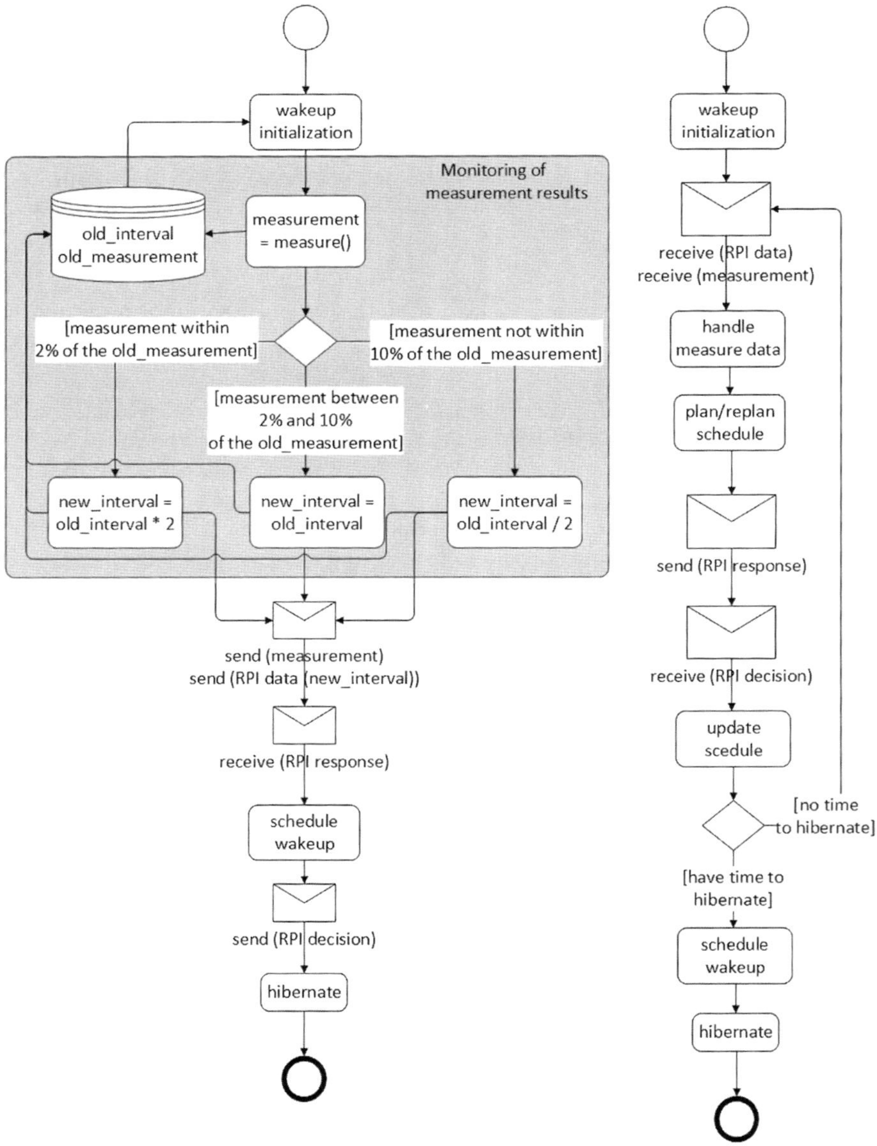

Fig. 2 M2M device and gateway agent activity diagram

discovered that consecutive measurements are same or similar, the interval (operating cycle) in which the M2M Device should wake up next increases. By implementing this approach, the M2M Device will consume less energy since the time it spends in hibernate mode increases. Furthermore, the M2M Device also communicates with M2M Gateway less frequently. On the other hand, if M2M Device

Agent discovers that measured values deviate from the older ones, the operating cycle decreases so that M2M Device Agent could obtain more measurements about the uncommon event.

Figure 2 shows activity diagram for M2M Device Agent on left-hand side and M2M Gateway Agent on right-hand side. Monitoring of measured values is executed just after wake-up, before exchanging RPI messages with M2M Gateway. Since we compare current and former measured values, the use of M2M Device's data store is obligatory in this scenario.

At the beginning of the operation, after wake-up from hibernate mode and performing initialization, the agent obtains measurement data from the sensor. After reading the measured values, it compares the old measured value saved in the data store with the current one. If the current measured value is within 2 % of the former one, the duration of the next operating cycle will be doubled. If the current measured value is between 2 % and 10 % of the former one, the operating cycle remains the same. Finally, if the current measured value is more than 10 % different than the former one, the duration of the operating cycle will be decreased by half.

After deciding on the duration of the next operating cycle, M2M Device informs M2M Gateway about that decision by sending RPI message. Then it waits to receive RPI response from M2M Gateway which contains information about the times when it has next wake-ups scheduled. Afterwards, M2M Device adjusts its wake-up time so that it fits to the M2M Gateway's schedule better, e.g. it should schedule the next wake-up closer to the already existing one so that M2M Gateway could serve this M2M Device and the other within the same operating cycle. Finally, the M2M Device sends that decision to M2M Gateway, schedules the next wake-up and goes to hibernate mode. After wake-up, M2M Gateway receives the initial schedule from a certain M2M Device along with the measurement. It plans the schedule and sends it to the M2M Device which decides whether it can adjust its own schedule to fit the M2M Gateway's schedule better. After receiving M2M Device's decision, M2M Gateway updates its schedule and goes to hibernate mode if it has no other tasks to perform in this operating cycle, or waits for messages from some other M2M Device.

5 Simulation of Energy Consumption

The purpose of the experiment was to determine the influence of on-mote data filtering on energy efficiency in context-aware M2M system. It is shown how data filtering approach described in Sect. 4 affects energy consumption of only one end M2M Device and of the entire system. To calculate energy consumption, the following equation was used:

$$E = P_{WM} \Delta t \tag{1}$$

where P_{WM} is the power used by Waspmote device, a sensor node implementation on which we evaluate our system, and Δt is the length of time during which we observe the consumption.

The power used by Waspmote device varies during operation. Each task from Fig. 2 uses different amount of power and lasts for a certain amount of time. In order to calculate the total energy consumption during a certain period of time, power consumption for each of the tasks and their durations were measured and analysed. The result is shown in Table 1.

Table 1 Power and time for each task during awake time in one operating cycle

State/task	Wake-up initialization	Measure	Process	Send +receive	Handle schedule	Send
Power consumption (mW)	70,13	141,12		260	141,12	258,95
Duration (ms)	80	50,56		464,1	1	255,3

To calculate total energy consumed during wake-up time in one operating cycle, energies consumed for each task need to be summed-up. When everything is multiplied and summed-up, it emerges that the total energy consumption for awake period of every operating cycle is 0.055 mWh. When using data filtering and when not applying that approach, the energy consumed during whole operating cycle is very similar because the energy consumed while staying in hibernate mode is negligible, since the device uses only 0.461 μW. However, the number of times when M2M Device wakes up is quite different. In the initial scenario setup of the measurement operating cycle is every 10 s in the case when data filtering is not used. In the case when data filtering is used, the initial operating cycle duration is also 10 s, but than it changes during operation. Measurement data which was used for that scenario comes from a temperature sensor situated in a room during the time of 24 h. Measurements were almost always within 2 % of the previous ones, so according to the algorithm, the intervals were being prolonged until exceeding the maximum interval limit which was set to 720 s. The intervals were being prolonged from 10 to 20, and then 40, 80, 160, 320 and 640 s. That was 7 wake-ups, and until the end of the day the intervals were almost always in the duration of 640 s, so it amounts to 133 wake-ups. In total, the M2M Device would wake up 140 times during one day. However, in 5 measurements the change in temperature was more than 2 %, so in the end it amounted to 145 wake-ups during the day. In the case when data filtering was not used, the M2M Device woke up every 10 s - totally 8640 times.

In the case when M2M Device uses data filtering approach and wakes up 145 times during 24 h, it consumes 8.05 mWh. On the other hand, when it does not use data filtering approach, and wakes up 8640 times, it consumes 479.20 mWh. It can be observed that the energy consumption is around 60 times lower when using data filtering approach. This is the case when data from temperature sensor located in an indoor space was used. The set up for the filtering algorithm should be updated according to the data values which are analysed and according to the requirements of the application which consumes the data.

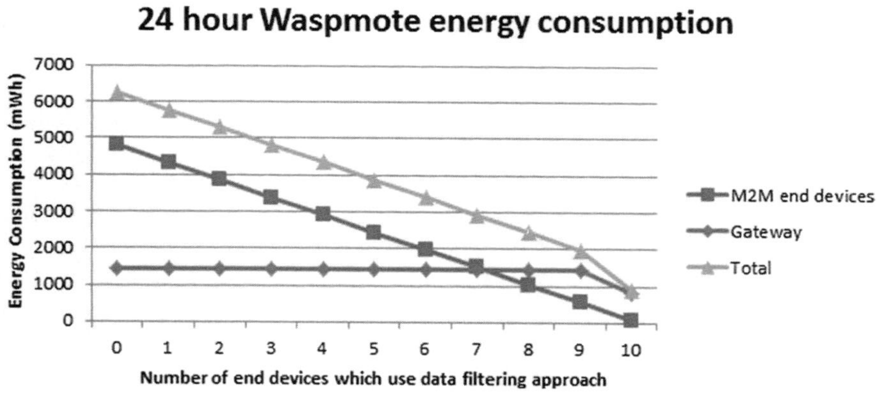

Fig. 3 Energy consumption for M2M system during 24 h

Total energy consumption in a system with ten M2M Devices and one M2M Gateway is shown in Fig. 3. If none of the M2M Devices uses filtering approach, the total energy consumption is around 4800 mWh. When number of M2M Devices which support filtering approach increases, and number of M2M Devices without filtering approach decreases, total energy consumption of M2M Devices decreases. When all M2M Devices use filtering approach, total consumption amounts to around 80 mWh. M2M Gateway which exchanges context-information in the form of RPI for machines uses the same amount of energy almost all the time because it has to stay available for M2M Devices which do not support filtering approach and therefore wake up every 10 s. Only in the case when all M2M Devices use data filtering approach, the consumption of the M2M Gateway decreases because in that case the M2M Gateway Agent can also adjust its operating cycle time to last longer than 10 s.

6 Conclusion

Energy efficiency is an important aspect that has to be taken into account when designing M2M systems since large number of devices are powered by batteries and do not have unlimited power supply. Different research activities are in motion in order to deal with this problem of energy scarcity in M2M systems. We proposed a data filtering approach which deals with repetitive data by increasing the interval in which data is forwarded from M2M Devices to M2M Gateway when results received from sensors are similar. By using this method, M2M Devices send less data, and can be in low-energy modes for a longer duration of time. When current measured value deviates from the old one, the transmission interval decreases so that M2M server receives more measurements in certain amount of time about the unusual event. The filtering approach is combined with the exchange of Rich Presence Information (RPI) which enables exchanging times when devices will be awake and available for communication. By using RPI exchange, the devices can synchronize easier to be awake when they are needed by other devices and they can stay longer in low-energy mode which improves energy efficiency of the entire system.

By using data filtering approach with RPI exchange, we showed that Libelium Waspmote devices used 60 times less energy during 24 h compared to the approach when only RPI exchange was used. Data filtering considerably improves energy efficiency of M2M Devices and, consequently, of the entire system, since they account for 90 % of all devices in the system where measurements were performed. When all M2M Devices in the system used data filtering approach, gateway consumption also decreased because operating cycles of M2M Devices increased, and M2M Gateway was also able to lengthen its own operating cycle.

When observing consumption of the entire system, it can be noticed that M2M Gateway is the device which has higher consumption than any M2M Device. Therefore, it would be useful to switch gateway capabilities between devices in the case when no device has unlimited power supply. For future work, we will focus on self-organization mechanisms that determine which device should assume liability for the tasks of a gateway based on the tasks which have to be performed in the system and battery state of all devices.

Acknowledgments This work was supported by strategic project "Energy Efficient M2M Device Communication" in cooperation with FTW - Forschungszentrum Telekommunikation Wien GmbH, Austria and by Croatian Science Foundation under the project 8065.

References

1. Song, J., Kunz, A., Schmidt, M., Szczytowski, P.: Connecting and managing m2m devices in the future internet. Mob. Netw. Appl. **19**(1), 4–17 (2014)
2. Pereira, C., Aguiar, A.: Towards efficient mobile m2m communications: survey and open challenges. Sensors **14**(10), 19582–19608 (2014)
3. Lu, R., Li, X., Liang, X., Shen, X., Lin, X.: Grs: the green, reliability, and security of emerging machine to machine communications. IEEE Commun. Mag. **49**(4), 28–35 (2011)
4. Chen, Y.K.: Challenges and opportunities of internet of things. In: Design Automation Conference (ASP-DAC), 2012 17th Asia and South Pacific, pp. 383–388 (2012)
5. Samsonov, S., Mueckenheim, J.: Power saving algorithm for static machine to machine smart meters. In: 2014 IEEE 34th International Conference on Electronics and Nanotechnology (ELNANO), pp. 474–477 (2014)
6. Theoleyr, F., Pang, A.C.: Internet of Things and M2M Communications. River Publishers (2013)
7. Home Gateway Initiative: Requirements For Wireless Home Area Networks (WHANs) Supporting Smart Home Services. Report, HGI-RD039 (2014)
8. Maracic, H., Miskovic, T., Kusek, M., Lovrek, I.: Context-aware multi-agent system in machine-to-machine communication. Procedia Comput. Sci. **35**(0), 241–250 (2014)
9. Oracle: Domains, VPDs and Vendor Access Solution. Report (2014)
10. Zoller, S., Vollmer, C., Wachtel, M., Steinmetz, R., Reinhardt, A.: Data filtering for wireless sensor networks using forecasting and value of information. In: 2013 IEEE 38th Conference on Local Computer Networks (LCN), pp. 441–449 (2013)

11. Papageorgiou, A., Schmidt, M., Song, J., Kami, N.: Smart m2m data filtering using domain-specific thresholds in domain-agnostic platforms. In: 2013 IEEE International Congress on Big Data (BigData Congress), pp. 286–293 (2013)
12. Costa, J., Miao, G.: Context-aware machine-to-machine communications. In: 2014 IEEE Conference on Computer Communications Workshops (INFOCOM WKSHPS), pp. 730–735 (2014)
13. Mathew, M., Weng, N.: Quality of information and energy efficiency optimization for sensor networks via adaptive sensing and transmitting. IEEE Sens. J. **14**(2), 341–348 (2014)
14. Hong, S.T., Oh, B.S., Chang, J.W.: A sampling-based data filtering scheme for reducing energy consumption in wireless sensor networks. In: 2011 IEEE Asia-Pacific on Services Computing Conference (APSCC), pp. 353–359 (2011)
15. Katusic, D., Skocir, P., Bojic, I., Kusek, M., Jezic, G., Desic, S., Huljenic, D.: Universal identification scheme in machine-to-machine systems. In: 2013 12th International Conference on Telecommunications (ConTEL), pp. 71–78 (2013)

Part II
Agent-Based Algorithms, Simulations and Organizations

Combining Process Simulation and Agent Organizational Structure Evaluation in Order to Analyze Disaster Response Plans

Nguyen Tuan Thanh Le, Chihab Hanachi, Serge Stinckwich and Tuong Vinh Ho

Abstract This paper shows how to simulate and evaluate disaster response plans and in particular the process and the organization set up in such situations. We consider, as a case study, the tsunami resolution plan of Ho Chi Minh City, Vietnam. We firstly examine the process model corresponding to this plan by defining three scenarios and analyzing simulations built on top of them. Then, we study the agent organizational structure involved in the plan by analyzing the role graph of actors and notably the power, coordination and control relations among them according to the Grossi framework. These evaluations provide recommendations to improve the response plan.

Keywords Agent organization evaluation · Crisis management · Process simulation · Role graph · Decision support system

1 Introduction

In crisis situations (tsunami or earthquake), coordination among the implied stakeholders (rescue teams and authorities) is of paramount importance to ease the efficient management and resolution of crises. Coordination may be supported by

N.T.T. Le (✉) · C. Hanachi
IRIT Laboratory (SMAC Team), Toulouse 1 University, Toulouse, France
e-mail: nguyen.le@irit.fr

C. Hanachi
e-mail: hanachi@univ-tlse1.fr

S. Stinckwich
UCBN & UMI UMMISCO 209 (IRD/UPMC), Toulouse, France
e-mail: serge.stinckwich@ird.fr

T.V. Ho
Institut Francophone International, Vietnam National University (VNU) & UMI UMMISCO 209 (IRD/UPMC), Hanoi, Vietnam
e-mail: ho.tuong.vinh@ifi.edu.vn

G. Jezic et al. (eds.), *Agent and Multi-Agent Systems: Technologies and Applications*, Smart Innovation, Systems and Technologies 38,
DOI 10.1007/978-3-319-19728-9_5

different related means such as plans, processes, organizational structures, shared artifacts (geographical maps), etc. [3].

Most often, coordination recommendations to manage crisis are available in a textual format defining the actors, their roles and their required interactions in the different steps of crisis life-cycle: mitigation, preparedness, response and recovery.

While coordination recommendations, in a textual format, are easy to manipulate by stakeholders, taken individually, they do not provide direct means to be analyzed, simulated, adapted, improved and may have various different interpretations, so difficult to manage in real time and in a distributed setting.

In [4], we propose an approach to transform a textual coordination plan into a formal process in order to have an accurate representation of the coordination, to reduce ambiguity and ease an efficient preparedness and resolution of tsunami at Ho Chi Minh City.

Formalizing coordination and producing models are a first step toward a better understanding and mastering of coordination. Then, it is also important to evaluate coordination models in order to provide recommendations to authority to help them improving coordination within resolution plans. Most of the time, authorities make real-world exercises to validate their plans but do not formally validate them. Unfortunately real-world exercises are not always possible (cost, impossibility to reproduce reality, etc.). Therefore simulation and formal validation become unavoidable.

Given these observations, it becomes useful to make formal evaluation of coordination models used during crisis situations. This is the approach followed in this work (see lifecycle of Fig. 1). Notably, our contribution consists in the definition of a framework to evaluate both the underlined process and the agent (actor) organization set up in a resolution plan. The two evaluation dimensions, process and organization, are complementary since the first one abstracts the coordinated behavior of the actors while the second abstracts the relationships (control, coordination, power …) between actors. Both are to be evaluated since they influence the efficiency, the robustness and the flexibility of the disaster response plans. Even if our

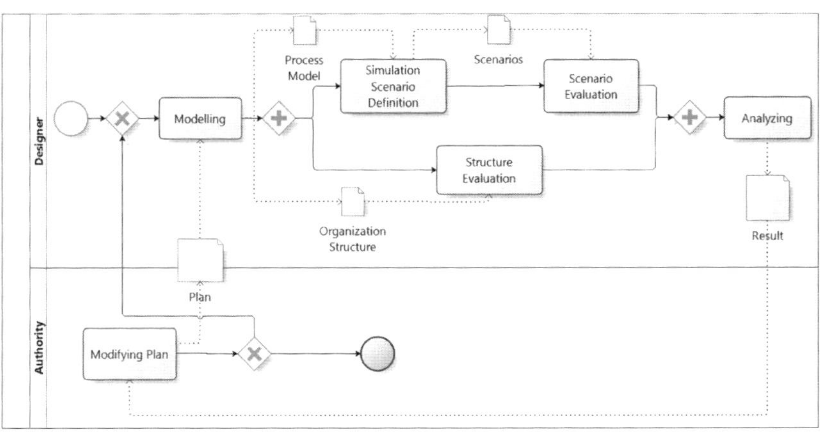

Fig. 1 Evaluation lifecycle of disaster response plans

work considers a concrete case study (i.e. the Ho Chi Minh City tsunami response plan), our approach is general enough to be applied to any crisis management plan.

The paper is organized as follows. We first recall the formal process model that we have proposed in [4] corresponding to the Ho Chi Minh City rescue plan. Related works about business process simulation and organizational structure evaluation are presented in Sect. 3. We then define three scenarios and analyze simulations built on top of them. Afterward, we evaluate the agent organizational structure involved in this plan by analyzing the role graph of the actors and notably the power, coordination and control relations among them according to the Grossi framework [8]. These evaluations provide recommendations to improve the response plan. Finally, we discuss the results and conclude our work.

2 Background

Response plans used during crisis situations involve the interactions of many actors and tasks organized in a flowchart of activities with interleaving decision points, that can be roughly be seen as a specific business process. We would like to apply business process techniques in crisis management. Therefore in [4], we have presented a process-based model to analyze coordination activities extracted from tsunami response plan proposed by People's Committee of Ho Chi Minh City. This conceptual model (Fig. 2), described with a Business Process Model and Notation (BPMN)

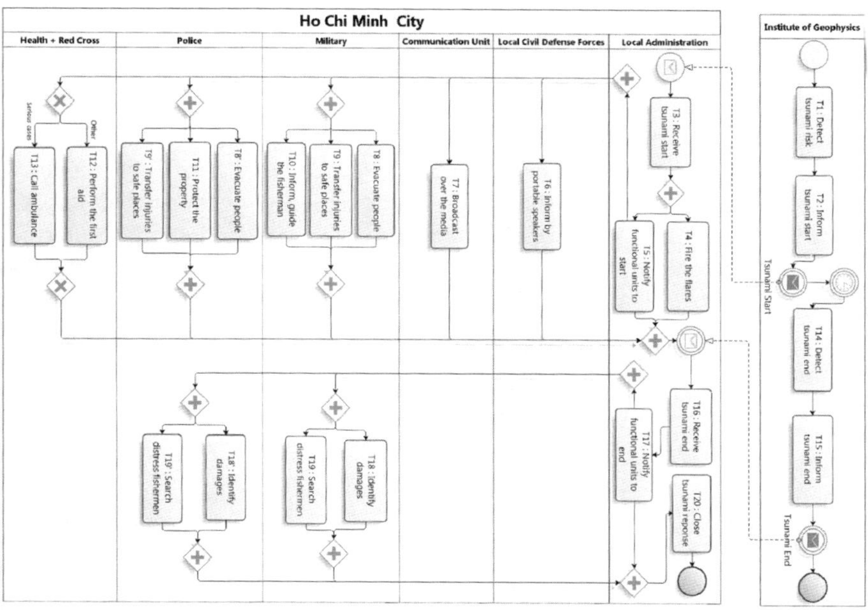

Fig. 2 Conceptual model of tsunami response plan proposed by Ho Chi Minh City

diagram, has been built by analyzing an official textual plan provided by the suitable authorities.

We can identify in the model above seven organizations (represented by lanes) involved with their flow of tasks and mutual interactions. In BPMN, a task (like *T1: Detect tsunami risk*) is represented by a rounded-corner rectangle. Several control structures are possible to coordinate the different tasks: sequence (arrow), parallelism (diamond including "+") or alternatives (diamond with "X"). We can notice, in Fig. 2, that *Military* and *Police* organizations are supposed to perform tasks in parallel. In this case, each organization members should be distributed over the parallel tasks according to a given policy (proportional distribution, distribution according to the importance given to each task). The *Health & Red Cross* organization has to choose to carry out only one task among two possible ones.

This model has been transformed and executed within a workflow system, namely Yet Another Workflow Language[1] (YAWL), to demonstrate the feasibility of managing the plan in a distributed setting. However, this transformation not only dropped lots of details of our conceptual model, but also did not provide process simulation functions, notably what-if simulation and performance analysis, useful for decision makers in charge of defining and updating plans. We will provide later in Sect. 4 a more elaborate model by having more realistic scenarios and organizational structure evaluations, that will allow more complex analysis of rescue plans.

3 Related Works

This section will situate our contribution according to three complementary points of view: coordination in Multi-Agent Systems (MAS), simulation of discrete event systems and organizational perspective.

The problem of coordinating the behaviour of MASs has been regularly addressed [2]. A coordination model is useful in crisis context since it helps in supporting interdependence between stakeholders, the achievement of common goals (e.g. saving victims), and the sharing of resources (vehicles, food, houses for victims, ...) and competencies (medical, carriers, ...). A coordination model can exploit and/or combine different techniques: (1) organizational structuring (2) contracting (3) negotiating (4) planning (5) shared artefacts. We follow in this paper a process-oriented technique which can be considered as a combination of plans within an organizational structure. The advantage of process-oriented coordination is to provide visibility on the whole crisis evolution: past, present and future activities and their relationships. [1] proposes a very detailed review of process management systems supporting disaster response scenarios. However, one main drawback of these systems is to support the real time managing of the crisis, while we consider the whole life cycle of the process and in this paper the simulation and validation steps.

[1]http://www.yawlfoundation.org/.

From a simulation point of view, a computer-based simulation of processes can be done following a *discrete-event simulation*, where a crisis evolution can be represented as a sequence of events. This approach has been applied successfully in workflow and business processes [5]. Process simulation helps to identify the bottlenecks in the flow of tasks and then optimize them with alternative ones or find out the better resource management solution. Rozinat et al. in [6] proposed an approach by analyzing the event logs (in structured format), then extracting automatically the useful information about: (1) control flow, (2) decision point, (3) performance, and (4) roles. Using these information, the authors constructed a four-facets simulation model and simulated it with a Petri nets tool, namely CPN.[2] Unlike [6], our model is created from an *unstructured textual guideline* so we cannot use an event miner such as ProM[3] tool to extract automatically the useful information. In our case, we have observed manually the necessary information by studying the textual plan, extracting the actors, their activities, and finally designed a corresponding conceptual model due to our comprehension.

In [7], the authors combined three types of information to generate a more accurate simulation model: (1) design information used to form model structure, (2) historic information (event logs) used to set model parameters (such as arrival rate, processing time) and (3) state information used to initialize the model. In our work, we have only used the design information to create our simulation model. We then added the necessary parameters like resource quantity, time constraints extracted from the official textual plan to the model.

From an organizational point view, Grossi et al. proposed in [8] a framework to evaluate the organizational structure based on a role graph with three dimensions: power, coordination and control. They introduced the concepts and the equations involved into the evaluation. Using these equations, we compared our results with the standard values proposed by Grossi in order to assess the robustness, the flexibility and the efficiency of our organization.

The novelty of our work is to evaluate resolution plans through a formal representation and to consider both process and organizational aspects at the same time and in a coherent framework.

4 Rescue Plans Assessment by Process Simulation

In this section, we will describe how to evaluate a rescue plan by using business process simulations. In order to perform these simulations, a conceptual model (such as Fig. 2) is not sufficient. Therefore, we need to add extra information (quantity of resources, time constraints) that will allow us to define more accurate scenarios.

[2]http://cpntools.org/.

[3]http://www.processmining.org/prom/start.

4.1 Definition of Simulation Parameters

Related to business process essence, we consider four input parameters as follows [9]: (1) the *Arrival process* expresses the arrival rate of new cases (i.e., process instances); (2) the *Probabilities for choices* indicates the probability of selecting one task to perform among several alternative tasks at a time; (3) the *Service time* expresses the required time for a task to complete its work; and (4) the *Number of human resources* specifies the kind of mobilized organizations and their quantity, as well as the allocated resources of tasks.

These four parameters are insufficient in our context. Indeed, BPMN simulation lacks some notions such as the actors' capacities and the priorities or the important factors of tasks. Hence, we have defined the notion of *importance factor* of a task T as an evaluation number of the importance of this task regarding its capacity in term of rescues or good salvage. The more this factor is high, the more its task can save persons or goods. Hence, we must pay attention to it since it influences the crisis resolution performance. This factor will be used in our context for allocating suitably the resources to parallel tasks, even if its use could be generalized to all types of tasks. As we will demonstrate it, taking into account this new notion will improve the overall performance of our process model.

To tune the *arrival process* and *service time* parameters, we could apply different kind of distributions such as Poisson distribution, Duration distribution, Normal distribution, Triangular distribution, etc.

Different from a typical business process as flight ticket booking, whose arrival rate is frequent (time distance between two customers' request is small), in crisis and disaster context, we do not meet the full queue or resource conflict problem. In our simulation, we have set the *arrival process* parameter to one, because we consider only one tsunami situation at a time.

We have set the *probabilities for choices* (in number between 0 and 1) of alternative tasks and the *importance factors* (in percent) for parallel tasks as shown in Table 1. We allocate resources to tasks in the order of their importance: important tasks are first served with the maximum resources according to their needs.

Table 1 Probabilities for choices (PC) of alternative tasks and importance factors (IF) of parallel tasks

Tasks	PC	Tasks	IF
T12/T13	0.8/0.2	**T4/T5**	40/60
		T18/T19	30/70
		T18'/T19'	70/30
		T8/T9/T10	70/20/10
		T8'/T9'/T11	10/10/80

We have also applied a *Duration distribution* for all tasks' *service time*, as shown in Table 2. We assumed that the time span of a tsunami is three hours.

Table 2 Service time (ST) of all tasks in tsunami response plan

Task	ST	Task	ST	Task	ST	Task	ST	Task	ST	Task	ST	Task	ST	Task	ST
T1	10 m	**T2**	15 m	**T3**	10 m	**T4**	30 m	**T5**	30 m	**T6**	1 h	**T7**	30 m	**T8**	3 h
T8'	3 h	**T9**	3 h	**T9'**	3 h	**T10**	3 h	**T11**	3 h	**T12**	3 h	**T13**	30 m	**T14**	10 m
T15	15 m	**T16**	10 m	**T17**	30 m	**T18**	1 h	**T18'**	1 h	**T19**	1 h	**T19'**	1 h	**T20**	30 m

Furthermore, we have modeled seven roles (or actors) with their corresponding acronym: Institute of Geophysics (abbr. IG), Local Administration (LA), Military (M), Police (P), Local Civil Defense Forces (LCDF), Communication Unit (CU), and Health & Red Cross (HR). The total *number of human resources* for each role is shown in Table 3. For the clarity purpose, we did not take into account other mobilized non-human materials such as the transport means (e.g., ambulances, fire trucks, canoes, etc.), or the machines (e.g., sprayer epidemic prevention machine, GPS machine, etc.).

Table 3 Human resources mobilized in our tsunami response plan

Resource	Quantity	Resource	Quantity
Institute of geophysics	5	Military	6836
Local administration	160	Communication unit	170
Local civil defense forces	6700	Police	3700
Health and red cross	2600		

Our expected outputs of the process simulation are two-fold: (a) the *Time use* representing the total time consumed by our tsunami response process, as well as the average time, the average waiting time, the minimum or maximum time for each task; and (b) the *Resource use* depicting the distribution of resources occupied by each actor.

Practically, we use Bizagi tool[4] to model and simulate our case study.

4.2 Definition of Scenarios

Following [10], we could define a scenario by four components: the purpose, the content, the form and the cycle. Regarding the purpose, crisis management simulation aims at answering the two following questions: (a) how could we allocate efficiently the human resources to tasks? and (b) what is the best resources allocation strategy? The content and the form of our scenarios are defined by the tasks' services time (in minutes), the number of mobilized actors (in positive integer values) and the probabilities for alternative tasks (in number) as well as the importance factor (in percent).

To demonstrate the efficiency of the *importance factor* notion, we have fixed the *arrival process*, the *probabilities for choices*, and the *service time* parameters. In a nutshell, we have shifted only the number of human resources allocated to tasks leading to the three scenarios:

[4]http://help.bizagi.com/processmodeler/en/index.html?simulation_levels.htm.

- *Scenario 1*: We name it *full-resource scenario*. For each task, we allocate to it the maximum number of human resources dedicated to it without considering any other aspects.
- *Scenario 2*: We call it *importance-focus scenario*. It is based on a percentage distribution of human resources allocated to each parallel and alternative task. These percentages are stated by the designer according to the importance factors or the probabilities for choices which he/she gives to each parallel or alternative task, respectively. We allocate a maximum value of human resources to all the other tasks.
- *Scenario 3*: It could be also called *all-equal scenario*. For parallel and alternative tasks, the same number of human resources is allocated without regarding to the probabilities for choices or the importance factors of tasks. The others tasks are allocated a maximum value.

The number of human resources allocated to each task for the three previous scenarios are shown in Table 4.

Table 4 Number of human resources allocated to tasks in the three scenarios

Task	Scen. 1	Scen. 2	Scen. 3	Task	Scen. 1	Scen. 2	Scen. 3
T1	5	5	5	**T14**	5	5	5
T2	5	5	5	**T15**	5	5	5
T3	160	160	160	**T16**	160	160	160
T4	160	64	80	**T17**	160	160	160
T5	160	96	80	**T20**	160	160	160
T6	6700	6700	6700	**T7**	170	170	170
T8	6836	4785	2278	**T8’**	3700	370	1233
T9	6836	1367	2278	**T9’**	3700	370	1233
T10	6836	683	2278	**T11**	3700	2960	1233
T18	6836	2050	3418	**T18’**	3700	2590	1850
T19	6836	4785	3418	**T19’**	3700	1110	1850
T12	2600	2080	1300	**T13**	2600	520	1300

4.3 Simulation & Analysis of Three Scenarios

We compare the different scenarios through the utilization rate of the resources. Figure 3 depicts the *resource utilization* (in percent) of each actor after the what-if simulation. As we see, scenario 1 (*full-resource scenario*) spends more human resources than others for parallel tasks performed by *Military* or *Police*. Otherwise for the actors having only ordered tasks, scenario 1 consumes the less human resources. Furthermore, except for the actor *Health & Red Cross* (in which we have an exclusive choice between two tasks: *T12* and *T13*), we observe that the resource utilization of scenario 2 (*importance-focus scenario*) and scenario 3 (*all-equal scenario*) are identical. For *Health & Red Cross* actor which has an alternative way, the resource utilization of *importance-focus scenario* is more efficient than the *all-equal scenario*.

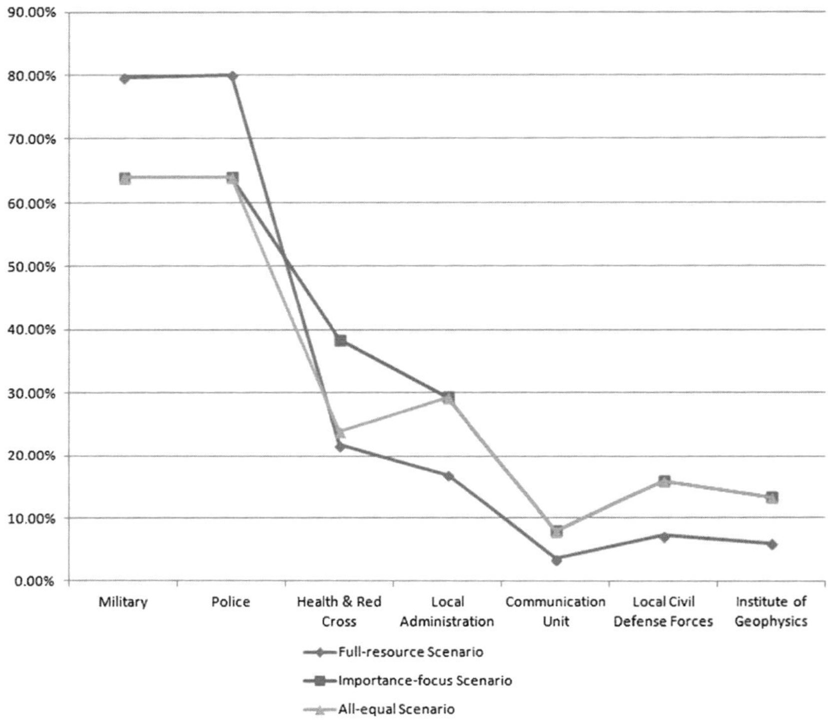

Fig. 3 Utilization of human resources corresponding to three scenarios

We finally have computed the average of resource utilization of all actors as shown in Table 5. The best strategy is the *Importance-focus scenario*.

Table 5 Comparing the average of resources used in three scenarios

	M	P	HR	LA	CU	LCDF	IG	Average
Scen. 1	79.69 %	80.00 %	21.82 %	16.97 %	3.64 %	7.27 %	6.06 %	30.78 %
Scen. 2	63.99 %	64.00 %	38.40 %	29.33 %	8.00 %	16.00 %	13.33 %	*33.29 %*
Scen. 3	63.99 %	63.99 %	24.00 %	29.33 %	8.00 %	16.00 %	13.33 %	31.23 %

5 Rescue Plans Assessment by Agent Organization Analysis

In this section, we evaluate the rescue plan organizational structure by using the framework provided by Grossi and al. [8]. This framework allows us to assess the *robustness*, *flexibility* and *efficiency* of our organization by using the *power*, *coordination* and *control*[5] relations between each pair of roles.

Grossi et al. state that: (a) the *robustness* means the stability of an organization in the case of anticipated risks; (b) the *flexibility* is the capacity of an organization to adapt to the environment changes; and (c) the *efficiency* refers to the amount of resources used by the organization to perform its tasks.

In our case, we will show that the structure organization is efficient and sufficiently flexible but not enough robust. Obviously, it is not possible to maximize simultaneously all criteria [8]. Since our organization is devoted to the disaster response, thus we would like to focus on the amount of resources used by tasks (the efficiency).

As Grossi's proposal, evaluating an organizational structure involves three steps: (1) building a role graph of the organization based on the three dimensions (*power*, *coordination*, *control*); (2) measuring specific properties of the organizational structure according to a set of formulas; (3) finally, comparing the obtained results with the optimum values proposed by Grossi in order to evaluate the qualities (*robustness*, *flexibility* and *efficiency*) of the organization.

5.1 Building the Role Graph

According to three dimensions described above, we have built the role graph corresponding to our organizational model (seven roles) as seen in Fig. 4. Each node corresponds to an organization while an arc corresponds to the relationship between two organizations. We can identify three types of relationships: power, coordination and control.

5.2 Computing the Metrics

Based on the role graph above, we have implemented isolation metrics (*completeness, connectedness, economy, unilaterality, univocity, flatness*) and interaction metrics (*detour, overlap, incover, outcover* and *chain*) as proposed by Grossi.

---

[5]The *power* dimension defines the task delegation pattern; the *coordination* dimension concerns the flow of knowledge within the organization; and the *control* dimension between agent A and agent B means that agent A has to monitor agent B's activities and possibly take over the unaccomplished tasks of agent B.

5.3 Measuring the Qualities

In order to evaluate criteria of our organization, we have compared our results (right-hand table) with the proposed optimum values (left-hand table) in Tables 6, 7 and 8.

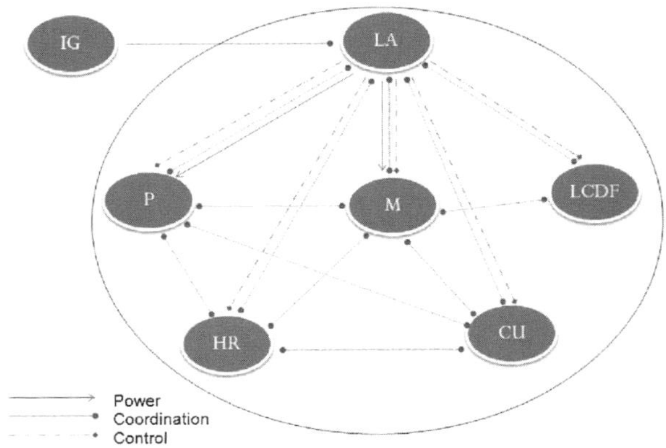

Fig. 4 Role graph of the tsunami response plan

Table 6 shows the organizational structure robustness of the rescue plans. We have three over twelve optimum metrics: $Connectedness_{Coord}$, $Overlap_{Coord-Pow}$ and $Chain_{Contr-Pow}$. The variation of our results with standard values is above average (0.54), so we can conclude that the organization is not robust enough.

Table 6 Organization robustness (on the right) versus standard values (on the left)

$Completeness_{Coord}$	1	25/42	$Overlap_{Coord-Pow}$	1	1
$Connectedness_{Coord}$	1	1	$Chain_{Contr-Pow}$	1	1
$Univocity_{Pow}$	0	1	$Chain_{Contr-Coord}$	1	0
$Unilaterality_{Coord}$	0	1/25	$InCover_{Contr-Coord}$	1	0
$Univocity_{Contr}$	0	1	$OutCover_{Pow-Contr}$	1	2/5
$Flatness_{Contr}$	0	1	$OutCover_{Pow-Coord}$	1	0

Table 7 shows how flexible is the organizational structure. We have two over six optimum metrics: $Chain_{contr-pow}$ and $Connectedness_{coord}$. The variation of our results with standard values is below average (0.33), thus our organization is sufficiently flexible.

Table 7 Organization flexibility (on the right) versus standard values (on the left)

$Completeness_{Pow}$	0	1/3	$Completeness_{Coord}$	1	25/42
$Connectedness_{Pow}$	0	1	$Connectedness_{Coord}$	1	1
$Chain_{Contr-Pow}$	1	1	$OutCover_{Pow-Contr}$	1	2/5

Table 8 depicts the efficiency of our organizational structure. We have six over ten optimum metrics: $Economy_{Pow}$, $Overlap_{Coord-Pow}$, $Unilaterality_{Pow}$, $Univocity_{Pow}$, $Economy_{Contr}$ and $Overlap_{Contr-Pow}$. The variation of our results with standard values is small (0.193), so our organization is quite efficient.

Table 8 Organization efficiency (on the right) versus standard values (on the left)

$Completeness_{Pow}$	1	1/3	$Unilaterality_{Pows}$	1	1
$Economy_{Pow}$	1	1	$Univocity_{Pow}$	1	1
$Economy_{Coord}$	1	17/36	$Economy_{Contr}$	1	1
$Overlap_{Coord-Pow}$	1	1	$Overlap_{Contr-Pow}$	1	1
$Overlap_{Pow-Coord}$	1	2/25	$Overlap_{Pow-Contr}$	1	2/5

6 Conclusion

In this paper, we have introduced two complementary evaluations of disaster management plans: process and organization evaluations. The process evaluation helps to identify the best allocation strategy of human resources according to the distribution rules of resources over the tasks. We have defined in our work three scenarios corresponding to three different distribution policies. In our case study, the best one corresponds to the “importance focus” i.e. allocating to tasks a number of resources based on its importance factor. In addition, the agent organizational structure evaluation assesses three criteria of our organization: robustness, flexibility and efficiency. In our case study, we have a flexible and efficient organization due to the fact that the roles are well connected while retaining a minimal of symmetric and redundant links. Even if we have a “quite good organization”, it remains not robust. An optimal robustness would require a complete connectivity between all nodes. This property is useful to guarantee the plan continuity in case where some resources are destroyed and a role is for example no more represented.

Thanks to this approach, we can provide interesting recommendations to improve crisis management plans. In our case study, two recommendations can be provided to the authorities of Ho Chi Minh City: (1) preferring the “importance focus” strategy and (2) improving the robustness in case of high-risk situations.

In the future research, we would like to establish a bridge between the discrete-event (process-oriented) simulation and the agent-based simulation by implementing a transformation from our business process model to a concrete multi-agent based model. The actors would be figured as the agents and the flow of tasks (coordination) would be distributed among agents. While the process-oriented simulation provided an aggregate vision of the plan thanks to roles, agent simulation would help to have an agent-centered view where different agents could play a same role with different behavior (e.g. following BDI architecture).

References

1. Franke, J., Charoy, F., Ulmer, C.: Handling Conflicts in Autonomous Coordination of Distributed Collaborative Activities, pp. 319–326, WETICE (2011)
2. Lesser, V., Corkill, D.: Challenges for multi-agent coordination theory based on empirical observations. In: Lomuscia, A., Scerri, P., Bazzan, A., Huhns, M. (eds.) Proceedings of 13th International Conference on Autonomous Agents and Multiagent Systems, IFAAMAS, pp. 1157–1160. In Challenge in Vision Special Track (2014)
3. Franke, J.: Coordination of distributed activities in dynamic situations. The case of inter-organizational crisis management. Ph.D. dissertation, University of Nancy, France (2011)
4. Thanh, L.N.T., Hanachi, C., Stinckwich, S., Vinh, H.T.: Representing, simulating and analysing Ho Chi Minh City Tsunami Plan by means of process models. In: ISCRAM (Information Systems for Crisis Response and Management) Vietnam 2013 Conference (2013)
5. Dugdale, J., Bellamine-Ben Saoud, N., Pavard, B., Pallamin, N.: Simulation and emergency management. In: van de Walle, B., Turroff, M., Hiltz, S.R. (eds.) Information Systems for Emergency Management, Chapter 10 (2010)
6. Rozinat, A., Mans, R.S., Song, M., Van der Aalst, W.M.: Discovering simulation models. In: Information Systems, vol. 34, no. 3, pp. 305–327, Elsevier (2009)
7. Rozinat, A., Wynn, M.T., Van der Aalst, W.M., ter Hofstede, A.H.M., Fidge, C.J.: Workflow simulation for operational decision support using design, historic and state information. In: Business Process Management, pp. 196–211. Springer (2008)
8. Grossi, D., Dignum, F., Dignum, V., Dastani, M., Royakkers, L.: Structural aspects of the evaluation of agent organizations. In: Coordination, Organizations, Institutions, and Norms in Agent Systems II, pp. 3–18. Springer (2007)
9. Van der Aalst, W.M., Nakatumba, J., Rozinat, A., Russell, N.: Business process simulation: how to get it right? In: BPM Center Report BPM-08-07, Citeseer (2008)
10. Rolland, C., Ben Achour, C., Cauvet, C., Ralyté, J., Sutcliffe, A., Maiden, N., Jarke, M., Haumer, P., Pohl, K., Dubois, E., Heymans, P.: A proposal for a scenario classification framework. In: Requirements Engineering Journal, vol. 3, no. 1, pp. 23–47. Springer (1998)
11. Longo, F.: Emergency simulation: state of the art and future research guidelines. In: SCS Modeling and Simulation Magazine, vol. 1 (2010)
12. Ahmad, A., Balet, O., Himmelstein, J., Boin, A., Schaap, M., Brivio, P., Ganovelli, F., Gobbetti, E., Pintore, G., de la Rivière, J.B.: Interactive simulation technology for crisis management and training: The INDIGO Project. In: Proceedings of the 9th International ISCRAM Conference (2012)

Evaluation of Organization in Multiagent Systems for Fault Detection and Isolation

Faten Ben Hmida, Wided Lejouad Chaari and Moncef Tagina

Abstract In the last decades MultiAgent Systems (MAS) have seen a growth of interest, not only from academics but also from industrialists. They have been in particular used to model and analyze the structure and dynamics of complex systems. One of the major uses of MAS in the industrial domain is diagnosis as they provide a good support for decision-making and seem especially helpful to detect and localize faults in a flexible manner. However, the lack of evaluation methods of MAS is an obstacle to the expansion of their use in such real world applications. In an attempt to bring MAS closer to industry, an evaluation of multiagent application for fault detection and isolation is carried out and presented in this paper. The used evaluation method focuses on organization as it is a very important characteristic of distributed systems in general and MAS in particular because of their social dimension. In this work, two multiagent models representing different diagnosis architectures are presented, evaluated and compared according to the proposed method.

Keywords Multiagent systems · Evaluation · Measurement · Graph theory · Organization · Diagnosis · Fault · Detection · Isolation

---

F.B. Hmida (✉) · W.L. Chaari · M. Tagina
COSMOS Laboratory, ENSI, National School of Computer Sciences, University of Manouba, Manouba, Tunisia
e-mail: faten.benhmida@ensi-uma.tn

W.L. Chaari
e-mail: wided.chaari@ensi-uma.tn

M. Tagina
e-mail: moncef.tagina@ensi-uma.tn

G. Jezic et al. (eds.), *Agent and Multi-Agent Systems: Technologies and Applications*, Smart Innovation, Systems and Technologies 38,
DOI 10.1007/978-3-319-19728-9_6

1 Introduction

MultiAgent Systems (MAS) are being increasingly used in industry where most environments are dynamic, complex and uncertain. Potential applications cover multiple areas such as control, diagnosis, transport, manufacturing and logistics. MAS provide a natural way to design and implement efficient distributed industrial systems as well as to design monitoring and decision support systems.

The literature review pointed out the early interest of researchers to develop MAS for industrial systems diagnosis and monitoring [1–3]. However, for such demanding and critical applications, achieving best solutions cannot be done without efficient evaluation tools. Analysis and comparison of the different alternatives for a system's architecture, for example, should be guided by adapted evaluation methods and through relevant criteria. In multiagent technology such tools and methods are still lacking, especially at the application level, even efforts are being provided to deal with this issue [4–6].

In fact, evaluating design methodologies and development platforms and tools is undeniably of great interest but remains insufficient to make a full coverage of multiagent systems evaluation. It becomes more and more necessary to provide means for evaluating and comparing multiagent applications. That is what we address through our research works which focus on agent-oriented applications evaluation based on specific characteristics especially structural ones [7, 8]. In this paper we are interested in organization. Several metrics to characterize it are inspired from graph theory and used to evaluate and compare two versions of a multiagent application for fault detection and isolation.

The paper is organized as follows: Sect. 2 presents the particularities of organization in MAS. In Sect. 3, the adopted approach of organization evaluation in MAS is described. Section 4 presents the several versions of the diagnosis multiagent application which are evaluated and compared. The results and interpretations are also exposed in the same section. A conclusion and a look at future work are presented in Sect. 5.

2 Organization Evaluation Method

A multiagent system is a society of agents in mutual interaction sharing tasks, resources and knowledge. The several interaction patterns in a multiagent system engender complex organizational structures and interrelated acquaintances networks [9]. Organization is a very important characteristic and a basic concept of multiagent systems as it transposes their social dimension [9–11]. The evaluation of organization is a very interesting issue; it allows developing an understanding of the multiagent system and its structure. In [9] the author defines the possible ways of analysing MAS organizations. There is functional analysis which aims to identify and describe the functions that the several entities of an organization are supposed to realize. From this prospect, organization is viewed as a set of roles and relations between those roles. There is also structural analysis which focuses on the interactions between agents trying to make sense to their complex interrelations and explain the resulting organizational structures.

In this work, great importance is accorded to the structural aspect of organizations, that's why the chosen representative model rely on graphs which expressively shapes the multiagent structure by drawing its entities and their relationships. Based on this model, the evaluation of organization is done thanks to specific metrics of graph theory which are presented in the following paragraph.

The following organization evaluation method is part of a global evaluation approach which consists in observing, modeling and measuring functional characteristics of MAS [7, 12].

The proposed solution consists in modeling the interaction network of the MAS through an oriented graph where the nodes represent the agents and the arcs represent the communication links between these agents. The arc direction is that of sending messages and each arc is weighted by the number of exchanged messages. The properties of this graph are then evaluated according to graph theory.

2.1 Degree Distribution

The degree of a node is the number of connections it has to other nodes. Here Ki designates the degree of a node i. In a graph, nodes may have different degrees. The degree distribution P(K) is defined as the fraction of nodes in the graph with degree K:

$$P(K) = \frac{N_K}{N} \tag{1}$$

with N_K the number of nodes having a degree K and N the total number of nodes.

Different types of networks have different characteristic degree distributions P (K). The shape of the degree distribution function allows distinguishing two main classes of networks. Homogeneous networks are characterized by a degree distribution that is concentrated around the degree mean value < K > given by:

$$<K> = \sum K \times P(K) \tag{2}$$

Heterogeneous networks are characterized by a degree distribution that generally follows a power law which is the case of most networks in the real world. This means that, in such networks, there is a large majority of nodes having low degree but a small number having high degree that greatly exceeds the average.

Those nodes are thought to be special and having a specific function in their networks.

2.2 Assortativity

Assortativity reflects the tendency of some nodes to be connected to other nodes having similar degrees. Assortativity is often examined in terms of correlation between nodes degrees. A simple characterization of the correlations between degrees of neighbor nodes is determined by the neighborhood degree Knn,i which expresses the average neighbors degree of a given node i.

$$K_{nn,i} = \frac{1}{K_i} \sum_j K_j \tag{3}$$

Based on this measure, degree correlation is given by the mean neighbourhood degree of nodes with K degree.

$$K_{nn}(K) = \frac{1}{N_K} \sum_{i/K_i = K} K_{nn,i} \tag{4}$$

The behaviour of $K_{nn}(K)$ defines two main classes of networks. We talk about assortative network when $K_{nn}(K)$ increases with K. This is the case of social networks when high degree nodes are preferentially associated with other high degree nodes. And we talk about disassortative network when $K_{nn}(K)$ decreases with K, such as in hierarchic networks where high degree nodes are connected to many nodes with low degree.

2.3 Centrality

Centrality determines the importance of a node within the network and how much it is influent. This metric is inspired from social networks analysis. Many terms are used to measure centrality such as degree centrality. Cd(i) is the degree centrality of node i.

$$Cd(i) = \frac{Ki}{(N-1)} \tag{5}$$

Where Ki is the degree of the node i and N is the number of nodes in the network.

2.4 Hierarchy

We talk about hierarchy in organizations if relations between the organization entities are determined by authority. In oriented graphs, hierarchy is defined as follows:

$$Hierarchy = 1 - \left[\frac{V}{MaxV}\right] \tag{6}$$

With V the number of pairs of nodes symmetrically connected and MaxV is the total number of pairs of nodes [13]. This measure is interesting to express structural hierarchy of the network but does not reflect hierarchy degree in multiagent systems as it does not consider semantic meaning of a link. For this reason, we propose to examine the nature of the links to determine whether there is any power relationship. The best way to do that is to pick up the nature of the message, and to deduce the existence of power relationship if the message is an order. The proposed measure for hierarchy is:

$$Hierarchy = \frac{OrdV}{MaxV} \tag{7}$$

Where OrdV is the number of nodes pairs related by order messages translating a hierarchic relationship.

2.5 Leadership

In a multiagent system several agents can come together into organization structures called groups which may be teams or coalitions. In both cases, there can be an agent having the special capability of coordinating the actions of the other agents. This agent is called “leader” [10]. While in a team the leader represents a decision authority, in a coalition the leader does not have a real decisional power but is rather a point of information centralization. To pick up such agents using graph formalism, a group is assimilated to a star sub-graph, with the leader as a central node surrounded by the other group agents [10]. A star graph Sn of order n is a graph on n nodes with one node having the degree n-1 and the other nodes having the degree 1 [14]. To recognize the leaders in a multiagent system, we have to identify the star sub-graphs in the whole representative graph.

2.6 Measures Summary and Signification

The following Table 1 summarizes the used measures, the way they are computed and the possible interpretations which can be concluded from such measures.

Table 1 Summary of measures and possible interpretations

Measure	Interpretation indications
Degree distribution $P(K) = \frac{N_K}{N}$	Nature of nodes:
	Homogeneous: egalitarian/balanced relationships
	Heterogeneous: authority/unbalanced relationships
Neighbourhood degree $K_{nn,i}$	Nature of the immediate neighbourhood and its membership degree:
	Strong cohesion with the rest of the MAS
	Low cohesion/isolation
Degree correlation K_{nn}	Nature of the MAS:
	Assortative
	Disassortative
Centrality C(d)	Dominance:
	Resources control
	Skills detention
	Decision-making
Hierarchy	Organization topology
	Hierarchy (simple or uniform)
	Holarchy
	Federation
Leadership	Authority:
	Hierarchical Superiority
	Synchronization
	Arbitration

3 Application to Fault Detection and Isolation

3.1 Evaluated Agent-Based Application

The evaluation method presented above was applied on two versions of a multi-agent diagnosis application. It is a multiagent application designed in order to detect and localize failures affecting an industrial system [15]. Its functioning principle can be summed up as follows: an analytical model describes the normal functioning way of the system. To diagnose the system, a test of the coherence between the observations done on the real system and the analytical model is performed through a set of residuals or fail indicators. The studied multiagent application of diagnosis is composed of different autonomous agents; each of them is responsible of a specific task and belongs to one of the following categories:

- *Detection agents*: they take the measures of the system variables to be supervised and compute the corresponding residuals. Then they send results to localization agents. Two detection methods are used: binary logic and fuzzy logic.
- *Localization agents*: they retrieve the different residuals from the detection agents, to localize the failure using different methods. The used localisation methods are binary logic, fuzzy logic, Probabilistic Neural Network (PNN) and Multi-layer Perceptron (MLP). For the two latter cases localization agents don't rely on detection agents since these methods allow both detection and localization of failures.
- *Interface agent*: it coordinates the other agents' processing by controlling the order of messages exchange and ensures communication with the application user. Interface agent sends requests to detection and localization agents to start diagnosis and display results to the user.

In addition to the detection, localization and interface agents, in the second version of the diagnosis application, a new agent DIAG is added which is an arbitrator agent that takes the diagnosis results from the several localization agents and determine the correct diagnosis if there is a conflict. It tells the several localization agents to send their credibility degree which is the number of times they emitted a good diagnosis in the past. Then it compares the different values received and takes the final diagnosis decision which corresponds to the agent having the highest credibility degree.

All these agents cooperate and communicate through message exchange to ensure the diagnosis process.

3.2 Results and Interpretations

In the following paragraphs, the results of the organization evaluation carried out on the two versions of the multiagent application for fault detection and isolation are presented and discussed. Visualization of the resulting graphs helps characterizing and comparing the several diagnosis architectures. Figure 1 illustrates the graph corresponding to the first version of the diagnosis MAS.

The graph presented in Fig. 2 highlights the presence of the agent DIAG which is an arbitrator agent. There is lot of communication added by the inclusion of this agent.

We notice that the detection agents DET1-5 and DETFL do not communicate with each other. They rely on the localisation agents LOCBN and LOCFL which are solicited more than the other agents. This is due to the fact that while detecting a fault, the detection agents send the residual values to the localisation agents whereas the agent INTER sends messages and does not receive any. This is because it is an

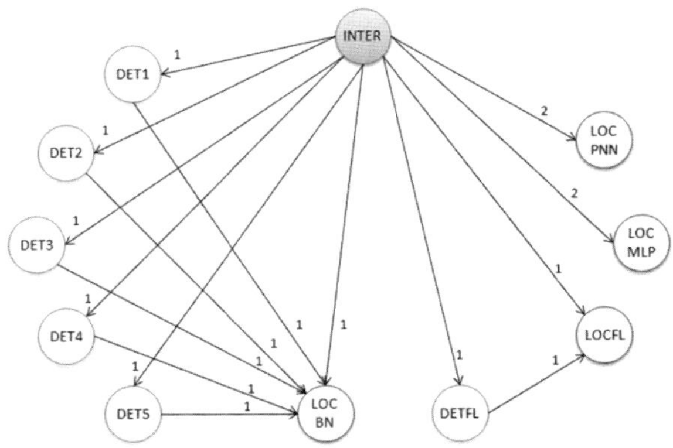

Fig. 1 Generated graph of the first agent-based diagnosis application

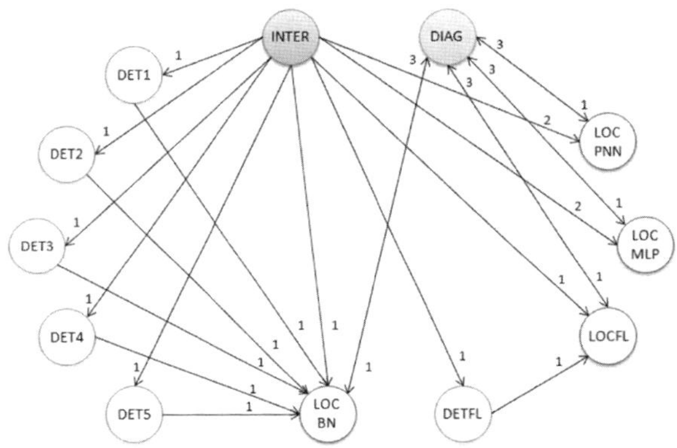

Fig. 2 Generated graph of the second agent-based diagnosis application

organiser agent; it transmits to the detection agents a diagnosis request and informs the localisation agents to receive the residual values from the detection agents. In the following we try to give some interpretations according to the obtained evaluation results. We start by examining the obtained values of the several used measures.

Figure 3 shows the several agents' degrees in the two versions of the diagnosis application.

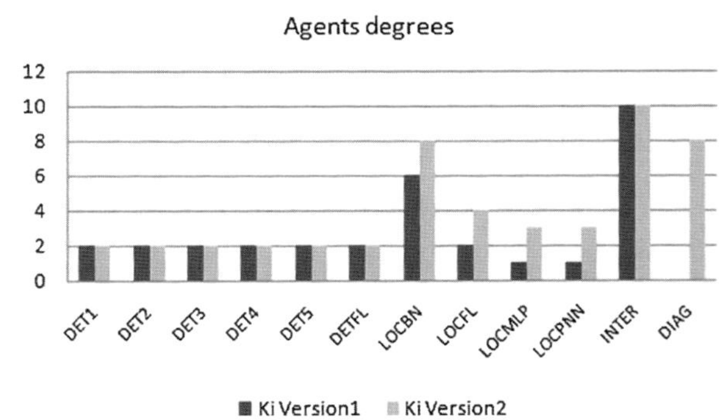

Fig. 3 Agents' degrees in the two application's versions

Two categories of agents can be distinguished:

- Agents whose degrees didn't change from a version to another such as detection and interface agents.
- Agents whose degrees have increased due to the addition of an arbitrator agent such as localization agents.

This is due to the fact that localization agents are those that emit diagnosis conclusions and are the more likely to be in conflict situations which explains resorting to an arbitrator agent.

Figure 4 illustrates the aspect of degrees distributions corresponding to the two versions of the diagnosis application.

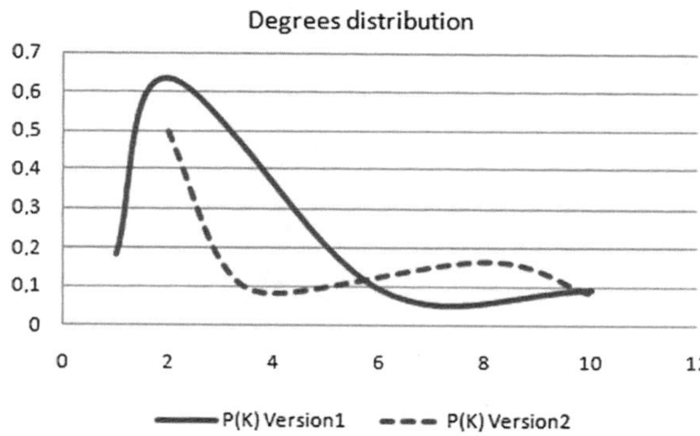

Fig. 4 Degrees distribution in the two application's versions

The obtained curves correspond to heterogeneous graphs; which means that there is no significant mean value around which the degrees are concentrated. In both cases, there is a majority of nodes with low degrees and a small number of nodes with high degrees exceeding the average. We also notice that the difference between the number of agents with a low degree and that of agents with a significant degree is much more pronounced in the first version of the application (without arbitration).

This is also what we see through Fig. 5 which shows the degrees correlation for both versions of the diagnosis application.

The correlation expresses the graph's assortativity degree. Both architectures are disassortatives. The disassortativity is a characteristic of hierarchical systems in which agents of significant degree tend to be associated to agents of lesser degree. We notice that in the case of diagnosis without arbitration, assortativity is slightly different compared to the case of diagnosis with arbitration. In the latter, the disassortativity is accentuated by the addition of the arbitrator agent whose degree is important compared to other agents. Figure 6 shows the centrality degrees of the different agents for both versions of diagnosis application.

We notice through these results that the centrality degrees of detection agents decreased by introducing the arbitrator agent and centrality degrees of localization

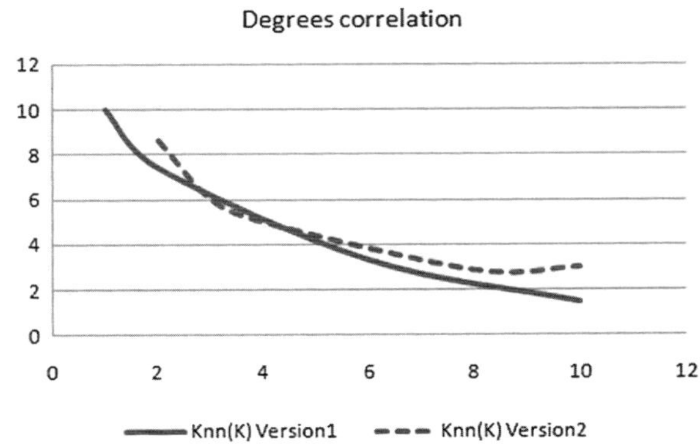

Fig. 5 Degrees correlation in the two application's versions

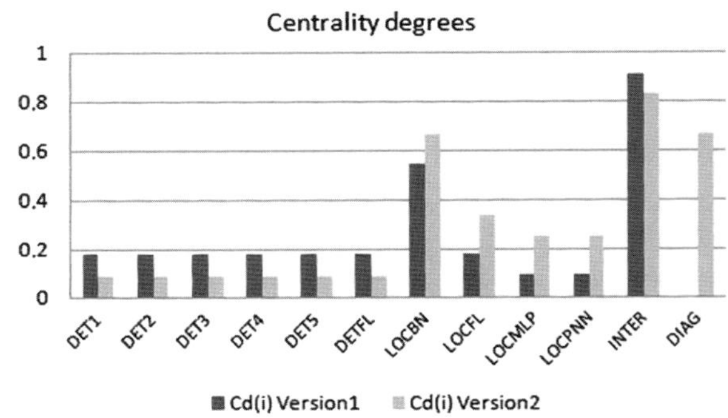

Fig. 6 Centrality degrees distribution in the two application's versions

agents have instead increased. This is explained by the fact that localization agents are connected to the arbitrator agent; which had increased the centralization within this acquaintance of the MAS.

The introduction of the arbitrator agent didn't change significantly the organizational structure of the MAS. The communication induced by adding the agent DIAG is a useful one since this will help the user in the decision making process. In fact, in the case of conflicts, the user has not to choose which diagnosis is the correct one; he will rely on the arbitrator agent which will take the decision instead of him according to the information retrieved from each localisation agent.

4 Conclusion and Future Work

In this paper we presented a multi-agent model as well as an appropriate evaluation method to analyse and compare different multiagent diagnosis architectures for fault detection and isolation. The evaluation approach focuses on the organization as it is an important characteristic in MAS. It is based on measures and indices stemming from graph theory. It should be noted that the purpose of the evaluation is not only to analyse and compare different MAS from a technological point of view. It aims also to help industrialists making choices between various architectural alternatives of the fault detection and isolation system. The evaluation allows industrialists to have a clear idea on the different diagnosis architectures specificities. In extension to this study, our future work includes two points. The first consists in extending the test-case diagnosis multiagent application in order to move to a larger scale and thus reconsider the evaluation results. The second point is to focus on the evaluated system's dynamics and the evolution of the proposed measures over time. For this purpose, the proposal of an evolving graph model instead of a static one is actually addressed.

References

1. Jennings, N.R., Corera, J.M., Laresgoiti, I.: Developing industrial multi-agent systems. In: ICMAS, pp. 423–430 (1995)
2. Albert, M., Längle, T., Wörn, H., Capobianco, M., Brighenti, A.: Multi-agent systems for industrial diagnostics. In: Proceedings of 5th IFAC Symposium on Fault Detection, Supervision and Safety of Technical Processes, pp. 483–488 (2003)
3. Pěchouček, M., Mařík, V.: Industrial deployment of multi-agent technologies: review and selected case studies. Auton. Agent. Multi-Agent Syst. **17**(3), 397–431 (2008)
4. Bonnet, G., Tessier, C.: Evaluer un système multiagent physique-Retour sur expérience. In: JFSMA, pp. 13–22 (2008)
5. Kaddoum, E., Gleizes, M. P., George, J. P., Glize, P., Picard, G.: Analyse des critères d'évaluation de systèmes multi-agents adaptatifs. In: JFSMA, pp. 123–132 (2009)
6. Bouzouita, K., Chaari, W. L., Tagina, M.: Utility-based approach to represent agents' conversational preferences. In: Information Processing and Management of Uncertainty in Knowledge-Based Systems, pp. 454–463. Springer International Publishing (2014)
7. Ben Hmida, F., Lejouad Chaari, W., Tagina, M.: Graph theory to evaluate communication in industrial multiagent systems. Int. J. Intell. Inf. Database Syst. **5**(4), 361–388 (2011)
8. Ben Hmida, F., Lejouad Chaari, W., Dupas, R., Séguy, A.: Evaluation of communication in multiagent systems for supply chain planning and control. In: ICTAI 14, pp. 408–412 (2014)
9. Jacques, F.: Multi-Agent Systems. An Introduction to Distributed Artificial Intelligence, Addison Wesley (1999)
10. Legras, F.: Dynamic organization of teams of autonomous vehicles by overhearing, Doctoral dissertation, SUPAERO, University of Toulouse, France (2003)
11. Boussebough, I.: Dynamically adaptive multi-agent systems, Ph.D. dissertation, University Mentouri, Constantine, Algeria (2011)
12. Ben Hmida, F.: Performance evaluation of multiagent systems, Ph.D. dissertation, University of Manouba, Tunisia, University of Bordeaux, France (2013)
13. Krackhardt, D.: Graph theoretical dimensions of informal organizations. Comput. Organ. Theory **89**(112), 123–140 (1994)
14. Harary, F.: Graph Theory, pp. 17–18. Addison-Wesley, Reading (1994)
15. Bouabdallah, B.S., Saddem, R., Tagina, M.: A Multiagent Architecture for Fault Detection and Isolation. In: 1st International Workshop on Applications with Artificial Intelligence. Patras (2007)

The Application of Co-evolutionary Genetic Programming and TD(1) Reinforcement Learning in Large-Scale Strategy Game VCMI

Łukasz Wilisowski and Rafał Dreżewski

Abstract VCMI is a new, open-source project that could become one of the biggest testing platform for modern AI algorithms in the future. Its complex environment and turn-based gameplay make it a perfect system for any AI driven solution. It also has a large community of active players which improves the testability of target algorithms. This paper explores VCMI's environment and tries to assess its complexity by providing a base solution for battle handling problem using two global optimization algorithms: Co-Evolution of Genetic Programming Trees and TD(1) algorithm with Back Propagation neural network. Both algorithms have been used in VCMI to evolve battle strategies through a fully autonomous learning process. Finally, the obtained strategies have been tested against existing solutions and compared with players' best tactics.

Keywords Genetic programming · Neural networks · Strategy games

1 Introduction

Artificial Intelligence is used in many domains of everyday life. One of the most interesting AI environments are computer games, where developers can use advanced AI algorithms to implement intelligent computer opponents. Any player can challenge such an opponent—this improves the knowledge base and assessment of the target solution.

Unfortunately, most open-source games, which are used as testing platforms for modern AI algorithms, are known only to AI researchers and are not popular among ordinary players. Few projects, like Civilization IV and Starcraft AI Competition, have been introduced to fill that gap. Now, the VCMI project comes to that place as well.

Ł. Wilisowski · R. Dreżewski (✉)
Department of Computer Science, AGH University of Science and Technology, Kraków, Poland
e-mail: drezew@agh.edu.pl

G. Jezic et al. (eds.), *Agent and Multi-Agent Systems: Technologies and Applications*, Smart Innovation, Systems and Technologies 38,
DOI 10.1007/978-3-319-19728-9_7

VCMI (http://vcmi.eu) is an open source project aiming to recreate Heroes of Might and Magic III game. HoMM3 has been a very successful release and is one of the most popular strategic games ever. What differentiates VCMI from other similar projects is huge knowledge base of best tactics and great number of active players, putting it right next to Starcraft challenge. With its unique environmental model and complex, turn-based gameplay, VCMI has the potential of becoming one of the most interesting and challenging AI environments.

2 Related Work

This section presents a general description of machine learning algorithms used in AI modules for strategic games. We mainly focus on reinforcement learning and evolutionary algorithms because our preliminary solution for VCMI is based on them. Few modern computational methods are briefly mentioned as well, as a part of research for future improvements and application of other algorithms in the presented environment.

Reinforcement learning algorithms are widely used in computational intelligence. They have been successfully applied to many AI environments, with additional constraints depending on complexity of defined model. Sutton's and Barto's paper [25] contains a general description of reinforcement learning. The examples of standard RL implementations can be found in [2, 5, 20]. In [28] Wender applied Q Learning and Sarsa algorithms, both with and without eligibility traces, to learn a city placement task in Civilization IV game.

Introduced first in [4] and later used by Ekker to play 5x5 Go game in [12], TD(μ) algorithm tries to separate changes caused by bad play from the changes caused by bad evaluation function. The algorithm is based on TD(λ) presented by Sutton in [26]. Ekker used TD(μ) combined with Resilient Back Propagation neural network to play Go with great effects.

Similar approach involving combination of different reinforcement learning algorithms and neutral networks was used in Othello game [23]. Described in [27] Minimax Q Learning is a modification of standard Q Learning algorithm which returns function of state/action/action triples instead of approximating function of state/action pairs. In this algorithm both actions of an agent and actions of its opponent are modeled. Another example of Minimax Q Learning was presented in [18].

Two other, related algorithms are worth mentioning here. R-max [6] is model-based reinforcement learning algorithm which can attain near-optimal average award in polynomial time. AWESOME, introduced in [8], is an algorithm that plays optimally against stationary opponents and converge to Nash equilibrium in finite number of steps.

Evolutionary algorithms are also widely used in the gaming AI. For player-opponent environments with competing individuals like VCMI, a co-evolutionary approach is usually employed. Co-Evolutionary algorithms have been well described in [29] and their agent-based variants were described in [9–11]. In [21] competitive co-evolutionary algorithm combined with different variants of Hall of Fame mechanism to evolve computer bots for RobotWars game. In [16] Fitnessless Co-Evolution, a new competitive co-evolutionary approach that lacks an explicit fitness function was introduced. It was successfully used this algorithm in the AntWars game.

There have been many publications related to AI algorithms in games in the last few years. Most of them focused on improving standard Reinforcement Learning and evolutionary algorithms or combining them with other methods to obtain better results. There were also introduced new algorithms like Fast Adaptive Learner (FAL) [13]—an algorithm which is highly reactive to opponent's strategy switches. There have been also few interesting adaptations of known computational methods, like for example swarm intelligence, into gaming environments. An example of such work would be an application of Ant Colony Optimization in Lemmings game [15].

Recently, one of the most popular algorithms is Monte-Carlo Tree Search (MCTS)—a method first used in Go and later applied with great effects to other AI environments. In [24] an application of MCTS algorithm in 7 Wonders game was presented. In [22] the algorithm has been used to solve traveling salesman problem. Some MCTS hybrid solutions have been introduced as well. In [3] a promising MCTS-Minimax hybrid algorithm was proposed. In [19] MCTS was combined with evolutionary techniques to optimize algorithm's performance. A complete study of MCTS methods can be found in [7].

3 VCMI Game

VCMI, which mirrors HoMM3's mechanics, is a strategic turn-based game that is focused on two major activities: developing player's kingdom (macro organization) and conducting battles (micro organization). This paper presents an implementation of Battle AI for VCMI. A short summary of battle mechanics will be presented now. A complete game documentation can be found at [1].

Fig. 1 Example screenshot from VCMI battle

Battles are held on hexagonal grid (Fig. 1). Two opposing sides represent two competitive game agents with armies composed of number of specific creature units.

Each creature has its own characteristics like attack, defense, speed and damage. The unit can also possess some special abilities. Additionally, the hero commanding an army influences the current battle state, both directly and indirectly. From this perspective VCMI battlefield resembles chess board with two players and opposing figures fighting each other, in much richer environment of larger scope.

Detailed battle mechanics are based on inflicting damage upon opponent units. During each turn, each unit has few available actions to make: moving, attacking, waiting, defending or using special ability if available. Both physical and magical actions can potentially inflict damage on the battle creatures. The win condition is met when all opponent units are killed.

4 Move Evaluation Function

One of the paper's main motivations was implementing Battle AI that could imitate human thinking process of conducting battle, especially concerning employment of advanced combat strategies. In order to achieve that, two additional assumptions had to be made. First, it was decided that the AI engine should use domain-specific knowledge of advanced tactics and important factors of the game. Second, instead of analyzing raw battle data, all relevant information should be presented on a higher abstraction level, based on human perception of the game environment.

Based on these assumptions an opponent model has been created. The model employed a greedy strategy, which could be summarized as follows: in each turn the player is provided with a list of candidate actions and he chooses the most beneficial one according to current evaluation function. The candidate actions are chosen from the whole space of possible moves based on standard, known game tactics.

Usually, the evaluation function is used to assess the utility of current battle state. However, this approach has been abandoned here because it requires assessment of all battle data constituting the selected environment state, which is a non-trivial task. Instead, is has been decided to use function analyzing actions only. In this approach the function evaluates only action results represented as a list of observable changes in the battle state, with few additional constraints. First, only relevant battle information is provided. Second, representation of data is supported with specific domain knowledge about game mechanics. Third, the list is short, elements atomic, relatively simple and quickly computed.

Table 1 Estimated action results

% of player units in queue	% of opponent units reaching	Damage inflicted*
% of player units reaching	% of opponent units in range	Damage received*
% of player units in range	% of protected shooters	Enemy retaliation taken
% of opponent units killed	% of released shooters	Is waiting
% of opponent units in queue	% of blocked shooters	Is dead

* computed using special formula

The list of all selected action results is presented in Table 1. The list has been chosen arbitrarily, which means that it is prone to improvements and reselection under further investigation.

5 Implemented Algorithms

The first algorithm used during learning phase was **co-evolution of genetic programming trees**. The idea was to create initial population of solutions (function approximations) and evolve them. In order to do it a competitive co-evolutionary algorithm has been implemented. Its pseudocode is shown in Alg. 1.

Algorithm 1 Genetic co-evolution (competitive)

```
Initialize population P
while end condition is not met do
  EVALUATE(P)                                   *(competitively)
  parents ← SELECT(P)
  childs ← RECOMBINE(parents)
  childs ← MUTATE(childs)
  P ← REPLACE(childs)
end while
Return best individual in P
```

The selected algorithm evaluated individuals on the basis of the number of battles won against other individuals using two-way K-Random-Opponents selection method. In this method every individual in the population is paired up with K opponents (other than itself) picked at random with replacement. Fitness values of both individuals are updated. In VCMI context pairing two individuals meant conducting one battle between two heroes of equal armies controlled by selected AI strategies. No auxiliary memory mechanism such as Hall-of-Fame has been used.

To represent population individuals strongly-typed genetic programming trees were used. They were based on Koza's canonical paper [17], with few additional constraints. First, it has been assumed that GP Tree should recognize two basic data types: logical Boolean and numeric floating point Double. All action results, in the form of high-level battle observations mentioned earlier, have been added to the tree as terminal nodes with appropriate types.

After that, a set of non-terminal function nodes has been added to the tree. All functions were equivalents of mathematical and logical operators—*not*(), *and*(), *or*(), *if*(), *add*(), *sub*(), *mul*(), *isEqual*(), *isGreater*()—with appropriate arguments and return types. They are all very straightforward and do not require further clarification.

Finally, evolved GP trees (mathematical expressions representing obtained evaluation functions) have been moved into VCMI battle engine. During the battle current action results have been dynamically computed and submitted to the provided mathematical expression, which returned the scalar utility value of currently evaluated action.

The second applied algorithm was **temporal difference algorithm with back propagation neural network**. Standard implementation of RL algorithms in VCMI would be very complicated due to a complexity of battle states representation. Usually, single game state should contain information about the whole observable environment at the certain time step. In our created opponent model only relevant action results have been currently interpreted by the game engine. To meet these requirements a TD(1) algorithm has been chosen.

TD(1) is a version of TD(λ) algorithm [25] with $\lambda = 1$. TD(1) is generally used for all undiscounted tasks where the observed reward is distributed equally to all previous states. This algorithm provides good generalization of Monte Carlo stochastic methods to online and incremental environments. The pseudocode of its standard version has been shown in Alg. 2.

In VCMI TD(1) algorithm has been customized to update function only when observing definite reward (after winning the battle). Updating function with each action could be added only when implementing full states representation. No intermediate rewards have been used during learning phase, but they could be added in the future if appropriate unit fitness function was provided.

Algorithm used ϵ-greedy policy for environmental exploration. Each strategy (move evaluation function) has been represented by a neural network due to continuous state-space of VCMI battle environment. The network has been learned using Back Propagation algorithm. Related battle observations normalized to [0,1] range constituted for network inputs. The network output returned an action utility value.

Algorithm 2 TD(1) Algorithm

Initialize function $V(s)$
γ—discount factor, α—learning rate
for all episodes **do**
  $s \leftarrow initialstate$
  repeat
    Use policy π to go to the next state s'
    Observe reward r
    $\delta \leftarrow r + \gamma V(s') - V(s)$
    for all previous states s **do**
      $V(s) \leftarrow V(s) + \alpha\delta$
    end for
    $s \leftarrow s'$
  until *finalstate* reached
end for

Regarding training opponents, two common approaches have been investigated. First of them was learning from so called self-play. Two reinforcement learning agents shared the same neural network updated with moves of the winning player. The second approach involved providing specific computer opponent to learn from. For this purpose a simple training AI has been implemented. The first approach will be referenced here as TD Self-Play and the second one as TD Challanged.

6 Experiments and Results

6.1 Learning Environment

Learning environment has been set up to mirror testing environment (VCMI) with one major difference: all random factors have been replaced with deterministic functions. For both environments all spells and external bonuses have been disabled and the role of commanding heroes maximally reduced.

All algorithms have been trained using one battle setup. Battle arena map designed specifically for tournament battles was taken from official HoMM3 competition website. The map included eight heroes from different towns with large and balanced armies—perfect for matching opponent strategies. Four of these setups have been chosen. The first of them (L—Fig. 2a) has been used throughout whole learning phase as training data. The other three setups (T1, T2 and T3—Fig. 2b–d) were additionally used during testing phase to provide experimental variance.

Fig. 2 Learning (a) and testing (b, c, d) setups

Two computer opponents have been implemented for testing algorithms in adapted learning environment. The first of them was DummyAI, which was based on simple, aggressive strategy. When current unit had some opponents in range, computer chose one of them and attacked it. If there were no enemies in current unit range, computer waited or moved towards random opponent unit if waiting was not possible. The second computer bot, RandomAI, was implemented to test algorithms against completely random player. Each algorithm played exactly one hundred games against both opponents in four possible setup configurations defined earlier, with half of the games played normally and the other half played with switched sides.

Additionally, a ChallengableAI has been implemented as an auxiliary opponent to be used during training phase of TD Challanged reinforcement learning algorithm. It was a modification of DummyAI containing additional probabilistic parameter of choosing completely random action. This allowed TD Challanged to win some initial battles and learn. Otherwise algorithm would not be able to bootstrap from losing strategy state. The probabilistic factor was initially set to 0.5 and gradually decreased to zero over half of the number of games played.

6.2 Testing Environment (VCMI)

The obtained strategies were then moved to VCMI game and tested against two original AI modules: StupidAI and BattleAI.

StupidAI is basic implementation of combat algorithm. It makes decisions based on maximizing difference between action payoff (inflicted damage) and action cost (received damage). It is very simple algorithm, but proves to be effective against non-experienced human opponents. StupidAI is used in VCMI as a standard battle algorithm, due to its clarity and reliability. BattleAI improves all-around performance of StupidAI. It provides better units placement and adds other functionalities like spell handling and management of creatures special abilities. It evaluates the utility of the current move based on complex predefined criteria. BattleAI module is still developed and adjusted by VCMI team according to players' best tactics.

No standard benchmarks have been used for presented solutions. This is because a new environment required a new kind of benchmark set up. For this purpose a baseline configuration has been used with arbitrary number of games to provide a foundation for universal testing platform in the future.

6.3 Algorithms Configuration

The GP Co-Evolution has been implemented with the use of ECJ library (http://cs.gmu.edu/~eclab/projects/ecj/). The following parameters have been set for the algorithm. Trees were created using ramped half-and-half method with maximum height of 6. Individuals have been evaluated with K-Random two-way algorithm of size 6. Single elimination tournament has also been tested, but did not yield satisfactory results. Standard reproduction and crossover ECJ pipelines were used with probability of 0.9 and 0.1 respectively. For node selection probability terminals have been given the value of 0.1 and non-terminals of 0.9.

The best co-evolutionary strategy emerged after 308 generations from initial population of 1000 individuals. That makes almost one million games played during algorithm's evaluation phase. Algorithm ran for 32 hours on a standard desktop machine with two-core processor. For providing algorithm variance another strategy has been evolved from smaller population of 600 individuals. After 223 generations, it turned out to be less effective than the original one in all battle setups except for T3, where it showed some improvement.

The following parameters have been set for TD(1) algorithm. Its policy epsilon value has been set initially to 0.3 for TD Challenged and 0.5 for TD Self-Play. It has been linearly decreased with every episode until reaching 0. Both lambda and discount factors have been set to 1.0. Move evaluation function has been represented by neural network with 15 input neurons, 4 hidden layers with 60 inputs each and single output neuron. The number of hidden layers and their neurons have been chosen arbitrarily based on multiple experiments. Also, it has been observed that less than 4 hidden layers resulted in deteriorated learning process. BP learning rate has been set to 0.01 for TD Challenged and 0.001 for TD Self-Play. Momentum has been set to 0.9.

TD Challenged ran for 10000 and TD Self-Play for 100000 episodes. Learning rates of both TD algorithms have been set to 1, because of already defined BP learning rates.

6.4 Results

Reinforcement learning algorithms obtained more than satisfactory results, given the nature and complexity of defined problem. Learning from basic stationary opponent yielded very good results against known opponent in learning environment. Also, the algorithm retained its efficiency against other functional agent on similar level (StupidAI). That allows to assume that, when provided with appropriate opponent and larger training data, algorithm would be able to successfully learn more advanced and efficient strategies.

TD Self-Play obtained good generalization of evaluation function, achieving even better results than TD Challenged in universal environment. However, unlike the TD Challenged, there is no guarantee that TD Self-Play would perform better in different learning conditions due to known temporal difference limitations of learning from self-play, explained in multiple publications.

GP Co-Evolution turned out to be a very efficient solution. It performed much better than both reinforcement learning algorithms and managed to outplay even the most demanding opponents in complex, non-deterministic VCMI environment.

Also, a few human involving battles were played against the algorithms in order to observe their behaviors exhibited during the battle. It has been observed that GP Co-Evolution employed some advanced combat strategies like unit positioning, blocking tactics, waiting and attacking. The RL algorithms pursued more simple, greedy strategies, though still effective against basic opponents.

Table 2 A number of games won against DummyAI and RandomAI in learning environment. 100 games have been played in each setup

	vs DummyAI				vs RandomAI			
Algorithm	L	T1	T2	T3	L	T1	T2	T3
GP Co-evolution	100	100	100	57	100	100	100	98
TD Challenged	100	0	100	100	97	68	99	61
TD Self-Play	53	0	45	100	96	70	100	82

Table 3 A number of games won against StupidAI and BattleAI in VCMI game environment. 100 games have been played in each setup

	vs StupidAI				vs BattleAI			
Algorithm	L	T1	T2	T3	L	T1	T2	T3
GP Co-evolution	100	97	100	63	99	96	91	41
TD Challenged	82	56	94	4	6	2	0	0
TD Self-Play	81	48	96	23	19	13	0	0

The detailed test results are presented in Table 2 and 3. It can be concluded that the deteriorated algorithms performance in T1 and T3 setups can be attributed to unknown test environment with creature-specific contributing factors. The decreased efficiency in VCMI, especially against the BattleAI, is believed to be a result of high complexity of the opponent and the random factors in the game.

7 Further Improvements

There are several ways in which performance of applied solutions could be improved. The most natural improvement would require increasing the size of candidate actions list, returned by enclosing AI functional entity. If the opponent model learn how to properly analyze candidate moves, it could be safely assumed that the final performance of the algorithm would mainly depend on the number of presented move options.

One prototype of such solution has been implemented, where algorithm additionally evaluated all available attack positions. That alteration increased GP Co-Evolution performance in T3 setup by 12 %, which is a very good result. With more computational resources, even whole candidate moves space could be analyzed. The alternative would be to include auxiliary local search methods to optimize suggestions of possible moves [14].

Another enhancement concept involves providing larger training data. Original algorithms have been trained with only one army setup due to required time constraints of learning phase. However, with more advanced computational node or cluster of nodes, algorithm could be trained with many possible army setups instead of only one. That could be very beneficial to final results of the algorithm. Some other specific RL related improvements could be recommended as well.

First, TD Challenged algorithm could be trained against more demanding, intelligent opponent. Second, neural network parameters could be obtained using evolutionary genetic algorithm, which is a common practice. Also, some other types of neural networks could be applied. Finally, intermediate RL rewards could be introduced to improve convergence of TD algorithms.

Also, some advanced learning mechanism could be used to obtain a generic unit fitness function returning a scalar value of current unit's strength. An accurate estimation of such function would provide additional information about analyzed actions and thus improve the algorithm's knowledge about the game environment. The fitness function would also facilitate battle states representation and it could be used with other AI modules like spell casting in order to optimize effects of the chosen strategy.

8 Summary and Conclusions

VCMI is a very interesting project. It provides a complex, demanding environment that can be analyzed by AI algorithms. Its turn-based gameplay with clear and structured rules is perfect for employing advanced solutions requiring more computational time. With its huge knowledge base, active players community and easily extensible architecture, VCMI becomes a perfect platform for testing AI solutions.

Conducted experiments showed potential benefits resulting from implementing global optimization algorithms in turn-based strategic games. The similar solutions can be used in all systems and decision engines that are based on environmental data and require behavior optimization. The efficiency of optimization algorithms will depend mostly on target system's complexity and defined environment model. System complexity can be simplified by the proper definition and reduction of the original problem. Environmental model should be supported with domain-specific knowledge and higher abstraction layer of analyzed data.

This paper provides an universal evaluation platform for future algorithms and can be treated as a starting point for further research. Such research should include analysis of all proposed improvements and assessment of other, innovative, state-of-the-art learning mechanisms in VCMI environment, with emphasis put on providing complete solution for the whole scope of battle AI. It is believed that further research on related issues will lead in future to significant development of artificial intelligence in complex strategic games and other similar systems.

Acknowledgments This research was partially supported by Polish Ministry of Science and Higher Education under AGH University of Science and Technology, Faculty of Computer Science, Electronics and Telecommunications statutory project.

References

1. Vcmi documentation. http://wiki.vcmi.eu/index.php?title=Game_mechanics
2. Amato, C., Shani, G.: High-level reinforcement learning in strategy games. In: van der Hoek, W. et al. (ed.) AAMAS. pp. 75–82. IFAAMAS (2010)
3. Baier, H., Winands, M.: Monte-carlo tree search and minimax hybrids with heuristic evaluation functions. In: Cazenave, T., Winands, M., Björnsson, Y. (eds.) Computer Games, Communications in Computer and Information Science, vol. 504, pp. 45–63. Springer International Publishing, Berlin (2014)

4. Beal, D.: Learn from your opponent—but what if he/she/it knows less than you? In: Retschitzki, J., Haddab-Zubel, R. (eds.) Step by Step : Proceedings of the 4th Colloquium Board Games in Academia. pp. 123–132. Editions Universitaires (2002)
5. Björnsson, Y., Hafsteinsson, V., Jóhannsson, Á., Jónsson, E.: Efficient use of reinforcement learning in a computer game. In: Computer Games: Artificial Intellignece, Design and Education (CGAIDE'04), pp. 379–383 (2004)
6. Brafman, R.I., Tennenholtz, M.: R-MAX—a general polynomial time algorithm for near-optimal reinforcement learning. J. Mach. Learn. Res. **3**, 213–231 (2002)
7. Browne, C., Powley, E.J., Whitehouse, D., Lucas, S.M., Cowling, P.I., Rohlfshagen, P., Tavener, S., Perez, D., Samothrakis, S., Colton, S.: A survey of monte carlo tree search methods. IEEE Trans. Comput. Intellig. AI Games **4**(1), 1–43 (2012)
8. Conitzer, V., Sandholm, T.: Awesome: a general multiagent learning algorithm that converges in self-play and learns a best response against stationary opponents. In: Fawcett, T., Mishra, N. (eds.) ICML. pp. 83–90. AAAI Press (2003)
9. Dreżewski, R., Siwik, L.: Multi-objective optimization using co-evolutionary multi-agent system with host-parasite mechanism. In: Alexandrov, V.N., van Albada, G.D., Sloot, P.M.A., Dongarra, J. (eds.) Computational Science—ICCS 2006. LNCS, vol. 3993, pp. 871–878. Springer, Berlin (2006)
10. Dreżewski, R., Siwik, L.: Multi-objective optimization technique based on co-evolutionary interactions in multi-agent system. In: Giacobini, M., et al. (eds.) Applications of Evolutionary Computing, EvoWorkshops 2007. LNCS, vol. 4448, pp. 179–188. Springer, Berlin (2007)
11. Dreżewski, R., Siwik, L.: Agent-based co-operative co-evolutionary algorithm for multi-objective optimization. In: Rutkowski, L., Tadeusiewicz, R., Zadeh, L.A., Zurada, J.M. (eds.) Artificial Intelligence and Soft Computing—ICAISC 2008. LNCS, vol. 5097, pp. 388–397. Springer, Berlin (2008)
12. Ekker, R.J.: Reinforcement Learning and Games. Master's thesis, University of Groningen (2003)
13. Elidrisi, M., Johnson, N., Gini, M.: Fast learning against adaptive adversarial opponents. In: AAMAS'12: Proceedings of the 11th International Conference on Autonomous Agents and Multiagent Systems (2012)
14. García-Martínez, C., Lozano, M.: Local search based on genetic algorithms. In: Siarry, P., Michalewicz, Z. (eds.) Advances in Metaheuristics for Hard Optimization. Natural Computing Series, pp. 199–221. Springer, Berlin (2008)
15. González-Pardo, A., Palero, F., Camacho, D.: Micro and macro lemmings simulations based on ants colonies. In: Esparcia-Alcázar, A.I., Mora, A.M. (eds.) Applications of Evolutionary Computation. Lecture Notes in Computer Science, pp. 337–348. Springer, Berlin (2014)
16. Jaskowski, W.: Algorithms for test-based problems. Ph.D. thesis, Institute of Computing Science, Poznan University of Technology, Poznan, Poland (2011)
17. Koza, J.R.: The genetic programming paradigm: genetically breeding populations of computer programs to solve problems. In: Soucek, B. (ed.) Dynamic, Genetic, and Chaotic Programming, pp. 203–321. Wiley, New York (1992)
18. Littman, M.L.: Markov games as a framework for multi-agent reinforcement learning. In: Proceedings of the 11th International Conference on Machine Learning (ML-94), pp. 157–163. Morgan Kaufman, New Brunswick (1994)
19. Lucas, S., Samothrakis, S., Pérez, D.: Fast evolutionary adaptation for monte carlo tree search. In: Esparcia-Alcázar, A.I., Mora, A.M. (eds.) Applications of Evolutionary Computation. Lecture Notes in Computer Science, pp. 349–360. Springer, Berlin (2014)
20. McPartland, M., Gallagher, M.: Learning to be a bot: reinforcement learning in shooter games. In: Darken, C., Mateas, M. (eds.) AIIDE. The AAAI Press, Menlo Park (2008)
21. Nogueira, M., Cotta, C., Fernández-Leiva, A.J.: An analysis of hall-of-fame strategies in competitive coevolutionary algorithms for self-learning in RTS games. In: Nicosia, G., Pardalos, P.M. (eds.) LION. Lecture Notes in Computer Science, vol. 7997, pp. 174–188. Springer, Berlin (2013)

22. Perez, D., Rohlfshagen, P., Lucas, S.: Monte-carlo tree search for the physical travelling salesman problem. In: Di Chio, C., et al. (eds.) Applications of Evolutionary Computation. Lecture Notes in Computer Science, vol. 7248, pp. 255–264. Springer, Berlin (2012)
23. van der Ree, M., Wiering, M.: Reinforcement learning in the game of Othello: learning against a fixed opponent and learning from self-play. In: IEEE Symposium on Adaptive Dynamic Programming and Reinforcement Learning (ADPRL), 2013, pp. 108–115 (2013)
24. Robilliard, D., Fonlupt, C., Teytaud, F.: Monte-carlo tree search for the game of "7 wonders". In: Cazenave, T., Winands, M., Björnsson, Y. (eds.) Computer Games, Communications in Computer and Information Science, vol. 504, pp. 64–77. Springer International Publishing, Basel (2014)
25. Sutton, R., Barto, A.: Reinforcement Learning: An Introduction. MIT Press, Cambridge (1998)
26. Tanner, B., Sutton, R.S.: Td(lambda) networks: temporal-difference networks with eligibility traces. In: De Raedt, L., Wrobel, S. (eds.) ICML. ACM International Conference Proceeding Series, vol. 119, pp. 888–895. ACM (2005)
27. Uther, W., Veloso, M.: Adversarial reinforcement learning. Technical report. In: Proceedings of the AAAI Fall Symposium on Model Directed Autonomous Systems (1997)
28. Wender, S., Watson, I.: Using reinforcement learning for city site selection in the turn-based strategy game civilization iv. In: Hingston, P., Barone, L. (eds.) CIG. pp. 372–377. IEEE (2008)
29. Wiegand, P.R.: An analysis of cooperative coevolutionary algorithms. Ph.D. thesis, George Mason University (2003)

Agent-Based Neuro-Evolution Algorithm

Rafał Dreżewski, Krzysztof Cetnarowicz, Grzegorz Dziuban, Szymon Martynuska and Aleksander Byrski

Abstract Neural networks are nowadays widely used across many different areas, both enterprise and science related. Regardless of the field in which they are applied, optimization of their structure is usually needed. During the last decades many methods for this procedure have been proposed, among them techniques based on evolutionary algorithms. In this paper a new algorithm for optimization of neural networks architecture based on multi-agent evolutionary approach is proposed. Also results of preliminary experiments aimed at comparing the proposed technique to already existing ones are presented.

Keywords Agent-based evolutionary algorithms · Neural networks · Prediction

1 Introduction

In the still growing range of numeric problems that require the use of large amounts of computational power, a significant area is taken by those, which are unsolvable or too costly in terms of resources for classical approaches. Among them there are tasks related to prediction, economical trends, weather forecasting, image recognition or exploring large data warehouses in search for hidden patterns.

Amid widely applied solutions for such challenging problems, artificial neural networks (ANN [7]) are, without a doubt, one of the most important. What gives an artificial neural network its advantage over classical approaches in the case of numerical problems is its distinctive ability to learn and adapt, which is loosely based on the way an actual neural network works.

R. Dreżewski (✉) · K. Cetnarowicz · G. Dziuban · S. Martynuska · A. Byrski
Department of Computer Science, AGH University of Science and Technology, Krakow, Poland
e-mail: drezew@agh.edu.pl

G. Jezic et al. (eds.), *Agent and Multi-Agent Systems: Technologies and Applications*, Smart Innovation, Systems and Technologies 38,
DOI 10.1007/978-3-319-19728-9_8

No matter what approach to constructing neural networks we assume, the main issue in the procedure is to teach the NN to process input data in desired way. The whole point of creating a neural network is to avoid forcing user to find the equations for calculating the problem and interaction should be reduced to minimum [10].

For any problem defined, which can be solved using neural networks, we can point out unlimited number of different network structures that can provide us with valid output. These layouts vary greatly in complexity and computational power/time needed to reach the solution. Due to a huge number of possible network configurations, manual process of refining the neural network architecture requires many experiments and may not necessarily lead to the best solution. That is why several methods of automatic optimization of neural network architecture were invented.

Among the most known algorithms for automatic pruning are Magnitude Based Pruning, Optimal Brain Surgeon, Optimal Brain Damage and Optimal Brain Surgeon with Nodes [9]. From the construction procedures we can highlight the following methods: Cascade Correlation [17] and Flex Net [13].

Each of the methods mentioned above can be considered a "classical" neural network optimization technique. The main advantage of such methods is undoubtedly their relative simplicity. Unfortunately, as the size of networks grow, their convergence and overall performance declines greatly. Good NN optimization algorithm must have two major characteristic—complexity growing very slowly with the increase of network architecture and an ability to search for the solution very effectively in both wide, and narrow space.

In this paper we propose a new agent-based approach. We will also present results of preliminary experiments aimed at comparing proposed agent-based approach to "classical" evolutionary algorithm used for neural network optimization.

2 Evolutionary Neural Network Optimization

Evolutionary neural networks (ENN) are an alternative approach to neural network optimization, in which the search for a desirable neural architecture is made by an evolutionary algorithm [12, 19]. It is a very flexible solution, but it requires some time to work properly. This method can be characterized as a "black-box" approach (compare [15]) in which we ignore internal structure, or semantics, of created solution, but we can evaluate its correctness and quality. Similar algorithms have proven to be especially helpful in the case of complex problems, e.g. NP-complete class, for which classical approaches give no satisfactory results [20].

The term "evolutionary algorithm" is a very wide one and describes a group of different techniques. One of them are co-evolutionary algorithms in which fitness of an individual is computed on the basis of its interactions with other individuals present in the population.

In order to encode the structure of the network into a chromosome generally two approaches are used—direct encoding and indirect encoding. In direct encoding longer chromosomes are needed, and indirect encoding suffers from noisy fitness evaluation [19].

During the evolution process not only connections of NN can be evolved. The same process can operate on weights, but it is much easier to evolve the structure of connections, leaving the search for the values of weights to the network training algorithm.

NEAT System belongs to a class of evolutionary algorithms, which change weights and the structure of neural network [16].

NeuroAGE was designed for weight and network structure evolution controlling complex control systems [6].

Another approach to the neuro-evolution was proposed in [14]. In this paper a Symbiotic Adaptive Neuroevolution System (SANE) for solving controlling problems using neural networks was used. Instead of evolving the whole network, SANE carries out operations on populations of nodes in the hidden layer. Unlike the previous described methods, SANE focuses only on evolution of neuron weights, working on predetermined topology.

Evolution of neural network architecture can be performed in multi-agent system. Taking advantage of autonomy, parallel actions and other agency features [18] the system can become a versatile alternative for classical sequential algorithms. Considering prediction of time series [1], when the signal to be predicted is very complicated (e.g. when different trends change over time or temporal variations in relationships between particular sequences are present) the idea of a multi-agent predicting system may be introduced, in which each agent may perform an analysis of incoming data and give predictions [2]. In the case of classification systems, many agent can treat the processed data in a different way, sometimes repeating themselves, sometimes applying completely different points of view (different transforms, preprocessing algorithms etc.)

3 Proposed Multi-agent Approach to Neuro-Evolution

Both NeuroAGE and NEAT system alike present classical approach to neuro- evolution problem. They each focus on developing topology and adapting weights of connections through evolution of networks treated as a coherent whole. In the SANE project it was proved, that better results can be achieved using separated optimization of individual parts of the given problem, by treating neurons as independent, cooperating, and the same time competing, entities. As a result, the population automatically preserves its diversity and is resistant to the premature convergence.

Another problem is the size of genotype growing in response to more complex network architectures. Genotype of a single individual must store information about subsequent connections. Again, in this scenario, SANE algorithm performs better, because genotypes of schemes and neurons have fixed size.

Proposed Multi-Agent Distributed SANE System (MADNESS) is an extension of ideas proposed in SANE system. MADNESS utilizes similar symbiotic approach to evolution of individual neurons but realizes it using agent-based system.

Unlike its predecessor, system has the ability to modify topology of networks. However, maintaining two populations was abandoned in favor of free transfer of neurons between tested networks.

3.1 Distributed Agent Model

In MADNESS neurons were implemented as agents independent from each other and placed inside an environment providing resources needed for survival and access to necessary information. Each agent is given by the environment a certain amount of resource ("life energy") at the start of his life. When this energy reaches zero, the individual dies and is removed from population. The environment checks periodically the quality of all networks and chooses the best neurons, which are given a certain amount of energy from the pool.

System is based on distributed model, thus enabling parallel existence of several independent environments (Fig. 1). A central node exists in this model, responsible for control of the whole experiment and processing nodes. Each node creates an individual, independent instance of environment, and initializes it with a random population of agents.

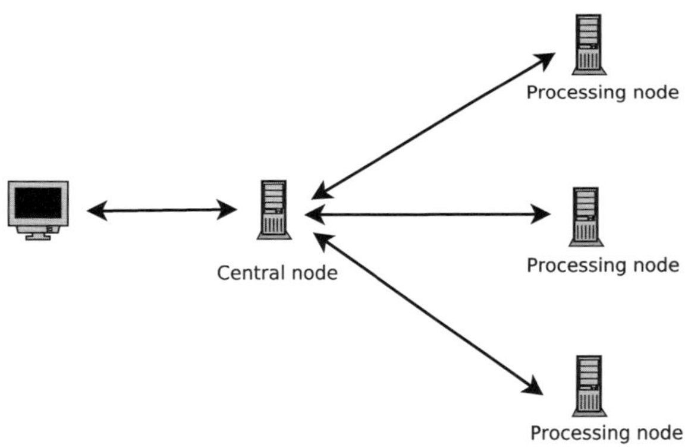

Fig. 1 Proposed system architecture

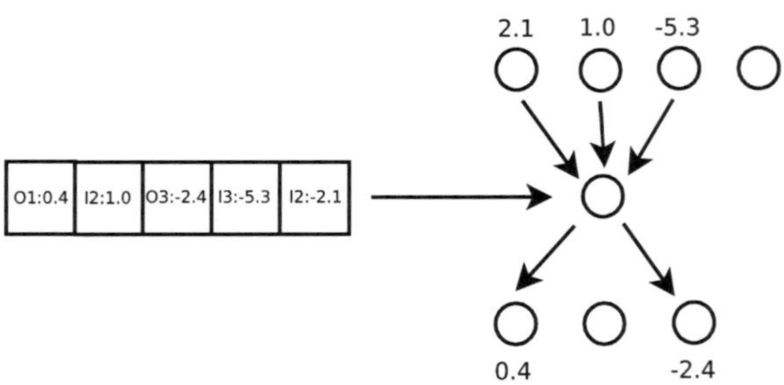

Fig. 2 A hidden layer neuron and its genotype

During the evolution central node controls and synchronizes the work of computing nodes. It also enables the transfer of neurons between environments. Distributing the processing on multiple independent environments significantly improves the algorithm performance. Thus, looking through a larger solution area is possible and the risk of clustering around a local minimum is reduced. An individual from one population, after migration and transferring to another one, has a significant possibility of boosting evolution in his new environment.

3.2 Evolutionary Operations

Genotype of a neuron is organized similarly to the one in SANE system. It is a sequence of 40-bit definitions of connections with neurons from input and output layers. Every definition begins with eight bits identifying neuron on the other side of connection. This eight bits can be simply treated as an unsigned integer from the range [0–255]. If its value is smaller than 128 then target neuron belongs to input layer, otherwise it comes from output layer. The result of modulo operation coming from division of this value and the number of neurons in target layer is the position of the neuron. After the first eight bits the next 32 come, which encode a floating point number serving as weight. An example genotype and a corresponding phenotype is presented in Fig. 2.

Mutation is done with the probability 0.001 and as an effect reverses a bit value. This operation is performed for every individual in each cycle.

For the neuron, that signals the readiness for reproduction and has enough energy, a partner is chosen from remaining individuals, which also are ready for the reproduction. The process itself is performed, similarly as in the SANE system, through one-point division of both parents' genotypes. However, copying one of the parents was completely abandoned. Instead of this, individuals created from joined parts of genotypes are only created. Newly created offspring is given a certain portion of life energy transferred from the parents.

3.3 Individuals Life Cycle

Each of environments is at the beginning initialized with an identical number of individuals, divided randomly between the same number of networks. Number of networks maintained in a given environment is constant for the duration of simulation. Individuals posses randomly generated genotypes of configurable length that has to be a multiple of number 40, and one chosen by user from few predefined activation functions. To each of them manager of resources allocates equal starting amount of life energy. Environment is then ready to start the simulation.

Computations proceeds as follows. First, each neuron undergoes the mutation according to the rules described above. Next, quality of all networks is measured. Values from an experimental set are provided to their input layers and their output is compared with expected result. On this basis their adaptation factor—fitness function value is evaluated. Individuals with the best qualities receive energy from resource manager from available pool and the others lose adequate amount of energy that is returned to the energy pool.

In the notification phase every individual decides about undertaking one of the predefined actions. The desire to perform an action is reported to the environment, which at the end performs all requests, making the impression that all actions were done at the same time.

Migration involves changing environment by a neuron that requested it. From such an individual all energy is taken away. For the purpose of maintaining good performance of the system, all migrating individuals are buffered and sent to the central node together, at the end of the computation cycle. The central node redirects them to the environment possessing the smallest number of individuals.

Changing the network operation causes one neuron to be transferred between hidden layers of two networks. There are multiple strategies for designating target network—strategy of the smallest network, in which neuron is assigned to the network with the least neuron count in hidden layer, strategy of best network, where target network becomes the one with the biggest fitness function value, strategy of worst network and random strategy, which is simply a random choice from all available networks. Probability of undertaking an action of changing network is changed dynamically according to energy gains in the last part of evaluation phase. If a neuron was given some energy, then probability is decreased, and if it was taken from him, probability increases. Change of probability value takes place in accordance to logarithmic function.

Swapping places operation is done by exchanging networks between two neurons connected by parent-child relation. It can be activated only by an individual that already has some descendants. Its result is a transfer of child node to the hidden layer of parent and vice versa. This is dictated by the fact that the progeny is, as a product of crossover, potentially a better individual than his parent.

The last but maybe most important, action is reproduction. As it should be performed only by the individuals with a very good fitness value, represented by amount of life energy accumulated, the minimal threshold of energy needed for reproduction should be adequately high. For every individual that is ready for reproduction a partner is chosen. Thus formed couples are subjected to crossover operation, performed according to rules described above, resulting in creating two new individuals per pair. Energetic costs are paid only by neuron which initialized reproduction—this energy is distributed between children.

Computations end when every environment executes a predefined number of steps. The whole process remains constantly under control of the central node.

3.4 SANE Implementation

In order to verify the functionality of MADNESS system it was necessary to confront its results with those obtained from its predecessor—the SANE system. However, for the comparison to be reliable it was necessary to maintain a certain level of similarity in the construction of both systems, with particular emphasis on technology used for their implementation.

All utilized SANE algorithms were implemented to, as possible, strictly imitate the original project [14]. This section focuses on modifications which were done in comparison to prototype and aspects that were not discussed in the original document.

The size of a single neuron genotype was modified. Taking into account the inner representation of floating-point numbers in the used programming language, the number of genes representing weights of connections was increased from 16 to 32. Because of the way in which target neuron is encoded, it is possible to place two or more connections to the same endpoint in one genotype. In such a case they are replaced in phenotype by one connection with weight which value is the sum of all substituted weights. Genetic operations (mutation and crossover) are executed exactly as in the case of regular neuron.

Similarly to MADNESS system, activation function used by neurons from hidden layer and output layer can be configured. Basic SANE module architecture is presented in Fig. 3. The value of fitness function for neuron is calculated as the sum of fitness values from five best network schemes which this particular neuron co-creates.

Together with the neuron population, also the scheme population is maintained. Every individual that belongs to it is built according to a particular prototype. All schemes are subjects to identical, two-phase mutation. Crossover is the effect of one-point division of parent genotypes, but there is no copying of complete genotype from one parent—both descendants are the result of swapping genes. The way of choosing schemes for reproduction and the operation of selecting individuals for next generation are the same as in the case of neurons.

The algorithm works as follows. From all the schemes neural networks are constructed, and to each of them input signal from the test set for the given problem is provided. Basing on the results from calculations, the fitness values for individuals from scheme population are calculated. Having all the networks evaluated, neurons can be given their respective fitness value. Next, a neuron reproduction phase is executed, in which parents are selected and assigned to pairs. From this couples new individuals are created. From current population worst individuals are removed and all new neurons join the group, founding new generation. At the end, selected genotypes are subjected to mutation. This phase is repeated for network scheme population. The whole procedure is done in cycles until the stop condition (executing a predefined number of steps or reaching by one of the individuals the necessary fitness level) is fulfilled.

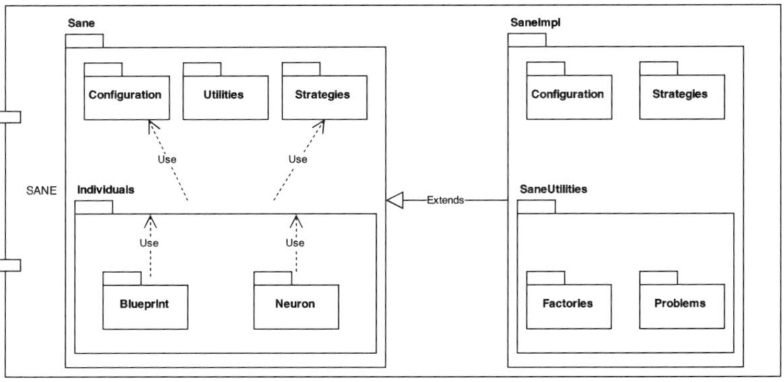

Fig. 3 Sane module architecture

4 Experimental Results

System was subjected to series of experiments. A set of problems solvable by neural networks was selected and for each of them the quality of results obtained from SANE and MADNESS modules were compared. Additionally, a number of strategies for allocation of neurons in networks were tested:

1. Steady/Uniform—neurons were added to smallest network in population, keeping a balanced distribution.
2. Random.
3. To worst—neurons added to the network with worst fitness.
4. To best—neurons added to the network with best fitness.

Each of the mentioned experiments was performed multiple times, which enabled the possibility of simple parameters evolution. Only results with the best fitness were taken into account during modules evaluation.

Presented preliminary experimental results have been obtained for multi-layered perceptrons [8] applied to the following problems:

- classification of simple bitmap image—the network had 9 neurons in the input layer and had to classify 4 patterns,
- prediction of a popular chaotic benchmark Mackey-Glass time series [11]—170 neurons in the input layer, one neuron in the output layer.

The parameters of the system for the problem of bitmap classification are given in Table 1.

The other parameters of the system were as follows: network population size was 50, hidden layer size was 10, size of neuron population was 50, size of genotype was 440, mutation probability was 0.001, percent of awarded individuals was 30, total amount of resources was 300, initial amount of resource was 5 and penalty value was 1.

Table 1 The values of parameters (classification)

	Prob.	Min. energy	Energ. cost
Death	N/A	N/A	N/A
Addition	0.5	5	1
Removal	0.01	N/A	N/A
Inheritance	0.1	N/A	N/A
Reproduction	1	10	6
Mutation	1	0	0

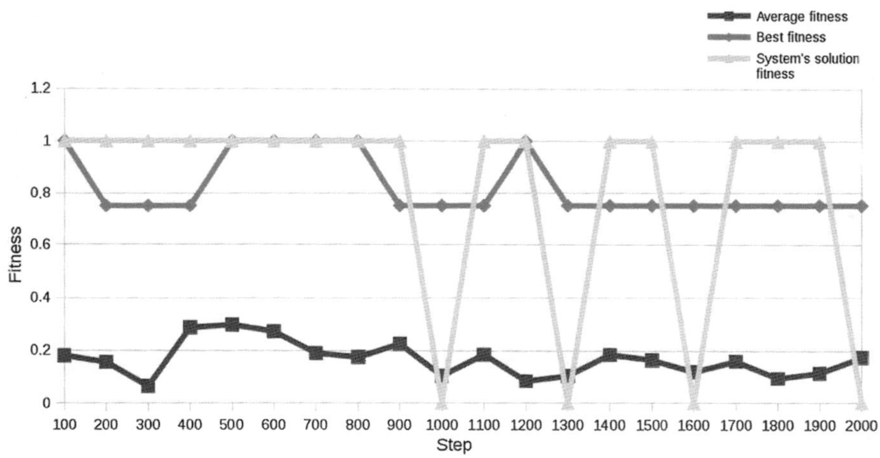

Fig. 4 SANE bitmap classification results

Figure 4 shows the results of the reference SANE system applied to bitmap classification. The results of MADNESS system obtained for different strategies of neuron assignment are shown in Fig. 5a–c. It seems, that the proposed MADNESS system produced slightly worse results than referenced SANE systems for the problem of simple bitmap classification.

The parameters of the system for Mackey-Glass time series prediction problem are given in Table 2. The presented results were obtained for prediction range 20.

Table 2 The values of parameters (prediction)

	Prob.	Min. energy	Energ. cost
Death	N/A	N/A	N/A
Addition	0.2	5	1
Removal	0.01	N/A	N/A
Inheritance	0.1	N/A	N/A
Reproduction	1	10	6
Mutation	1	0	0

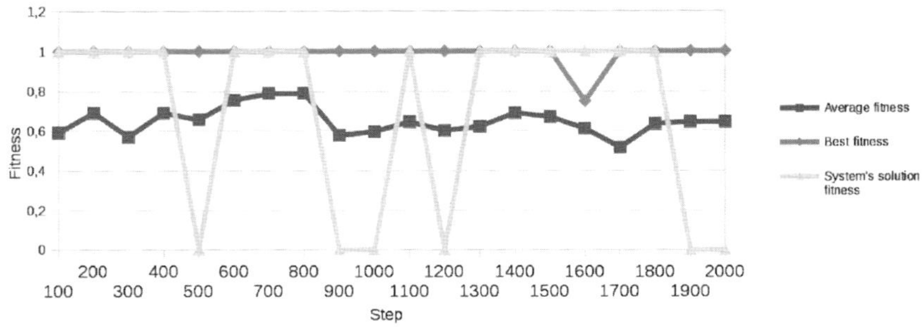

(a) Uniform strategy

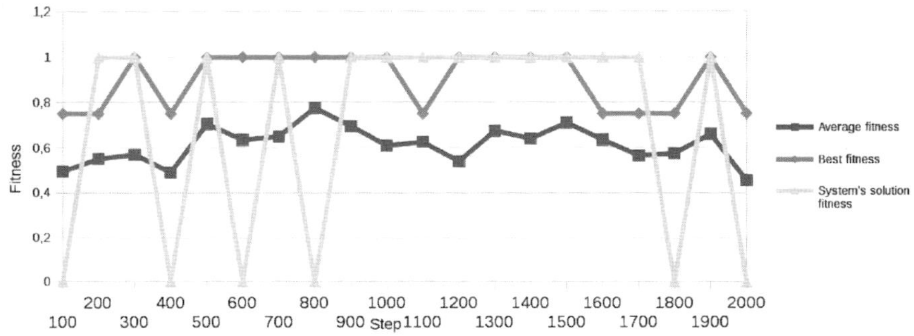

(b) "To the worst"strategy

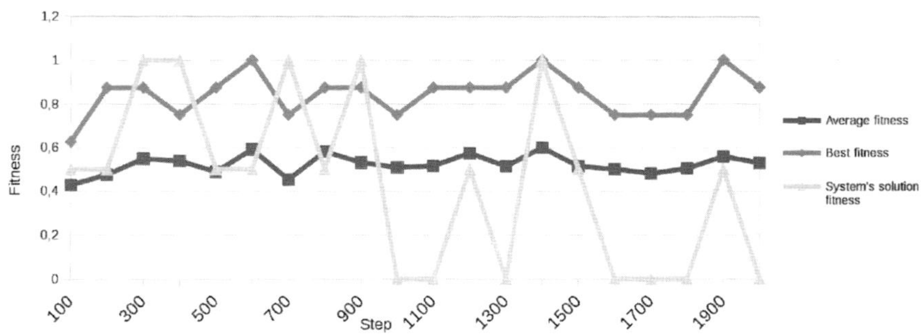

(c) Random strategy

Fig. 5 MADNESS bitmap classification results

The other parameters for the system were as follows: network population size was 100, hidden layer size was 10, size of neuron population was 100, size of genotype was 1000, mutation probability was 0.001, percent of awarded individuals was 30, total amount of resources was 700, initial amount of resource was 5 and penalty value was 1.

Mackey-Glass series used in experiments is an approximate solution of the following differential equation:

$$\frac{dx}{dt} = \beta \frac{x_\tau}{1 + x_\tau^n} - \gamma x, \gamma, \beta, n > 0 \tag{1}$$

with the following values of parameters: $\beta = 0.1, \gamma = 0.2, n = 10, \tau = 17$.

Figure 6a shows the results of prediction obtained with SANE system (blue—average fitness, red—best fitness and yellow—system's solution fitness), while in Fig. 6b the actual prediction that may be observed (blue—original time series, red—predicted time series). In the similar way, the prediction results obtained with MADNESS system are presented in Fig. 7a, b. The prediction results are more or less equivalent in the means of prediction accuracy, however it is easy to see, that the diversity of solutions obtained with the use of MADNESS system is much lower than in the case of SANE system (compare Figs. 6a and 7a).

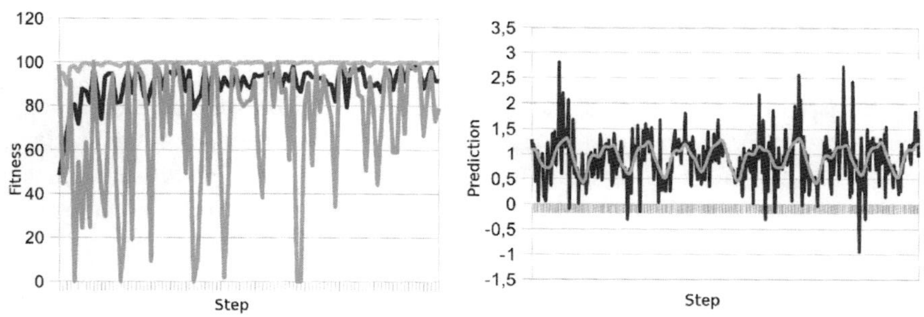

(a) Prediction results (b) Original and predicted time series

Fig. 6 SANE: Mackey-Glass series prediction results

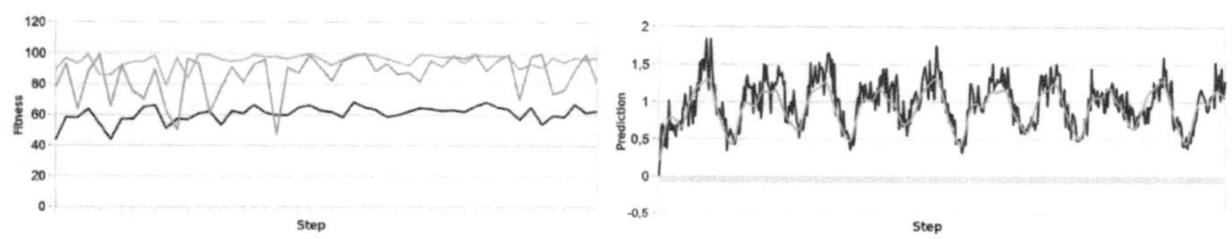

(a) Prediction results (b) Original and predicted time series

Fig. 7 MADNESS: Mackey-Glass series prediction results

4.1 Efficiency of SANE and MADNESS

Promising results were obtained when examining efficiency of the two examined systems. Features of MADNESS agent-based approach (lack of global control, varying number of neural individuals) allowed to significantly decrease the computation time (see Table 3).

Table 3 Computation time (in ms/100 iterations) for different number of individuals

	50	100	200	500
SANE	266	590	780	1780
MADNESS	98	210	390	130

The number of neural individuals was the same in the beginning of computation, but it could be decreased in the case of MADNESS system, hence the possibility of speeding-up the computation in an intelligent way. The presented results have been obtained on PC dual-core computer.

5 Concluding Remarks

In the passed years in which neural network based solutions were utilized, few distinctive groups of optimization algorithms emerged. The first wave—classical iterative algorithms had the advantage of being fairly simple and straightforward, but they soon proved to be far too limited for further use. Vacant spot for new optimization techniques was soon to be filled by genetic algorithms. These kind of methods were more flexible and provided many workarounds for problems previously categorized as unsolvable.

Nonetheless, their maximum potential was also reached in time. Seeing the more and more obvious limitations of GA, some of new implementations of optimization techniques started to be based on multi-agent approach.

In this paper the new agent-based evolutionary approach was presented. Preliminary results are very promising, especially for bitmap classification problem. Average, and for some cases best, fitness is significantly higher than when using pure SANE algorithm. Computation time needed to utilize the system is within reasonable boundaries but of course further experiments and analysis will be needed.

Future research will include performance improvements of the proposed agent-based algorithm. Also new techniques based on the general agent-based neuro-evolution approach would be proposed—for example different variants of agent-based co-evolution, such as cooperative [5] and competitive [4] approach, as well as techniques based on sexual selection mechanism [3].

Acknowledgments This research was partially supported by Polish Ministry of Science and Higher Education under AGH University of Science and Technology, Faculty of Computer Science, Electronics and Telecommunications statutory project.

References

1. Byrski, A., Kisiel-Dorohinicki, M., Nawarecki, E.: Agent-based evolution of neural network architecture. In: Hamza, M. (ed.) Proceedings of the IASTED International Symposium: Applied Informatics. IASTED/ACTA Press (2002)
2. Cetnarowicz, K., Kisiel-Dorohinicki, M., Nawarecki, E.: The application of evolution process in multi-agent world (MAW) to the prediction system. In: Tokoro, M. (ed.) Proceedings of the 2nd International Conference on Multi-Agent Systems (ICMAS'96). AAAI Press (1996)
3. Dreżewski, R., Cetnarowicz, K.: Sexual selection mechanism for agent-based evolutionary computation. In: Shi, Y., van Albada, G.D., Dongarra, J., Sloot, P.M.A. (eds.) Computational Science—ICCS 2007, 7th International Conference, Beijing, China, May 27–30, 2007, Proceedings, Part II. LNCS, vol. 4488, pp. 920–927. Springer, Berlin (2007)
4. Dreżewski, R., Siwik, L.: Multi-objective optimization technique based on co-evolutionary interactions in multi-agent system. In: Giacobini, M., et al. (eds.) Applications of Evolutinary Computing, EvoWorkshops 2007: EvoCoMnet, EvoFIN, EvoIASP, EvoINTERACTION, EvoMUSART, EvoSTOC and EvoTransLog, Valencia, Spain, April 11–13, 2007, Proceedings. LNCS, vol. 4448, pp. 179–188. Springer, Berlin (2007)
5. Dreżewski, R., Siwik, L.: Agent-based co-operative co-evolutionary algorithm for multi-objective optimization. In: Rutkowski, L., Tadeusiewicz, R., Zadeh, L.A., Zurada, J.M. (eds.) Artificial Intelligence and Soft Computing—ICAISC 2008, 9th International Conference, Zakopane, Poland, June 22–26, 2008, Proceedings. LNCS, vol. 5097, pp. 388–397. Springer, Berlin (2008)
6. Dürr, P., Mattiussi, C., Floreano, D.: Neuroevolution with analog genetic encoding. In: Parallel Problem Solving from Nature—PPSN IX, Procedings, pp. 671–680 (2006)
7. Floreano, D., Mattiussi, C.: Bio-inspired Artificial Intelligence. MIT Press (2008)
8. Haykin, S.: Neural Networks: A Comprehensive Foundation. Prentice Hall (1999)
9. Kavzoglu, T., Vieira, C.A.O.: An analysis of artificial neural network pruning algorithms in relation to land cover classification accuracy. In: Proceedings of the Remote Sensing Society Student Conference, pp. 53–58. Oxford University Press (1998)
10. Kriesel, D.: A brief introduction to neural networks (2007). http://www.dkriesel.com
11. Mackey, M., Glass, L.: Oscillation and chaos in physiological control systems. Science **197**, 287–289 (1977)
12. Mitchell, M.: An Introduction to Genetic Algorithms. MIT Press (1998)
13. Mohraz, K., Protzel, P.: Flexnet a flexible neural network construction algorithm. In: Proceedings of the 4th European Symposium on Artifical Neural Networks (ESANN '96), pp. 111–116 (1996)
14. Moriarty, D.E., Miikkulainen, R.: Forming neural networks through efficient and adaptive coevolution. Evol. Comput. **5**, 373–399 (1997)
15. Precup, R.E., David, R.C., Petriu, E.M., Preitl, S., Paul, A.S.: Gravitational search algorithm-based tuning of fuzzy control systems with a reduced parametric sensitivity. In: Soft Computing in Industrial Applications, pp. 141–150. Springer (2011)
16. Stanley, K.O., Miikkulainen, R.: Evolving neural networks through augmenting topologies. Evol. Comput. **10**(2), 99–127 (2002)
17. Ragg, T., Braun, H., Landsberg, H.: A comparative study of neural network optimization techniques. In: 13th International Conference On Machine Learning: Workshop Proceedings On Evolutionary Computing And Machine Learning, pp. 111–118. Springer (1997)
18. Wooldridge, M.: Introduction to Multiagent Systems. Wiley, New York (2001)

19. Yao, X., Liu, Y.: Evolving artificial neural networks through evolutionary programming. In: Fogel, L.J., et al. (ed.) Evolutionary Programming V: Proceedings of the 5th Annual Conference on Evolutionary Programming. MIT Press (1996)
20. Zăvoianu, A.C., Bramerdorfer, G., Lughofer, E., Silber, S., Amrhein, W., Klement, E.P.: Hybridization of multi-objective evolutionary algorithms and artificial neural networks for optimizing the performance of electrical drives. Eng. Appl. Artif. Intell. **26**(8), 1781–1794 (2013)

Towards Designing Android Faces After Actual Humans

Evgenios Vlachos and Henrik Schärfe

Abstract Using their face as their prior affective interface, android robots and other agents embody emotional facial expressions, and convey messages on their identity, gender, age, race, and attractiveness. We are examining whether androids can convey emotionally relevant information via their static facial signals, just as humans do. Based on the fact that social information can be accurately identified from still images of nonexpressive unknown faces, a judgment paradigm was employed to discover, and compare the style of facial expressions of the Geminoid-DK android (modeled after an actual human) and its' Original (the actual human). The emotional judgments were achieved through an online survey with video-stimuli and questionnaires, following a forced-choice design. Analysis of the results indicated that the emotional judgments for the Geminoid-DK highly depend on the emotional judgments initially made for the Original, suggesting that androids inherit the same style of facial expression as their originals. Our findings support the case of designing android faces after specific actual persons who portray facial features that are familiar to the users, and also relevant to the notion of the robotic task, in order to increase the chance of sustaining a more emotional interaction.

Keywords Android robot • Facial expression • Emotion • Static signals • Social perception • Human-agent interaction

E. Vlachos (✉) · H. Schärfe
Department of Communication and Psychology, Aalborg University, Rendsburggade 14, 9000 Aalborg, Denmark
e-mail: evlachos@hum.aau.dk

H. Schärfe
e-mail: scharfe@hum.aau.dk

G. Jezic et al. (eds.), *Agent and Multi-Agent Systems: Technologies and Applications*, Smart Innovation, Systems and Technologies 38,
DOI 10.1007/978-3-319-19728-9_9

1 Introduction

How a face appears to be is a combined result of genetic factors, environmental and cultural moderations, and individual choices [1]. Faces convey messages via their four types of signals; *the static signals* which include the permanent aspects of the face, such as skin pigmentation, morphological/bone structure (i.e., jaw size), cartilage, fatty deposits, and size/shape/location of the facial features (mouth, nose, eyes, brows), *the slow signals* which include changes in the facial appearance due to ageing, for instance permanent wrinkles, or changes in muscle tone, *the artificial signals* such as cosmetics and plastic surgery, and *the rapid signals* which are temporary changes in facial appearance produced by momentary movement of facial muscles (also known as microexpressions) [2]. This paper is focused on the messages transmitted by the static and slow signals of an android robot, an artificial system designed with the goal of mimicking humans in their external appearance/shape, featuring human-like characteristics in its behavior, regarding motion, intelligence and interaction/communication patterns [3–6]. Androids take advantage of their anthropomorphic design to facilitate social interaction, and elicit social responses [7]. Using their face as their prior affective interface, and as an identification provider, androids and other agents embody emotional facial expressions, and convey messages on their identity, gender, age, race, and attractiveness through their static and slow facial signals [2, 8]. Related research on the rapid signals can be found in [9, 10].

Central ideas of P. Ekman and W. V. Friesen [2] were used to examine whether androids can convey emotionally relevant information via their static facial signals, just like humans. Considering the fact that socially relevant information (i.e., personality, sociosexuality, aggression, trust-worthiness) can be identified with accuracy in human faces from visible cues in neutral static images alone, as well as the fact that a large number of functional neuroimaging studies have used neutral faces as a baseline condition for comparing facial expressions, we resorted to using neutral images for our study [11–21]. A judgment paradigm at zero acquaintance ("*perceivers are given no opportunity to interact with targets who are strangers to them*" [22]) was employed to discover, and compare the style of facial expressions of the Geminoid-DK android (modeled after an actual human) and its' Original (the actual human). We want to discover the relation between the emotional judgments for the Geminoid-DK, and the emotional judgments initially made for the Original. *We hypothesize that the emotional judgments for the Geminoid-DK will depend on the emotional judgments initially made for the Original.* Our purpose is to persuade researchers to model android and agent faces after specific actual persons who portray facial features that are relevant to the notion of the robotic task, and also familiar to the users, in order to increase the chance of sustaining a more emotional Human-Agent Interaction (HAI). Hardly a product is appealing to everyone, hardly a human is sympathized by all of his/her acquaintances, and thus hardly a generic anthropomorphic agent would achieve a broad appeal.

Mimicking the facial characteristics of a real person is a typical approach when designing android robots today. However, most androids in the scientific literature are built for research purposes without anyone giving consideration to their faces matching any specific function. Notable instances are: Albert HUBO modeled after Albert Einstein, PKD-A after the novelist Philip K. Dick, Android Twin after the roboticist Zou Ren Ti, Repliee R1 after a five year old Japanese girl, "Rex, the bionic man" after the psychologist Bertolt Meyer, FACE android is based on a real subject, Bina48 after the co-founder of the Terasem Movement Foundation, Face Robot after the death mask of a human, EveR-2 is based on a Korean female, Geminoid-HI is built after its creator Prof. H. Ishiguro, and Geminoid-DK after Prof. H. Scharfe [23–32].

2 Facial Expressions

Facial expressions signal information related not only to the emotional states, but also to the disposition, and the behavioral intention of the interaction partners predicting their future actions [33–35]. Research on personality judgments from facial images indicates that a core accuracy indeed exists in social perception of faces [36], while in [37] it is showcased how a social outcome, such as an electoral success, can be accurately predicted through the brains ability to automatically categorize faces. The evaluation of novel faces possibly influencing the likelihood of social engagement with unfamiliar conspecifics is performed by the primate amygdala [38]. Even though the frontiers of Artificial Intelligence are constantly expanding, simulating the way this subcortical brain region works, and applying it to robots is not yet feasible.

2.1 Styles of Facial Expressions

Emotions can exist without facial expressions, and facial expressions can exist without congruent underlying emotional states, as they can be modified by voluntary muscle movements [39]. The individual's culture, gender, or family background imposes different unwritten codes governing the manner emotions are expressed [40]. The underlying reason for the phenomenon of attributing specific facial expressions to specific social contexts has been observed to be the intention to reveal less negative emotions, leading eventually to tighter bonds within the group [41, 42].

The need to control one's behavior through managing the appearance of a particular emotional expression appropriate for a particular situation in a certain context is described by societally defined rules called display rules [43]. Deeply ingrained habits about managing facial expressions, idiosyncratic to an individual, are developed and learned during childhood, or through a particular experience [44], resulting in a particular cast to someone's facial expressions leading to eight characteristic *styles* of *facial expressions* [2]: *the Withholders* have an unexpressive face, and rarely reveal any emotion, *the Revealers* are the opposite of the withholders, and cannot modulate their facial expressions - emotions are "written" all over their faces, *the Unwitting expressors* do not know what emotion their face is showing, *the Blanked expressors* whose faces look blank when they think they are showing an emotion, *the substitute expressors* substitute the appearance of an emotion for another without knowing that this is happening, *the frozen-affect expressors* always show a trace of one of the emotions in some part of the face when actually not feeling any emotion at all, *the ever-ready expressors* characteristically show one of the emotions at their first response to almost any event, and situation, and *the flooded-affect expressors* show one, or two emotions in a fairly definite way almost all the time -there is never a time when they are feeling neutral.

3 The Experiment

3.1 Stimuli

A Geminoid is a teleoperated robot built after an existing person, and developed as a communication medium to address several telepresence, and self-representation issues [32]. The facial expressions of the Geminoid can be programmed and evaluated by reference to the Original person [9]. Following the recommendations of P. Ekman and W. V. Friesen, the style of facial expression can be extracted through 5 judgments on neutral pictures [2]. We selected a facial expression from our database were the Original looked neutral, not experiencing any emotion, and looking calm/relaxed (Fig. 1/left). Then, we adjusted the values of the 12 pneumatic actuators of the Geminoid-DK to mimic that expression (Fig. 1/right). The stimuli was composed of two videos, each one depicting one of the selected neutral expressions of Fig. 1 in between two black frames, resembling the blinking of the eye. Each video lasted 7 s; the first 4 s informed the viewers about the briefness of the video, the 5th one was left blank, the 6th projected the image and the last one was left blank again. We also projected one more video in the beginning of the survey depicting the Original when surprised (having the answer already noted), serving as an example of the questionnaire process. The recognition rate of emotional expressions might be higher, and have less ambiguity when dynamic sequences are shown rather than still pictures [45], but we wanted to simulate a behavior analogous to interacting with the robot in real life. The movements of the Geminoid are mechanical, abrupt and the change of facial status takes a lot more than a micro-expression. The emotions revealed have almost zero onset and offset time, and are depicted in a position around the peak of the emotional display. Communication partners of the android cannot tell if the emotion is emerging, or if it is dissipating.

3.2 Design

According to the judgment-based approach, emotion can be recognized entirely out of context, while the judgments depend on the judges' past experience of that particular facial expression, either of his own face or of someone else's in conjunction with a revealing behavior [9, 46]. We launched an online questionnaire with video stimuli in order to attract judges. It was a within-subjects design (every user judged both videos) on zero acquaintance (any impact of the stimulus target can be attributed primarily to the physical features of the target). For further validation, the stimuli were tested against the Noldus Face Reader 5, a tool providing emotional assessments (six basic emotions, and the neutral one). Face reading software gives the ability to minutely analyze, and validate assumptions about both natural and artificial faces.

3.3 Procedure

Judges were prompted to answer the forced-choice type question "*What emotion do you think the face in the video is showing?*" by selecting from a list with the pre-determined six basic emotions (disgust, sadness, happiness, fear, anger, surprise) on a 2-point intensity scale; either the emotion existed, or not. The neutral choice was not included as people have a tendency to pick this answer when they are uncertain of a facial expression [2]. Judges could re-view the videos since our goal was not to test them whether they can recognize emotions, or train them to do so.

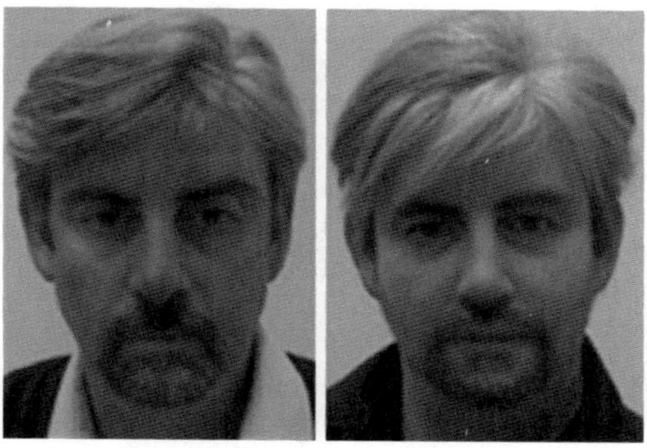

Fig. 1 Facial blueprints for the neutral face of the original (left), and the Geminoid-DK (right)

3.4 Participants

The judges were non - expert respondents, representing a group of people that resembled real world end-users. P. Ekman and W. V. Friesen state that with the assistance of five judges the results will be sound. All responses were anonymous, providing comfort to the judges to freely select an emotion from the predetermined list. We attracted the attention of 50 participants (34 females, and 16 males), who belonged mainly to the 21–30 years old age group.

4 Results and Statistical Analysis

4.1 Results

The results illustrated at Fig. 2 indicate that the strong majority of the judges (half, or more) named the emotion of Anger as the dominant one for the Original, with 36 judgments. There was little agreement upon the judgment for the Geminoid; almost every emotion term was used by one, or another judge. The 23 Sadness judgments as well as the 24 Anger judgments are less than the half (25), but still significantly high.

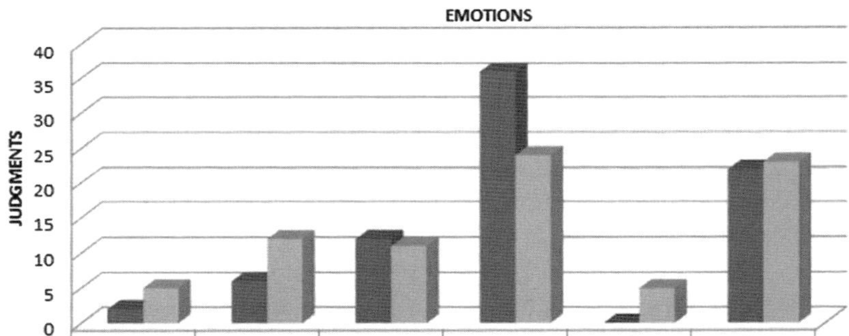

	Surprise	Fear	Disgust	Anger	Happiness	Sadness
Original	2	6	12	36	0	22
Geminoid-DK	5	12	11	24	5	23

Fig. 2 Emotion judgments for the neutral faces

The Noldus Face Reader 5 software tool provides results independent of human judgments. We loaded the same facial blueprints to the Face Reader, after having assigned a calibration to each participant; one to the Original, and one to the Geminoid-DK. By using the individual calibration method, the Face Reader can correct person specific biases towards a certain emotional facial expression. The calibration consists of a few seconds video with the participant maintaining a neutral mode while making mild facial expressions of emotion. As depicted in Fig. 3, both images were classified as Neutral (long horizontal bar in the Expression Intensity module), but the emotion of Anger was predominantly present (short horizontal bar) while the rest of the emotions remained on a zero level.

Fig. 3 Printscreens from the Noldus Face Reader 5 System featuring the Analysis Visualization (left), and the Expression Intensity (right) modules for the neutral faces of the original (up), and the Geminoid-DK (bottom)

4.2 Statistical Analysis

Regression analysis showed that the R value, which represents the Pearson Correlation (how strong the linear relationship is), is 0.939. This value indicates a very high degree of correlation between the emotion values of the Original and the Geminoid. The R-squared value, representing the coefficient of determination, equals to 0.882, indicating that the dependent variable "Geminoid-DK" can be explained to a high degree by the independent variable "Original". In other words, **the emotional judgments for the Geminoid-DK highly depend on the emotional judgments for the Original, thus our hypothesis is true**. The statistical significance of the regression model that was applied is 0.005 (p-value), less than 0.05, indicating that the model applied can statistically significantly predict the outcome variable. Details for the ANOVA and Regression Coefficients are depicted in Table 1. The residuals that are illustrated in Table 2 show the difference between the observed value of the dependent variable (Original) and the predicted value (Geminoid). The residual plot seems to have a fairly random pattern (values range around zero and seem normally distributed) which dictates that a linear model provides a decent fit to the data.

5 Discussion

The results indicated that the strong majority of the judges named the emotion of Anger as the dominant one for the Original (36 judgments). *This outcome suggests that Anger is an emotion often revealed through the face of the Original, and he could be regarded as a frozen-affect expressor for the emotion of Anger*, either because he maintains some element of that emotion in his face due to not totally relaxing his muscles when not feeling any emotion, or due to his static signals, and the morphology of his face (deep set eyes, and a low eyebrow). Differences in facial morphology could be the outcome of life long differences in expressiveness, but they could also be attributed to the fact that masculinity and anger expressions share perceptual space such that masculine faces tend to be perceived as angrier than non-masculine faces. Research by D. Vaughn Becker et al. shows that the "Spontaneous generation of a mental image of an emotional expression is likely to summon an associated gender: Angry faces are visualized as male" [47]. The same study reveals that neutral male faces, relative to neutral female ones, were more likely to be misidentified as angry and less likely to be identified as happy. The zero Happiness judgments for the original support this case. The 23 Sadness judgments for the Geminoid are justified, as Sadness is a frequent response to a neutral face [2]. However, the 24 Anger judgments combined with the results of the Face Reader software (Fig. 3) which indicate that Anger is an emotion predominantly present in the Geminoid, *suggest that the Geminoid-DK has a frozen- affect for the emotion of Anger to a slight degree. The overall results, thus, conclude in stating that the style of facial expressions of the Original, and part of his personal display rules have passed to the android robot.*

Table 1 ANOVA and Regression Coefficients tables

ANOVA						
	df	SS	MS	F (Ftest)	Significance F (Pvalue)	
Regression	1	311.5115789	311.51158	29.794214	0.005476	
Residual	4	41.82175439	10.455439			
Total	5	353.3333333				
	Coefficients	Standard Error	t Stat	P-value	Lower 95 %	Upper 95 %
Intercept	5.889122807	1.898035984	3.102746	0.036125	0.61933	11.15892
X Variable 1	0.572631579	0.104908173	5.458408	0.005476	0.28136	0.863903

On the occasion where indeed the judges misidentified the human model (connecting anger with masculinity), then they will act similarly when judging the robot by making the same "mistakes", and by repositing the same misconceptions, and impressions to the robot, as the statistical analysis indicated. In that case, our study is not affected at all. On the contrary, it would enhance our argument that it is possible to copy the style of facial expressions of a real human into a robot.

Table 2 Residual output and residual plot for the original

Observation	*Predicted Geminoid-DK*	*Residuals*
1	7,034385965	-2,03439
2	9,324912281	2,675088
3	12,76070175	-1,7607
4	26,50385965	-2,50386
5	5,889122807	-0,88912
6	18,48701754	4,512982

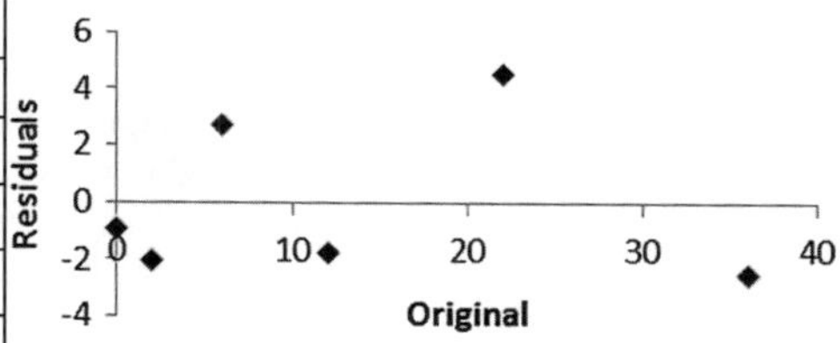

6 Conclusion

We (the humans) have adjusted our space, actions, and performed tasks according to our morphology, abilities, and limitations. Thus, the properties of a social agent should fit within these predetermined boundaries. For a successful HAI, the agent should meet the expectations of its interaction partners, satisfy the goals of the task, match its appearance, and behavior to the given situation, and, lastly, be overt. Agents should let their users know about what they are capable of doing, as well as of not doing, prior to HAI in order to avoid deceit, or attribution of false capabilities respectively [48]. We do not propose embodiment of anthropomorphic cues to all types of agents, not even to all the social ones. However, *if a certain task can be best accommodated by an android, then by modeling the androids' face after a specific human who portrays facial features that are familiar to the users, and relevant to the notion of the task the chances for prolonged and more meaningful HAI will probably be increased.* That, of course, still remains to be tested. By showing that the emotional judgments for the Geminoid-DK highly depend on the emotional judgments initially made for its Original, we suggest that androids inherit the same style of facial expression as the humans they are modeled after. This study is only a step towards designing android faces after actual humans. Future research plans include experimentation with more androids that are modeled after humans.

References

1. Young, A., Bruce, V.: Pictures at an exhibition: the science of the face. Psychologist **11**(3), 120–125 (1998)
2. Ekman, P, Friesen, W.V.: Unmasking the face – a guide to recognising emotions from facial expressions. Malor Books (2003)
3. Duffy, B.R.: Anthropomorphism and the social robot. Spec. Issue Soc. Interact. Robots Rob. Auton. Syst. **42**(3–4), 177–190 (2003)

4. Walters, M.L., et al.: Avoiding the uncanny valley – robot appearance, personality and consistency of behavior in an attention-seeking home scenario for a robot companion. Auton. Rob. **24**(2), 159–178 (2008)
5. MacDorman, K.F., Ishiguro, H.: Toward social mechanisms of android science. In: A CogSci 2005 Workshop, Interaction studies, vol. 7, no. 2, pp. 289–296 (2006)
6. Fukuda, T., et al.: How far is "Artificial Man"? In: IEEE Robotics and Automation Magazine, pp. 66–73. IEEE Press (2001)
7. Duffy, B.R.: Anthropomorphism and robotics. In: Proceedings of the Society for the study of Artificial Intelligence and the Simulation Behavior (AISB '02). Imperial College (2002)
8. Bar-Cohen, Y., Hanson, D., Marom, A.: How to make a humanlike robot. In: The Coming Robot Revolution-Expectations and Fears About Emerging Intelligent, Humanlike Machines, pp. 57–74. Springer (2009)
9. Vlachos, E., Schärfe, H.: Android emotions revealed. In: Proceedings of the International Conference on Social Robotics (ICSR 2012), LNAI 7621, pp. 56–65. Springer (2012)
10. Becker-Asano, C., Ishiguro, H.: Evaluating facial displays of emotion for the android robot Geminoid F. In: IEEE SSCI Workshop on Affective Computing *and* Intelligent, pp. 1–8. IEEE Press (2011)
11. Kramer, R.S.S., Ward, R.: Internal facial features are signals of personality and health. Q. J. Exp. Psychol. **63**(11), 2273–2287 (2010)
12. Little, A.C., Perrett, D.I.: Using composite images to assess accuracy in personality attribution to faces. Br. J. Psychol. **98**, 111–126 (2007)
13. Boothroyd, L.G., Jones, B.C., Burt, D.M., DeBruine, L.M., Perrett, D.I.: Facial correlates of sociosexuality. Evolution and Human Behavior **29**(3), 211–218 (2008)
14. Stirrat, M., Perrett, D.I.: Valid facial cues to cooperation and trust: male facial width and trustworthiness. Psychol. Sci. **21**(3), 349–354 (2010)
15. Carré, J.M., McCormick, C.M., Mondloch, C.J.: Facial structure is a reliable cue of aggressive behavior. Psychol. Sci. **20**(10), 1194–1198 (2009)
16. Sprengelmeyer, R., et al.: Loss of disgust Perception of faces and emotions in Huntington's disease. Brain **119**(5), 1647–1665 (1996)
17. Andersen, A.H., et. al.: Neural substrates of facial emotion processing using fMRI. Cogn. Brain Res. **11**(2), 213–226 (2001)
18. Pessoa, L., McKenna, M., Gutierrez, E., Ungerleider, L.G.: Neural processing of emotional faces requires attention. Proc. Natl. Acad. Sci. **99**(17), 11458–11463 (2002)
19. Kilts, C.D., Egan, G., Gideon, D.A., Ely, T.D., Hoffman, J.M.: Dissociable neural pathways are involved in the recognition of emotion in static and dynamic facial expressions. Neuroimage **18**(1), 156–168 (2003)
20. Phillips, M.L., et. al: A specific neural substrate for perceiving facial expressions of disgust. Nature **389**(6650), 495–498 (1997)
21. Lee, E., Kang, J.I., Park, I.H., Kim, J.J., An, S.K.: Is a neutral face really evaluated as being emotionally neutral? Psychiatry Res. **157**(1), 77–85 (2008)
22. Albright, L., Kenny, D.A., Malloy, T.E.: Consensus in personality judgments at zero acquaintance. J. Pers. Soc. Psychol. **55**(3), 387 (1988)
23. Oh, J.H., et al.: Design of android type humanoid robot Albert HUBO. In: IEEE/RSJ International Conference on Intelligent Robots and Systems, pp. 1428–1433. IEEE Press (2006)
24. Hanson, D., et al.: Upending the uncanny valley. In: Proceedings of the 20th National Conference on Artificial Intelligence (AAAI'05), vol. 4, pp. 1728–1729. MIT Press (2005)
25. IEEE spectrum. http://spectrum.ieee.org/automaton/robotics/robotics-software/roboticist_and_his_android_twi
26. Minato, T., Shimada, M., Ishiguro, H., Itakura, S.: Development of an android robot for studying human-robot interaction. In: Innovations in Applied Artificial Intelligence, pp. 424–434. Springer, Berlin, Heidelberg (2004)
27. Channel4. http://www.channel4.com/news/bionic-man-android-channel-4-science-museum

28. Pioggia, G., et al.: FACE: facial automaton for conveying emotions. Appl. Bion. Biomech. **1** (2), 91–100 (2004)
29. BINA 48. https://www.lifenaut.com/bina48
30. Hashimoto, T., Hitramatsu, S., Tsuji, T., Kobayasi, H.: Development of the face robot SAYA for rich facial expressions. In: SICE-ICASE 2006 International Joint Conference, pp. 5423–5428. IEEE Press (2006)
31. Ahn, H.S., et al: Development of an android for singing with facial expression. In: IECON 2011–37th Annual Conference on IEEE Industrial Electronics Society, pp. 104–109. IEEE Press (2011)
32. Vlachos, E., Schärfe, H.: The geminoid reality. In: HCI International 2013-Posters' Extended Abstracts, pp. 621–625. Springer, Berlin, Heidelberg (2013)
33. Hareli, S., Hess, U.: Introduction to the special section-the social signal value of emotions. Cogn. Emot. **26**(3), 385–389. Psychology Press (2012)
34. Scherer, K.R.: On the nature and function of emotion: a component process approach. In: Scherer et al., K.R. (eds.), Approaches to emotion, Hillsdale, pp. 293–317. Erlbaum (1984)
35. Keltner, D.: Facial expressions of emotion and personality. In: Magai et al., C. (eds). Handbook of emotion, adult development and aging, pp. 385–391. Academic Press Inc. (1996)
36. Penton-Voak, I.S., Pound, N., Little, A.C., Perrett, D.I.: Personality judgments from natural and composite facial images: more evidence for a "Kernel of Truth" in social perception". Soc. Cogn. **24**(5), 607–640 (2006)
37. Olivola, C.Y., Todorov, A.: Elected in 100 milliseconds: Appearance-based trait inferences and voting". J. Nonverb. Behav. **34**(2), 83–110. Springer (2010)
38. Todorov, A., Engell, A.: The role of the amygdala in implicit evaluation of emotionally neutral faces. Soc. Cogn. Affect. Neurosci. **3**(4), 303–312 (2008)
39. Ekman, P.: Facial expression and emotion. Am. Psychol. **48**(4), 376–379. American Psychological Association (1993)
40. Ekman, P.: Universals and cultural differences in facial expressions of emotion. In: Nebraska Symposium on Motivation, vol. 19, pp. 207–282. University of Nebraska Press (1972)
41. Saarni, C.: An observational study of children's attempts to monitor their expressive behavior. Child Dev. **55**, 1504–1513 (1984)
42. Garner, P.W.: The relations of emotional role taking, affective/moral attributions, and emotional display rule knowledge to low-income school-age children's social competence. J. Appl. Dev. Psychol. **17**, 19–36 (1996)
43. Reissland, N., Harris, P.: Children's use of display rules in pride-eliciting situations. Br. J. Dev. Psychol. **9**(3), 431–435 (1991)
44. Malatesta C.Z., Haviland, J. M.: Learning display rules: the socialization of emotion expression in infancy. Child Dev. **53**(4), 991–1003. Society for Research in Child Dev. (1982)
45. Edwards, K.: The face of time: temporal cues in facial expressions of emotion. Psychol. Sci. **9**(4), 270–276. American Psychological Society (1998)
46. Cohn, J. F.: Foundations of human computing: facial expression and emotion. In: Proceedings of the 8th International Conference on Multimodal Interfaces (ICMI2006), pp. 233–238 (2006)
47. Becker, D.V., Kenrick, D.T., Neuberg, S.L., Blackwell, K.C., Smith, D.M.: The confounded nature of angry men and happy women. J. Pers. Soc. Psychol. **92**(2), 179 (2007)
48. Vlachos, E., Schärfe, H.: Social robots as persuasive agents. In: Social Computing and Social Media, Lecture Notes in Computer Science, vol. 8531, pp. 277–284. Springer (2014)

Part III
IS: Agent-Based Modelling and Simulation

Multi-objective Optimization for Clustering Microarray Gene Expression Data - A Comparative Study

Muhammad Marwan Muhammad Fuad

Abstract Clustering is one of the main data mining tasks. It can be performed on a fuzzy or a crisp basis. Fuzzy clustering is widely-applied with microarray gene expression data as these data are usually uncertain and imprecise. There are several measures to evaluate the quality of clustering, but their performance is highly related to the dataset to which they are applied. In a previous work the authors proposed using a multi-objective genetic algorithm – based method, NSGA – II, to optimize two clustering validity measures simultaneously. In this paper we use another multi-objective optimizer, NSPSO, which is based on the particle swarm optimization algorithm, to solve the same problem. The experiments we conducted on two microarray gene expression data show that NSPSO is superior to NSGA-II in handling this problem.

Keywords Clustering • Microarray gene expression data • Multi-objective optimization

1 Introduction

Optimization problems are encountered in my scientific, economic, and engineering domains. Many of these problems are complex in nature and finding exact solutions for them requires considerable computational time and involves recruiting a large number of computing resources. In many cases we are satisfied with "acceptably" optimal solutions which can be obtained in a reasonable time.

M.M. Muhammad Fuad (✉)
Institutt for Kjemi, NorStruct, The University of Tromsø - the Arctic University of Norway, Forskningsparken 3, 9037 Tromsø, Norway
e-mail: marwan.fuad@uit.no

G. Jezic et al. (eds.), *Agent and Multi-Agent Systems: Technologies and Applications*, Smart Innovation, Systems and Technologies 38,
DOI 10.1007/978-3-319-19728-9_10

An optimization process can be defined as follows: Let $\vec{X} = [x_1, x_2, \ldots, x_{nbp}]$ (*nbp* is the number of parameters) be the candidate solution to the problem for which we are searching an optimal solution. Given a function $f: U \subseteq \mathbf{R}^{nbp} \to \mathbf{R}$, find the solution $\vec{X^*} = [x_1^*, x_2^*, \ldots, x_{nbp}^*]$ which satisfies $f\left(\vec{X^*}\right) \leq f\left(\vec{X}\right), \forall \vec{X} \in \mathrm{U}$. The function *f* is called the *fitness function* or the *objective function*.

Bio-inspired Optimization is a class of optimization algorithms which is based on the social behavior of insects, birds, fish and other animals. These algorithms model the collective intelligence of these living organisms to solve optimization problems. Bio-inspired optimization algorithms have been employed to solve different practical optimization problems for their robustness and ability to be implemented using distributed computing environments.

Many practical optimization problems have to satisfy several objectives simultaneously. These objectives could be conflicting or harmonious. This class of optimization problems is called *Multi-objective Optimization* (MOO). An *m*-dimensional MOO problem can be formulated as follows:

$$min\{f_1(\vec{X}), f_2(\vec{X}), \ldots, f_m(\vec{X})\}$$

Where $\vec{X} \in \mathbf{R}^{nbp}$

The optimal solution for MOO is not a single solution as for single-objective optimization problems, but a set of solutions defined as *Pareto optimal* solutions [1], also called a *non-dominated* solution. A solution is Pareto optimal if it is not possible to improve a given objective without deteriorating at least another objective.

In [2] the authors suggested applying a multi-objective optimization algorithm of fuzzy clustering using a genetic algorithm-based optimizer. In this paper we suggest using another optimizer based on the particle swam optimization algorithm to handle the same problem.

The rest of the paper is organized as follows; in Sect. 2 we present related background, in Sect. 3 we present the previous method and we introduce the new one that we suggest to handle this problem in Sect. 4. We conduct comparative experiments in Sect. 5, and we conclude in Sect. 6.

2 Background

Data mining, also called *knowledge discovery*, can be defined as the process of analyzing data in the aim of finding useful information. This process must be automatic or (more usually) semiautomatic. The patterns discovered must be meaningful in that they lead to some advantages. Data mining focuses mainly on discovering hidden information in large amounts of data.

As with other domains in information technology, several papers have proposed applying bio-inspired algorithms to solve data mining problems [3, 4].

Data mining performs several tasks. In the following we present a particular one which it related to the topic of this paper. This task is *clustering*. Clustering is the task of partitioning data into groups, or clusters, so that the objects within a cluster are similar to one another and dissimilar to the objects in other clusters [5]. In clustering there is no target variable. Instead, clustering algorithms seek to segment the entire dataset into relatively homogeneous subgroups or clusters [6].

In general, clustering methods can be grouped in two classes: *Partitioning Methods* and *Hierarchical Methods.*

2.1 Partitioning Methods

A partitioning method divides n data objects into k partitions or clusters. The basic partitioning methods typically adopt exclusive cluster separation. Some partitioning methods require that each object belong to exactly one cluster, whereas others relax this requirement. Figure 1 shows the general scheme of partitioning clustering methods.

The three main partitioning clustering methods are *k-means clustering*, *k-medoids clustering*, and *fuzzy c-means clustering*.

k-means clustering: In k-means clustering we have a set of n data points in d-dimensional space R^d and an integer k. The task is to determine a set of k points in R^d, the centroids, so as to minimize the mean distance from each data point to its nearest center [7]. Formally, the k-means clustering error can be measured by:

$$E_{k-means} = \sum_{i=1}^{k} \sum_{j=1}^{n_j} d\left(u_{ij}, c_i\right) \tag{1}$$

Where u_{ij} is the jth point in the ith cluster, and n_j is the number of points in that cluster. The quality of the k-means clustering increases as the error given in Eq. (1) decreases.

k-medoids clustering: This clustering method is a variation of the k-means clustering. In k-medoid clustering instead of taking the mean value of the objects in a cluster as a reference point, we can pick actual objects to represent the clusters using one representative object per cluster. Each remaining object is assigned to the cluster of which the representative object is the most similar. Partitioning is then performed based on the principle of minimizing the sum of dissimilarities between each object p and its corresponding representative object. The absolute error criterion is defined as:

Algorithm Partitioning-based Clustering

```
Require c the number of clusters, n data objects.

1. (Randomly) initialize the cluster centers

2. For each data object, and for each center, compute
      the distance between this data object and the
      cluster center, assign the data object to the
      closest center according to some predefined
      criterion.

3. Update the cluster centers.

4. Repeat steps 2-3 until convergence.
```

Fig. 1 Partitioning-based clustering

$$E_{k-medoids} = \sum_{i=1}^{k} \sum_{p \in C_i} d(p, o_i) \tag{2}$$

where o_i is the representative object of cluster C_i [7].

Fuzzy c-means clustering (FCM): FCM is based on the concept of *Partial Membership* where the allocation of the data points to the centers becomes a matter of degree – the higher the membership value, the stronger its bond to the cluster [8].

The objective of FCM clustering is to minimize the measure:

$$E_{fmc} = \sum_{p=1}^{c} \sum_{i=1}^{n} u_{pi}^{m} d^2 (z_p, x_i) \tag{3}$$

Where u is the fuzzy membership and m is fuzzy exponent. In the following we present a brief description of FCM [9]. FCM starts with random c initial cluster centers. In the next step, and at each iteration, the algorithm finds the fuzzy membership of each data object to every cluster using the following equation:

$$u_{pi} = \frac{1}{\sum_{j=1}^{c} \left(\frac{d(z_p, x_i)}{d(z_j, x_i)} \right)^{\frac{2}{m-1}}} \tag{4}$$

for $1 \le p \le c$, $1 \le i \le n$.

Based on the membership values, the cluster centers are recomputed using the following equation:

$$z_p = \frac{\sum_{i=1}^{n} (u_{pi})^m x_i}{\sum_{i=1}^{n} (u_{pi})^m} \tag{5}$$

$1 \leq p \leq c$

The algorithm continues until a stopping condition terminates it.

2.2 Hierarchical Methods

Hierarchical methods work by creating a decomposition of the data objects using some criterion. This is done using a bottom-up approach, or a top-down one.

The hierarchical clustering model lies in iteratively grouping objects using a particular method of linkage. Of this we mention: *single linkage* (nearest neighbor), *complete linkage* (furthest neighbor), and *average linkage* (unweighted pair-group average)/(weighted pair-group average) [10].

It is worth mentioning that clustering has also been the subject of bio-inspired optimization. In [11] we presented a paper that applies differential evolution, a popular bio-inspired optimization algorithm, to the problem of k-means clustering. We also studied the application of several optimizers to FCM in [12].

3 Fuzzy Clustering of Microarray Data Using Multi-objective Optimization

Gene microarray is used to simultaneously measure the expression levels for hundreds of thousands of genes. Gene microarray is a powerful tool to study the regulatory relationship between genes.

The huge amount of data generated by gene microarray experiments is explored to answer fundamental questions about gene functions and their interdependence [13].

Clustering has been used in microarray experiments to identify the sets of genes which share similar expression profiles. Genes that are similarly expressed are often co-regulated and involved in the same cellular processes. Therefore, clustering suggests functional relationships between groups of genes [9].

In [2] the authors suggested applying a multi-objective optimization algorithm of fuzzy clustering which simultaneously optimizes several fuzzy cluster validity measures. The performance of their proposed method was compared with other partitioning and hierarchical clustering methods for clustering gene expression data. The authors validate their method through experiments on *Yeast Sporulation* and

Human Fibroblasts Serum data. The optimization method they applied is *Non-dominated Sorting Genetic Algorithm II.*

3.1 Non-dominated Sorting Genetic Algorithm II (NSGA-II)

NSGA-II [14] is a widely used method for solving MOO problems. NSGA-II can be summarized as follows [15]: a random parent population is initialized. The population is sorted based on non-domination in two fronts, the first front being completely a non-dominant set in the current population and the second being dominated by the individuals in the first front only. This procedure continues until all the solutions in the initial population have been ranked. Each solution is assigned a rank equal to its non-domination level based on the front it belongs to. Individuals in the first front are assigned a fitness value of 1 whereas individuals in the second are assigned a fitness value of 2, and so on. In NSGA-II the authors introduce a new parameter which is the *crowding distance*. This parameter measures how close every individual is to its neighbors. The crowding distance is calculated for each individual of the population. Parents are selected from the population by using a binary tournament selection based on the rank and the crowding distance. An individual is selected if its rank is less than that of the other or if its crowding distance is greater than that of the other. The selected population generates offspring from crossover and mutation operators. The population with the current population and current offspring is sorted again based on non-domination and only the best *popSize* individuals are selected, where *popSize* is the population size. The selection is based on the rank and on the crowding distance on the last front. The process repeats to generate the subsequent generations *nGen*. □

4 A Non-dominated Sorting Particle Swarm Optimizer for Clustering Microarray Gene Expression Data

In this paper we address the multi-objective optimization problem presented in Sect. 3 using a highly competitive multi-objective optimizer which is an extension of the *Particle Swarm Optimization* (PSO) algorithm. But first we present a brief description of PSO.

PSO is a member of a family of naturally-inspired optimization algorithms called *Swarm Intelligence* (SI). PSO is inspired by the social behavior of some animals, such as bird flocking or fish schooling [16]. In PSO individuals, called *particles*, follow three rules: a- *Separation:* each particle avoids getting too close to its neighbors. b- *Alignment:* each particle steers towards the general heading of its neighbors, and c- *Cohesion:* each particle moves towards the average position of its neighbors.

PSO starts by initializing a swarm of *popSize* particles at random positions $\vec{X}_i^0$ and velocities $\vec{V}_i^0$ where $i \in \{1, .., popSize\}$. In the next step the fitness function of each position, and for each iteration, is evaluated. The positions $\vec{X}_i^{k+1}$ and velocities $\vec{V}_i^{k+1}$ are updated at time step $(k+1)$ according to the following formulae:

$$\vec{V}_i^{k+1} = \omega.\vec{V}_i^k + \varphi_G\left(\vec{G}^k - \vec{X}_i^k\right) + \varphi_L\left(\vec{L}_i^k - \vec{X}_i^k\right) \tag{6}$$

$$\vec{X}_i^{k+1} = \vec{X}_i^k + \vec{V}_i^k \tag{7}$$

where $\varphi_G = r_G.a_G$, $\varphi_L = r_L.a_L$, $r_G, r_L \to U(0, 1)$, $\omega, a_L, a_G \in \mathbf{R}$, $\vec{L}_i^k$ is the best position found by particle *i*, $\vec{G}^k$ is the global best position found by the whole swarm, ω is called the *inertia*, a_L is called the *local acceleration*, and a_G is called the *global acceleration*. The algorithm continues for a number of iterations. □

In [17] the author presented the *Non-dominated Sorting Particle Swarm Optimizer* (NSPSO) to solve MOO problems. NSPSO incorporates the main mechanisms of NSGA-II to the PSO algorithm. In NSPSO once a particle has updated its position, instead of comparing the new position only against the *pBest* position of the particle, all the *pBest* positions of the swarm and all the new positions recently obtained are combined in just one set of *popSize* solutions. Then, the algorithm selects the best solutions among them to form the next swarm (by means of a non-dominated sorting). This algorithm selects the leaders randomly from the leaders set among the best of them, based on two different mechanisms: a niche count and a nearest neighbor density estimator. This algorithm uses a mutation operator that is applied at each iteration step only to the particle with the smallest density estimator value (or the largest niche count) [18].

We compare in the following section the performance of NSPSO with that of NSGA-II for solving the problem of fuzzy clustering of microarray gene expression data.

5 Experiments

We conducted our experiments on the same datasets and the same multi-objective optimization problem used in [2].

5.1 The Data

Yeast sporulation data: This dataset consists of 6118 genes measured across 7 time points (0, 0.5, 2, 5, 7, 9 and 11.5 hours) during the sporulation process of budding yeast. The data are then log-transformed. This dataset is available at [19]. Among the 6118 genes, the genes whose expression levels did not change significantly during harvesting have been ignored from further analysis. This is determined by a threshold level of 1.6 for the root mean squares of the log2-transformed ratios. The resulting set consists of 474 genes [20].

Human fibroblasts serum data: This dataset contains the expression levels of 8613 human genes. The dataset has 13 dimensions corresponding to 12 time points (0, 0.25, 0.5, 1, 2, 4, 6, 8, 12, 16, 20 and 24 hours) and one unsynchronized sample. A subset of 517 genes whose expression levels changed substantially across the time points has been chosen. The data are then log2-transformed [20]. This dataset is available at [21].

5.2 The Protocol

There are different cluster validity measures in the literature. The performance of these measures depends on the dataset on which the clustering task is performed. Therefore it is advisable to use several validity measures simultaneously. In this paper, as in [2], we search for the optimal clustering of the data through a multi-objective optimization problem which simultaneously optimizes two objectives. In order to compare NSPSO with NSGA-II we use the same objectives used in [2] which are the E_{fmc} measure (Eq. 3) and the Xie-Beni (XB) separation measure [22] which is defined as:

$$XB = \frac{\sum_{p=1}^{c} \sum_{i=1}^{n} u_{pi}^2 \left\| x_i - z_p \right\|^2}{n.\ \underset{i,j}{min} \left\| z_p - z_i \right\|^2} \tag{8}$$

The objective of our optimization problem is to minimize the Xie- Beni and the E_{fmc} measures simultaneously.

In addition to comparing NSPSO with NSGA-II, we also compare it with other clustering methods that use a single criterion such as the three partitioning methods mentioned in Sect. 2.1 (k-means, k-medoids, FCM) and the three hierarchical methods presented in Sect. 2.2 (single linkage, complete linkage, average linkage).

In order to evaluate the performance of these different methods we use the *Silhouette index* (SI) [23] which is defined as follows [24]; given a dataset $\{X_1, X_2, \ldots, X_N\}$ which is partitioned into k clusters $\{C_1, C_2, \ldots, C_k\}$. The silhouette width of the ith vector in cluster C_j is given by:

$$s_i^j = \frac{b_i^j - a_i^j}{max\{b_i^j, a_i^j\}} \tag{9}$$

Where

$$a_i^j = \frac{1}{m_j - 1} \sum_{\substack{k=1 \\ k \neq i}}^{m_j} d\left(X_i^j, X_{k}^j,\right), i = 1, \ldots, m_j$$

And

$$b_i^j = \min_{\substack{n=1, \ldots, K \\ n \neq j}} \left\{ \frac{1}{m_n} \sum_{k=1}^{m_n} d\left(X_i^j, X_k^n\right) \right\}, i = 1, \ldots, m_j$$

The silhouette of cluster C_j is then defined as:

$$S_j = \frac{1}{m_j} \sum_{i=1}^{m_j} s_i^j \tag{10}$$

And the SI, the global Silhouette index, is given by:

$$SI = \frac{1}{K} \sum_{j=1}^{K} S_j \tag{11}$$

Greater values of SI indicate a better clustering quality.

Table 1 shows the results of our experiments. As we can see, the SI of NSPSO, and for both datasets tested, is higher than all the SI values of all the other methods. Particularly, the SI value using NSPSO is higher than the one obtained when using NSGA-II as an optimizer. This proves that NSPSO is more appropriate to solve this optimization problem.

Table 1 Comparison of the Silhouette index of datasets *Yeast Sporulation* and *Human Fibroblasts Serum* using different clustering methods

Clustering method		Dataset	
		Yeast sporulation	Human fibroblasts serum
Partitioning	k-means	0.573	0.324
	k-medoids	0.594	0.261
	FMC	0.588	0.330
Hierarchical	Single linkage	−0.491	−0.328
	Average linkage	0.501	0.298
	Complete linkage	0.439	0.278
NSGA-II		0.646	0.413
NSPSO		0.698	0.463

6 Conclusion

This paper compares the performance of two multi-objective optimization algorithms: non-dominated sorting genetic algorithm II (NSGA-II) and the non-dominated sorting particle swarm optimizer (NSPSO), in solving the problem of fuzzy clustering microarray gene expression data. The experiments we conducted on *Yeast Sporulation* and *Human Fibroblasts Serum* datasets using two objectives; fuzzy c-means (FMC) and Xie- Beni measures, which were evaluated by means of the Silhouette index (SI), show that NSPSO is more appropriate for processing this optimization problem as it gives a higher SI value for the two datasets tested.

References

1. El-Ghazali, T.: Metaheuristics: from design to implementation. Wiley (2009)
2. Mukhopadhyay, A., Maulik, U., Bandyopadhyay, S.: Multiobjective Evolutionary Approach to Fuzzy Clustering of Microarray Data, pp. 303–328. World Scientific, Singapore (2007)
3. Muhammad Fuad, M.M.: Differential evolution versus genetic algorithms: towards symbolic aggregate approximation of non-normalized time series. In: Sixteenth International Database Engineering & Applications Symposium– IDEAS'12, Prague, Czech Republic, Published by BytePress/ACM, 8–10 Aug 2012
4. Muhammad Fuad, M.M.: Using differential evolution to set weights to segments with different information content in the piecewise aggregate approximation. In: 16th International Conference on Knowledge-Based and Intelligent Information & Engineering Systems, KES 2012, San Sebastian, Spain, Published in *Frontiers of Artificial Intelligence and Applications* (FAIA), IOS Press, 10–12 Sept 2012
5. Han, J., Kamber, M., Pei, J.: Data mining: concepts and techniques, 3rd edn. Morgan Kaufmann (2011)
6. Larose, D.T.: Discovering Knowledge in Data: An Introduction to Data Mining. Wiley, New York (2005)
7. Kanungo, T., Netanyahu, N.S., Wu, A.Y.: An Efficient k-means clustering algorithm: analysis and implementation. IEEE Trans. Pattern Anal. Mach. Intell. **24**(7) (2002)
8. Krzysztof, J.C., Pedrycz, W., Swiniarski, R.W., Kurgan, L.A.: Data Mining: A Knowledge Discovery Approach. Springer-Verlag New York, Inc., Secaucus (2007)
9. Maulik, U., Bandyopadhyay, S., Mukhopadhyay, A.: Multiobjective Genetic Algorithms for Clustering: Applications in Data Mining and Bioinformatics. Springer (2011)
10. Gorunescu, F.: Data mining: Concepts. Blue Publishing House, Cluj-Napoca, Models (2006)
11. Muhammad Fuad, M.M.: Differential evolution-based weighted combination of distance metrics for *k*-means clustering. In: The 3rd International Conference on the Theory and Practice of Natural Computing – TPNC 2014, Granada, Spain. Published in Lecture Notes in Computer Science, vol. 8890, 9–11 Dec 2014
12. Muhammad Fuad, M.M.: On the application of bio-inspired optimization algorithms to fuzzy *c*-means clustering of time series. In: The 4th International Conference on Pattern Recognition Applications and Methods - ICPRAM 2015, Lisbon, Portugal. SCITEPRESS Digital Library, 10–12 Jan 2015
13. Lodhi, H., Muggleton, S.: Elements of Computational Systems Biology. Wiley, New York (2010)
14. Deb, K., Pratap, A., Agarwal, S., Meyarivan, T.: A fast and elitist multiobjective genetic algorithm: NSGA-II. IEEE Trans. Evol. Comput. (2002)

15. Ma, Q., Xu, D., Iv, P., Shi, Y.: Application of NSGA-II in Parameter Optimization of Extended State Observer. Challenges of Power Engineering and Environment (2007)
16. Haupt, R.L., Haupt, S.E.: Practical Genetic Algorithms with CD-ROM. Wiley-Interscience, (2004)
17. Li, X.: A Non-dominated Sorting particle swarm optimizer for multi-objective optimization. In: Proceedings of the Genetic and Evolutionary Computation Conference (GECCO'2003), Lecture Notes in Computer Science, vol. 2723, pp. 37–48. Springer (2003)
18. Reyes Sierra, M., Coello Coello, C.: Multi-objective particle swarm optimizers: a survey of the state-of-the-art. Int. J. Comput. Intell. Res. **2**(3), 287–308 (2006)
19. http://cmgm.stanford.edu/pbrown/sporulation
20. Maulik, U., Mukhopadhyay, A., and Bandyopadhyay, S.: Combining pareto-optimal clusters using supervised learning for identifying co-expressed Genes. BMC Bioinfor. **10**(27), 1–16, 20 Jan 2009
21. http://www.sciencemag.org/feature/data/984559.shl
22. Xie, X.L., Beni, G.: A validity measure for fuzzy clustering. IEEE Trans. Pattern Anal. Mach. Intell. **13**, 841–847 (1991)
23. Rousseeuw, P.J.: Silhouettes: a graphical aid to the interpretation and validation of cluster analysis. J. Comp. Appl. Math. **20**, 53–65 (1987)
24. Petrovic, S.: A comparison between the silhouette index and the davies-bouldin index in labelling IDS clusters. In: Proceedings of the 11th Nordic Workshop of Secure IT Systems (2006)

Group Control Strategy for a Fleet of Intelligent Vehicles-Agent Performing Monitoring

Viacheslav Abrosimov

Abstract The article considers the territory monitoring process by intelligent vehicles. The aim is to create an efficient algorithm for the iterative solution of Vehicle Routing Problem in view of danger situation on the monitoring objects as well as on the route between the objects. We introduce a group control strategy for a fleet of vehicles performing monitoring. Vehicles are considered as intelligent agents. Vehicles make optimal decisions on the route adjustment by means of negotiations on the principle of «an auction» when there are problems on the route. The route is built with the use of an ant algorithm taking into account the situation at the previous stages of the monitoring. The main advantages are that the strategy of intelligent control objects takes into account the danger of the situation on the object and a danger on the route between the objects.

Keywords Travelling salesman problem · Vehicle routing problem · Intelligent agent · Monitoring · Intellectual control vehicle · Collective control · Strategy

1 Introduction

It is known that the Travelling Salesman Problem-finding a shortest route on the graph, which once includes each node of the graph - belongs to the class of NP-hard problems and cannot be resolved by the search methods [1, 2]. Methods of discrete optimization, particularly the branch-and-cut method allowing the discovery of optimal or approximate solutions, are used to find specific solutions. Exact methods are often combined with the heuristic ones for searching of the routes [2].

V. Abrosimov (✉)
Moscow Aviation Institute -National Research University, 4 Volokolamskoe Shosse, Moscow 125993, Russian Federation
e-mail: avk787@gmail.com

G. Jezic et al. (eds.), *Agent and Multi-Agent Systems: Technologies and Applications*, Smart Innovation, Systems and Technologies 38,
DOI 10.1007/978-3-319-19728-9_11

A Travelling Salesman Problem for several salesmen– Vehicle Routing Problem (VRP) - is more interesting [3–9]. Each salesman must visit a number of cities and return to the initial city; the total way all salesmen have travelled must be minimum. In practice, the specified statements of the problems arise in such areas as servicing of territorially distributed objects of the oil and gas industry, flights around remote crews and airdropping of cargos, maintenance of engineering facilities, etc. The class of such problems includes but is not limited to the problems of logistics, traffic management, monitoring, etc. [4].

There are various variants of VRP that are used in a variety of application areas [5]. Some of them are: Multiple Depot Vehicle Routing Problem (MDVRP) – VRP with several depots, Vehicle Routing Problem with Time Windows (VRPTW) – VRP with the additional restriction that the time window is associated with each customer, Split Delivery Vehicle Routing Problem (SDVRP) – VRP in which the same customer can be served by different vehicles, Stochastic Vehicle Routing Problem (SVRP) – VRP in which one or several components of the problem are random, Vehicle Routing Problem with Pick-Up and Delivering (VRPPD) – VRP in which the possibility that customers return some commodities is contemplated, Vehicle Routing Problem with Satellite Facilities (VRPSF) – VRP in which satellite facilities are used to replenish vehicles during a route and so on. It is necessary to emphasize that all problem definitions do not include the necessity of iterative.

In the conventional statements it is supposed that distribution of the objects to vehicles takes place beforehand. During monitoring they function individually and independently from each other. However, in practice there is often a situation when some vehicles can not fulfill their tasks because of the various non-standard situations [6]. Appearance of a new objects, refinement of the cargo delivery time, failure of the vehicle, and evolving of emergencies on the route, etc. are possible directly in the course of the task fulfillment. Classical statement conditions are also upset: vehicles are initially not in the same city, have different resources, begin to move if a different time, etc. They can meet with various obstacles on the route and even with the prohibition on visiting of the city. The above situations significantly complicate the task. The vehicles become constrained in their strategies. They must collectively fulfill a common task and redistributing he routes when necessary.

This article considers the territory monitoring problems by intelligent vehicles. A specific feature consists in the inherent antagonism of the motion environment, which is a common situation in practice. We introduce a group control strategy for a fleet of vehicles performing monitoring. Intellectual Control Vehicles (ICV) considers as salesman. In practice, various vehicles with intellectual control systems can appear as ICV-aircrafts, helicopters, drones, on which the corresponding equipment for monitoring is established. The cities will be point as objects (in practice – settlements, remote bases, groups of people in distress, etc.) and extended objects (pipelines, impassable roads, forest areas, etc.), whose condition has to be inspected. Such practical problems often arise in the emergency conditions - fires, flooding, natural disasters, etc. The degree of emergency is variable and is evaluated by the level of its intensity, distribution and duration. Therefore, one of the most important conditions for resolution of such tasks is the necessity of its multiple solution in iterative regime. So the main our feature concerns the need for recurrent flying around points. A common situation in practice is when a series of vehicles still continue their tasks, but an additional fleet of vehicles must be allocated for surveying separate points and domains on a territory.

The route selection as a group of objects to be sequentially inspected and iterative repetition of such in section. Basic criterion in the first iteration is the minimization of the total flight time of all ICV, which in fact reflects the saving of fuel as a basic resource. In the next iterations, it is necessary to use another criterion – primarily inspection of those objects where (a) the intensity situation is maximal and (b) the intensity on the route to which is also maximal. Our analysis of the monitoring problems in these conditions shows that they have an essential likeness to the ant algorithms [7]. Indeed, the ants leave so-called «pheromones» – special odor marks scores – when passing the route; other ants more probably follow the way, von which there are more such pheromone marks. This is the base of ant algorithms for optimization of the optimum way search. For the second and subsequent iterations, the route is therefore dependent on the intensity degree of the object situations and routes to them in the previous iterations.

2 Control Object as an Intellectual Agent

Our analysis shows a possibility to use the methodology of multiagent intellectual systems for building a collective behavior strategy of several control objects when resolving a specific problem [8–13]. Intellectual agents act individually and independently process the information and interact with other agents to obtain solution of a collective problem [10, 11].

Our primary task is to endow all vehicles with the properties of an intelligent agent. In practice, a reasonable approach implies the following. First, construct a general solution on monitoring using a common algorithm implemented by the leading agent and assign motion routes to each vehicle. Second, allow each vehicle to make decisions depending on the current situation in the autonomous mode. A vehicle exchanges information with other vehicles, reports its current state, action strategy and the fulfillment degree of a group mission. A vehicle considers the goals and intentions of other vehicles. Imagine that a vehicle or several vehicles fail to perform the tasks of monitoring due to difficult conditions or the impact of an unfriendly environment. In this case, other vehicles can take new tasks if available resources are sufficient.

During the movement, the space-time situation can develop unfavorably for separate ICVs that will necessitate a refusal of the earlier resolved problems and redirecting to resolution of other problems or even return to the base. Acting as an agent, any ICV can receive the information on the necessity to solve additional problems, evaluate its capabilities, declare its readiness and capability to decide and undertake them on the request and/or instructions of other ICVs. Such tasks constitute the commitments of ICVs. Commitments can be also built up as a kind of hierarchical structure when some control hierarchy is established between ICVs; the commitments are then based on the subordination conditions.

In a standard Vehicle Routing Problem, the strategy consist only in the searching of an optimal route of the vehicles between the cities [5]. Let us complicate the problem with the following conditions:

- potential obstacles on the vehicles route to the objects up to the impossibility of moving to them;
- availability of an object inspection danger up to the strict prohibition on visiting it;
- possibility of refusal of one or several vehicles to fulfill the task;
- necessity of multiple going around the objects until a certain condition is met.

These conditions require consideration in iterative algorithms possible hazards and on-line interaction ICVs between themselves in order to collective task would perform successfully.

3 Strategy of an Intellectual Control Vehicle as an Agent in an N-Dimensional Vehicle Routing Problem

All specified limitations must be considered in the building of respective strategies as additional criteria of the route selection. The developed strategy of a standard ICV is realized in the form of the following sequence.

1. Definition of a distance matrix D_{sn} between all N objects.
2. Setting of input data in the form of a sequence $M_n\colon \{\mu_n\} \forall n \in N$ with the elements μ_n, describing the situation intensity of the n-th object, matrix M_{sr} of the emergency situation intensity, matrix degree of danger W_{sr} on the route between the objects s and r (for example because of air defense system).
3. Setting of input data in the form of a sequence $F_n\colon \{\varphi_n, \rho_n\} \forall n \in N$ with the elements describing the priority degree φ_n of monitoring an n-th object and danger ρ_n in the neighborhood of this object.
4. Generation of a route (sequence) of going around the monitoring objects in the χ-th monitoring cycle for each ICV $_i$ taking into account the matrixes M_{sr}, W_{sr} and sequences M_n and F_n:

$$M^{\chi}_{iH} = \{n_{i1}, n_{i2} \ldots n_{ih} \ldots, n_{iH_\chi}\} \forall n \in N, \forall i \in I \quad (1)$$

Upon the results of the solution of this problem, the objects are assigned to ICVs and the parameter α_i^n (object n is assigned in the course of distribution of the objects to separate ICVs of the whole group (target distribution) to ICV_i) is assigned with a value equal to one for each n. If $\alpha_i^n = 0$ at least for one object after target distribution, the structure or capabilities of the selected group of ICVs is insufficient for the fulfillment of the monitoring task. In this case, it is necessary to increase the power of the set I (expansion of the group with additional ICVs).

5. Simulation of the motion of $ICV_i \forall i \in I$ from the s-th to the next r-th objectat a speed v_i determined by the construction features of ICV_i.
6. Obtaining of the information μ_{sn} on the emergency situation intensity on the route and information μ_n on the emergency situation intensity at the n-th object in the course of movement by means of the surveillance equipment of ICVs.
7. Correction of the matrix M_{sr} element's values of the emergency situation intensity on the route by the values $\sigma\mu_{sr}^{\tau}$ with the additional information received for the period τ on the route connecting the objects n_s and n_r

$$\mu_{sr}^{t+\tau} = \mu_{sr}^{t} + \sigma\mu_{sr}^{\tau} \tag{2}$$

8. Correction of the matrix M_n element's values of the emergency situation in tensity at the n-th object by the values $\sigma\mu_n^{\tau}$ of the additional information received at the n-th at the moment of its monitoring

$$\mu_n^{t+\tau} = \mu_n^{t} \pm \sigma\mu_n^{\tau} \tag{3}$$

9. In the second and subsequent iterations for all objects r, to which the transition from the s-th object is possible, the objects for the route are selected using an analog of the ant algorithm of the following form

$$\Omega_{sr} = \left\{(\varphi_r * \mu_{rq}^{\varepsilon}) / \sum\nolimits_N (\varphi_n * \mu_{rq}^{\varepsilon})\right\} + (\pi_{sr}^{\lambda} \times \mu_{srq}^{\lambda}) / \sum\nolimits_{A^D} (\pi_{sr}^{\lambda} \times \mu_{srq}^{\lambda}) \tag{4}$$

where:

- Ω_{sr} probability of crossing from the object n_s to the object n_r;
- π_{sr}^{ε} inverse distance between the objects n_s and n_r;
- μ_{rq}^{ε} value of the emergency situation intensity at the object n_r in the q-th monitoring cycle;
- μ_{srq}^{ε} value of the emergency situation intensity on the route between the objects n_s and n_r determined in the q-th monitoring cycle;
- ε and λ empirical indexes for consideration of the problem solution specifics.

The first item considers the importance of the object φ_r and degree of its situation intensity μ_{rq}^{ε}, which stipulates the necessity of the monitoring of an object with the highest importance and worst situation. The second item "weighs" the distances to the objects and facilitates the selection of an object with the minimal distance to it taking into account the emergency situation intensity between the objects μ_{sr}^{λ}. All characteristics can be both enhanced and weekend with the empirical indexes ε and λ.

4 Model of Intellectual Control Vehicles Negotiations

Negotiations of agents are studied in many publications [8, 12, 13]. Among different types of agents cooperation (coordinated cooperation, simple cooperation, collective/individual competition for resources), we will consider simple cooperation. This implies experience integration among agents (i.e., task allocation and knowledge exchange), and additional measures on action coordination are unnecessary. In group monitoring, each agent possesses individual resources and offers them for participation in negotiations, since agents accomplish a group mission.

We have already introduced the notion of an unfriendly environment for vehicle motion. It should be distinguished from a negotiation environment. The work [8] separated out four settings of a negotiation environment, namely, the number of issues, the number of participants, environment state and negotiation relationship. In the territory monitoring problems, vehicles organize negotiations on a single issue (route correction if necessary). The number of participants is the number of vehicles in a fleet after deduction of vehicles that announced their task failure (multilateral participants). We will consider dynamic negotiations of agents: decision rules in negotiations are based on situation forecasting, negotiations run in the online regime and the current situation is continually refined in the course of motion. In the sequel, a negotiation relationship is treated as a solo: negotiations on route correction do not depend on other negotiations.

A virtual "electronic bulletin board" is used in this article to organize the interaction of ICVs as independent agents resolving a collective problem. To do so, a resource with a preset structure of the content, to which all ICVs can address both for obtaining and placement of the information accommodation in the on-line mode, is allocated in the Internet. Collective information of three types is posted hereon the electronic board: (a) current information on the real situation on the route of the objects reconnaissance, received from the surveillance equipment; (b) information on the impossibility of the task fulfillment by ICV_j and (c) announcement that ICV_i has residual resources for fulfilment of the task, including those of other ICVs refused their commitments.

Description of the state of the n-th inspected object at the current moment t is saved on the electronic board in the form of a sequence

$$n(t): \{x_n, y_n, \varphi_n, \mu_n, \mu_{sr}, \omega_{sr}, \rho_n, \chi_n, \chi_{rn}, \alpha_i^n\} \tag{5}$$

where

x_n, y_n	GPS-coordinates of the objects;
$\varphi_n \in [0.1]$	priority degree of monitoring of the n-th object;
$\omega_{sr} \in [0.1]$	flight danger degree on the route between the Objects n_s and n_r;
χ_n	indicator of the number of the monitoring (visits, fly-arounds, cargo airdropping, etc.) of the n-th object;
χ_{rn}	indicator of the number of the ICV flights on the monitoring route between the plants n_s and n_r
α_i^n	indicator that n-th object has examined already

Characteristics $x_n, y_n, \varphi_n, \mu_n, \mu_{sr}, \omega_n, \rho_n$ are used in the routing. Objects, in which the danger degree exceeds the predetermined ($\rho_n \geq \rho_n^{\text{fix}}$), are thus excluded from the consideration. If the flight danger degree between the objects n_s and n_r also exceeds the predetermined ($\omega_{sr} \geq \omega_{sr}^{fix}$), these objects are included in the route of ICV_i, having appropriate capabilities σ_i for motion in these conditions.

Description of the state of ICV_i at the current time t is saved on the electronic bulletin board in the form of a sequence, in which, apart from the route in the given iteration M_{iH}^{χ}, a list of the inspected M_{iH+}^{χ} and non-inspected M_{iH-}^{χ} objects, residual resource Res_i and attribute A_i of the current activity of ICV_i are stored:

$$ICV_i : \left\{i,\ M_{iH}^{\chi}, M_{iH+}^{\chi}, M_{iH-}^{\chi}, Res_i, A_i\right\} \tag{6}$$

Let ICV_p have technical or other problems, which do not allow the fulfillment of the task for monitoring of the object n_h, in the course of ICV movement on the routes at the time t_{fix}. In this case, ICV_p re-zeros the current activity attribute ($A_p := 0$), and index α_p^h of an object from the sequence M_{p-}^{χ} is assigned with a value equal to zero. Thus, a part of the objects $n_{p1}, n_{p2}, n_{ph}, \ldots .n_{pH}$ becomes unassigned to ICVs and is subject to a new distribution.

Negotiations of the remaining ICVs can be realized in this situation indifferent ways. The analysis shows [8, 12, 13] that the model of auctions is the most appropriate to the resolved problem. The algorithm of the auctions for incorporation of the new objects in the route has the following form.

1. For each $h \in H$ and for each ICV_i involved in monitoring process the value

$$\Delta F_i(n_h) = \min_j\{d(n_h, n_{ij}) + d(n_h, n_{i,j+1}) - d(n_{ij}, n_{i,j+1})\}/v_i\} \\ 1 \leq j \leq k-1 \tag{7}$$

is calculated. It represents the most effective conditions for incorporation of the object n_h in the route and reflects the minimum time, which is required to leave the route M_i for monitoring of the object n_h and return back to the same route. Object n_h is included in the route of that ICV_i, for which the following conditions are simultaneously perform:

$$\Delta F_i^*(n_h) = \min_i\{\Delta F_i(n_h)\} \text{ and } Res_i - \Delta F_i(n_h) \geq 0 \tag{8}$$

where Res_i is predictable resource margin of ICV_i after travelling the preset route.

2. If $\alpha_i^n = 1 \forall n \in N$ then a target distribution attribute is developed (all $\in N$ of the objects for monitoring are assigned to $i \in I$ by respective ICVs).

3. If $\alpha_i^n = 0$ at least for one $h \in N$, the process of target distribution proceeds according to the following algorithm:

(a) Function $\Delta F_i(n_h)$ and mismatch of the resources for achievement of the object n_h are calculated for each ICV_I

$$c_{ih} = res_i - \Delta F_i(n_h) \tag{9}$$

(b) ICV_f, for which $|c_{ih}|_f = \min_I |c_{ih}|$ is determined. If such minimum exists at least for one ICV_f, the object n_h is assigned to ICV_f and $\alpha_f^h := 1$.

5 Conclusions

In a standard N-dimensional salesman problem the strategy of the Vehicle Routing Problem consists in the searching of an optimal route of the vehicles movement between the cities only. However, in practice there is a significant amount of specific details: the cities differ in the importance; refusal of one or several vehicles to fulfill the task is possible. There are various dangerous, occasionally absolute obstacles on the vehicles route to the objects up to the impossibility of moving to them. Terrain monitoring problems require a multiple process of reconnaissance of the cities depending on the situations, which arise in the course of the objects inspection. The vehicles must thus exchange the information in the on-line mode on the retention of their full independence in decision-making to resolve a collective problem.

The group control strategy for a fleet of intelligent vehicles performing monitoring is developed in this article. Control vehicles are considered as intellectual agents. The problem of route generation using an analog of the ant algorithm is thus resolved in the on-line mode. When a problem arise (agent functioning disturbance or loss, limitations and problems on the route, etc.), the agents negotiate on a virtual site, revise the previously made decisions on the reconnaissance of the objects and update their routes.

At last we would like to emphasize the advantages of presented approach. First of all, it is the idea to endow all vehicles with the properties of an intelligent agent. Secondly, the strategy of intelligent control objects constructed based on the ant algorithm to incorporate new elements describing the key criteria taking into account during the monitoring of the territory: the danger of the situation on the object and a danger on the route between the objects. Thirdly, the procedure of negotiation agents and route redistribution on the basis of a virtual electronic board became very simple. And fourthly, procedure for solving the problem is constructed as an iterative procedure.

The developed algorithms are a strategy of collective behavior of a group of several control vehicles at the various importance of inspected objects, danger situation on the route and new problems arising during the monitoring.

Acknowledgments The reported study was partially supported by RFBR, research project No. 13-08-00721-a.

References

1. Rego, C., Gamboa, D., Glover, F., Osterman, C.: Traveling salesman problem heuristics: leading methods, implementations and latest advances. Eur. J. Oper. Res. **211**(3), 427–441 (2011)
2. Woeginger, G.J.: Exact algorithms for NP-hard problems: a survey. In: Combinatorial Optimization, Lecture notes in Computer Science, vol. 2570, pp. 185–207 Springer (2003)
3. The Vehicle Routing Problem: Latest Advances and New Challenges.: In: Golden, B.L., Raghavan, S., Wasil, E.A. (eds.) (2008)
4. Allan, L.: The Dynamic Vehicle Routing Problem. LYNGBY Printed by IMM, DTU Bookbinder Hans Meyer (2001)
5. Frank, T.:Applying monte carlo techniques to the capacitated vehicle routing problem. Master thesis, Computer Science Leiden Institute of Advanced Computer Science, Leiden University, The Netherlands (2010)
6. Jingquan, Li : Models and algorithms of real-time vehicle rescheduling problems Under Schedule Disruptions. Ph.D. thesis, The Graduate College of the University of Arizona (2006)
7. Dorigo, M., Gambardella, L.M.: Ant colonies for the traveling salesman problem. Bio. Syst. **43**, 73–81 (1997)
8. Ren, F.: Autonomous agent negotiations strategies in complex environment. Ph.D. thesis, School of Computer Science and Software Engineering, Faculty of Engineering, University of Wollongong (2010)
9. Miaomiao, D., Hua, Yi.: Research on multi-objective emergency logistics vehicle routing problem under constraint conditions. J. Ind. Eng. Manage. JIEM. **6**(1), 258–266 (2013)
10. Tan, K.C., Cheong, C.Y., Goh, C.K.: Solving multiobjective vehicle routing problem with stochastic demand via evolutionary computation. Eur. J. Oper. Res. **177**, 813–839 (2007)
11. Wooldridge, M., Jennings, N.R.: Intelligent agents: theory and practice. Knowl. Eng. Rev. **10** (2), 115–152 (1995)
12. Multiagent Systems: Algorithmic, Game-Theoretic, and Logical Foundations Hardcover. Cambridge University Press (2008)
13. Multiagent Systems: A modern approach to distributed artificial intelligence. In: Gerhard, W. (ed.) MIT Press (2000)

Economic Demand Functions in Simulation: Agent-Based Vs. Monte Carlo Approach

Roman Šperka

Abstract The aim of this paper is to compare an agent-based and Monte Carlo simulation of microeconomic demand functions. Marshallian demand function and Cobb-Douglas utility function are used in simulation experiments. The overall idea is to use these function as a core element in a seller-to-customer price negotiation in a trading company. Furthermore, formal model of negotiation is proposed and implemented to support the trading processes. The paper firstly presents some of the principles of agent-based and Monte Carlo simulation techniques, and demand function theory. Secondly, we present a formal model of demand functions negotiations. Lastly, we depict some of the simulation results in trading processes throughout one year of selling commodities to the consumers. The results obtained show that agent-based method is more suitable than Monte Carlo, and the demand functions could be used to predict the trading results of a company in some metrics.

Keywords Simulation · Agent-based · Monte carlo · Demand functions · Business process · Trading negotiation

1 Introduction

Business process management and simulation (BPMS) is often viewed as one of the technologies, which might help to rationalize the traditional execution of business processes in contemporary fluctuant business environment. The aim of BPMS is to lower the resources (financial, personal, time, etc.) needed for the day-to-day business praxis of companies [1–3]. Regardless of the area, which is company

R. Šperka (✉)
School of Business Administration in Karviná, Department of Informatics and Mathematics, Silesian University in Opava, Univerzitní Nám. 1934/3, 733 40 Karviná, Czech Republic
e-mail: sperka@opf.slu.cz

G. Jezic et al. (eds.), *Agent and Multi-Agent Systems: Technologies and Applications*, Smart Innovation, Systems and Technologies 38,
DOI 10.1007/978-3-319-19728-9_12

active in, it is used to find some reserves and to lower the costs. One of the possibilities to use BPMS is an automated way to execute the business processes – Workflow Management Systems (WfMS) [4]. We used more in-depth approach to the trading price negotiation to cover the seller-to-customer interaction and to use it further to simulate their behavior [5–10] in our partial research.

The approach introduced in this paper compares the simulation results achieved by two techniques. Firstly, we will use an agent-based method in the form of a multi-agent system to serve as a simulation platform for the seller-to-customer negotiation in a virtual trading company. Secondly, we will present a Monte Carlo simulation of the same phenomena.

The main idea concentrates around the negotiated price establishment. Here we used a demand function. The overall scenario comes from the research of Barnett [11]. He proposed the integration of real system models with the management models to work together in real time. The real system (e.g., ERP system) outputs proceed to the management system (e.g., simulation framework) to be used to investigate and to predict important company's metrics (KPIs – Key Performance Indicators). Actual and simulated metrics are compared and evaluated in a management module, which identifies the steps to take to respond in a manner, that drives the system metrics towards their desired values. Firstly, we used a generic control loop model of a company [8] and implemented a multi-agent simulation framework, which represents the management system. And secondly, we used this paradigm in a Monte Carlo simulation. This task was rather complex, therefore we took only a part of the model – trading processes, specifically the seller-to-customer negotiation.

The paper is structured as follows. Section 2 represents some of the theoretical incomes. A formal model is described in Sect. 3. The core of this section is the demand function definition. The simulation results are presented in Sect. 4.

2 Theoretical Background

Agent-based modeling and simulation (ABMS) provides some opportunities and benefits resulting from using multi-agent systems. ABMS should serve as a platform for simulations with the aim to investigate the consumers' behavior. Agent-based models are able to integrate individually differentiated types of consumer behavior. The agents are characterized by a distributed control and data organisation, which enables to represent complex decision processes with few specifications. There were published many scientific works in this area in the recent past. They concern in the analysis of companies' positioning and the impact on the consumer behavior [12–14]. Often discussed is the reception of the product by the market [15, 16], and the innovation diffusion [17–19]. More general deliberations on the ABMS in the investigating of consumer behavior show, e.g., [20, 21].

Monte Carlo simulation method (or Monte Carlo experiments) are a broad class of computational algorithms that rely on repeated random sampling to obtain numerical results; typically one runs simulations many times over in order to obtain the distribution of an unknown probabilistic entity.

The core problem to be solved in the business process of selling the commodities to the consumers while using the simulation, is the price negotiation. We used specific functions from the economic theory in this partial research. We built our experimental research on a demand function. In microeconomics, a consumer's Marshallian demand function (named after Alfred Marshall) specifies what the consumer would buy in each price and wealth situation [22], assuming it perfectly solves the utility maximization problem[1]. Marshallian demand is sometimes called Walrasian demand (named after Léon Walras) or uncompensated demand function instead, because the original Marshallian analysis ignored wealth effects[2] [23, 24]. We also used a Cobb-Douglas utility function and preferences stating that quantity demanded of each commodity does not depend on income, in fact the quantity demanded of each commodity is proportional to income [25]. The formal model is introduced in the next section.

3 Formal Model

The seller-to-customer negotiation is described, and the definition of a Marshallian demand curve is proposed in this section. Marshallian demand function is used during the contracting phase of agents' interaction. It serves to set up the limit price of the customer agent as an internal private parameter.

Only a part of the company's generic structure was implemented. This part consists of the sellers and the customers trading with commodities (e.g. tables, chairs, etc.). One stock item simplification is used in the implementation. Participants of the contracting business process in our multi-agent system are represented by the software agents - the seller and customer agents interacting in the course of the quotation, negotiation and contracting. There is an interaction between them. The behavior of the customer agent is characterized by the Marshallian demand function based on the Cobb-Douglas utility function.

At the beginning the disturbance agent analyses historical data – calculates average of sold amounts for whole historical year as the base for percentage calculation. Each period turn (here we assume a week), the customer agent decides whether to buy something. His decision is defined randomly. If the customer agent decides not to buy anything, his turn is over; otherwise he creates a sales request and sends it to his seller agent. Requested amount (generation is based on a normal distribution) is multiplied by disturbance percentage. Each turn disturbance agent

[1]In microeconomics, the utility maximization problem is a problem, which consumers face: "How should I spend my money in order to maximize my utility?" It is a type of optimal decision problem.

[2]The wealth effect is an economic term, referring to an increase (decrease) in spending that accompanies an increase (decrease) in perceived wealth.

calculates the percentage based on historical data and sends the average amount values to the customer agent. The seller agent answers with a proposal message (a certain quote starting with his maximal price: *limit price * 1.25*). This quote can be accepted by the customer agent or not.

The customer agents evaluate the quotes according to the demand function by calculating his maximal price. The Marshallian demand function was derived from Cobb-Douglas utility function and represents the quantity of the traded commodity as the relationship between customer's income and the price of the demanded commodity. If the price quoted is lower than the customer's price obtained as a result of the demand function, the quote is accepted. In the opposite case, the customer rejects the quote and a negotiation is started. The seller agent decreases the price to the average of the minimal limit price and the current price (in every iteration is getting effectively closer and closer to the minimal limit price), and resends the quote back to the customer. The message exchange repeats until there is an agreement or a reserved time passes.

Marshallian function specifies what would consumer buy at each specific price and income, assuming it perfectly solves utility maximization problem. For example: If there are two commodities and the specific consumer's utility function[3] is:

$$U(x_1, x_2) = x_1^{0.5} x_2^{0.5} U(x_1, x_2) = x_1^{0.5} x_2^{0.5} \tag{1}$$

Then the Marshallian demand function is a function of income and prices of commodities:

$$x(p_1, p_2, I) = \left(\frac{I}{2p_1}, \frac{I}{2p_2} \right) \tag{2}$$

Where I represents income and p_1 and p_2 are the prices of the commodities. In general, Cobb-Douglas utility function can be defined as:

$$U(x_1, x_2) = x_1^{\alpha} x_2^{1-\alpha} \tag{3}$$

The corresponding Marshallian demand function is:

$$x(p_1, p_2, I) = \left(\frac{\alpha I}{p_1}, \frac{(1-\alpha)I}{p_2} \right) \tag{4}$$

Only one commodity is calculated (which is traded by the simulated company) in the model. In this case –there are two commodity baskets using the Marshallian demand function, where one is represented by company traded one and the rest represents all alternative commodities, that customer can buy. So only x_1 is used

[3]Ratio of the same utility, consuming the commodities – Cobb-Douglas utility function.

sup-posing that utility ratio α is known and that for the rest of commodities the utility ratio is $(1\text{-}\alpha)$. Therefore the demand function looks like this:

$$X = \frac{\alpha I}{p} \quad (5)$$

where X represents amount of commodity, α is utility ratio, I is income and p is the price of the commodity. Customer's decision is described by retrieving the price from the demand function. We also include here the ability of the seller for increasing/decreasing the price according to his skills:

$$p = S\frac{\alpha I}{X} \quad (6)$$

This is the core formula, by which the customer decides if the quote is acceptable. The aforementioned parameters represent global simulation parameters set for each simulation experiment. Other global simulation parameters are:

I – customer's income – it's normal distributed value generated at the beginning and not being changed during the generation;

α – utility ratio – normal distributed value, which is generated for each customer each turn (week, while customers' preferences can change rapidly);

p – commodity price;

S – seller skills (ability to change price);

X – amount of commodity – normal distributed value generated, when customer decides to buy something.

Customer agents are organized in groups. Each group is being served by a specific seller agent. Their relationship is given; none of them can change the counterpart. Seller agent is responsible to the manager agent. Each turn, the manager agent gathers the data from all seller agents and stores KPIs of the company. The data is the result of the simulation and serves to understand the company behavior in a time – depending on the agents' decisions and behavior. The customer agents need to know some information about the market. This information is given by the informative agent. This agent is also responsible for the turn management and represents outside or controllable phenomena from the agents' perspective.

4 Experimental Results

At the start of simulation experiments phase several parameters were set. Agents list and their parameterization are listed in Table 1. This simulation session works with agent-based simulation.

The complex Monte Carlo parameterization is listed in Table 2 for the second simulation session. The purpose of these experiments was to compare the

Table 1 Parameterization of agent-based simulation. Source: own

Agent type	Agent count	Parameter name	Parameter value
Customer	500	Maximum disc. Turns	10
		Mean Quantity	40 m
		Quantity S. Deviation	32
		Mean income	600 EUR
		Income St. Deviation	10
		Mean utility Ratio	I.15
		Utility St. Deviation	0.2
Seller	25	Mean ability	1
		Ability St. Deviation	0.03
		Minimal price	0.36 EUR
Manager	1	Purchase price	0.17 EUR
Market info	1	Iterations count	52 weeks
Disturbance	1		

simulation results from the agent-based to the Monte Carlo simulation method. Nevertheless, we also aim to prove if the demand functions could serve as a core element in the seller-to-customer negotiation.

Table 2 Monte Carlo simulation parameterization. Source: own

Parameter name	Parameter value
Number of customers	10 000
Mean quantity	50 m
Quantity S. Deviation	29
Mean income	600 EUR
Income St. Deviation	10
Mean utility ratio	0.15
Utility St. Deviation	0.2
Mean ability	1
Ability St. Deviation	0.03
Mean sell price	3.15 EUR
Sell price St. Dev.	3.0
Purchase price	0.17 EUR
Iterations count	52 weeks

Agents were simulating one year – 52 weeks of interactions. As mentioned above – the manager agent was calculating the KPIs. In the case of a Monte Carlo method the simulation was done for each customer. The results were counted in four categories frequently used to describe the company's trading balance. The categories are: sold amount, income, costs, and gross profit. We unveil some gross profit values in the graphs bellow.

The commodity to be traded with was a UTP cable. Indeed, companies are dealing with a whole portfolio of products. In our simplification we concentrated only on one product and this was the UTP cable. Further, total gross profit was chosen as a representative KPI. Figure 1 contains the month sums of average gross profit for real and generated data for the agent-based simulation. Figure 2 represents the Monte Carlo simulation results for the same category. As can be seen from these figures, the result of simulations represents trend, which is quite similar to the real data.

Real data was taken from a Slovak anonymous company trading with PC components and supplies. The time series was discovered for the 2012 and the parameters of the simulations were set to mirror the situation on the market in that time.

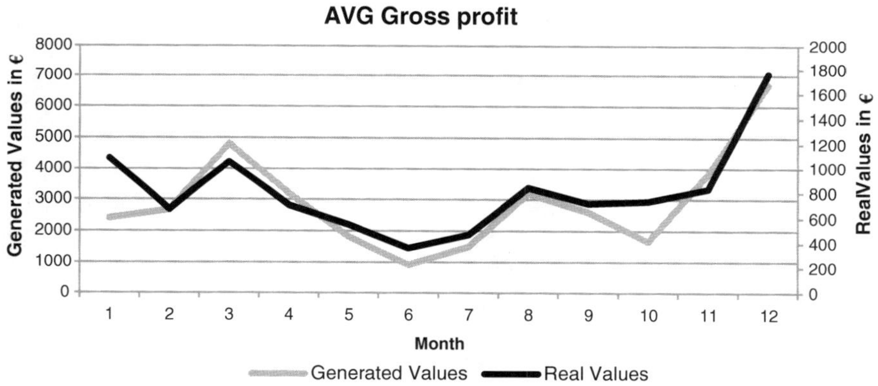

Fig. 1 Agent-based simulation. The generation values graph (AVG Gross Profit) – monthly (Source: own)

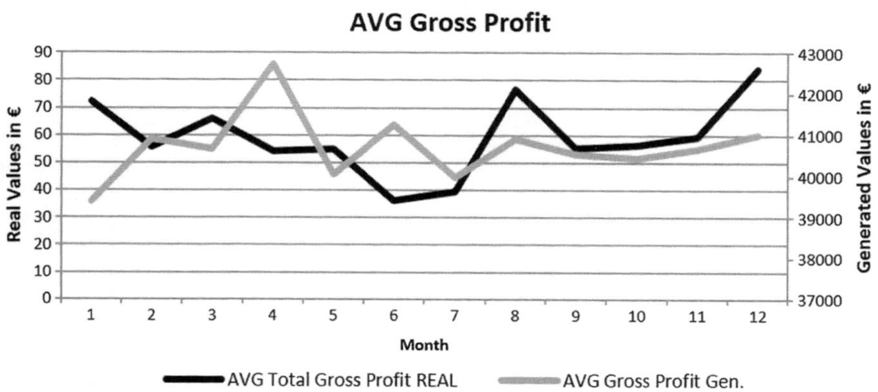

Fig. 2 Monte Carlo simulation. The generation values graph (AVG Gross Profit) – monthly (Source: own)

To discover the correlation between the real and generated AVG month price the correlation analysis was performed. The correlation coefficient for a total gross profit amount was 0.894 in the case of the agent-based simulation, which represents very strong correlation between real and generated data.

The correlation coefficient for a total gross profit amount was 0.693 in the case of the Monte Carlo simulation, which represents strong correlation between real and generated data. These results show that the demand functions could be used in further experiments to support the predictive purposes of decision making tools based on it.

5 Conclusion

The paper introduces an agent-based and a Monte Carlo simulation approach dealing with trading processes within a virtual company. The experiments were set to prove the idea, that microeconomic demand functions could be used as a core element in a seller-to-customer price negotiation. The overall idea is to use this approach to implement decision support models that could be connected to real management information systems in order to serve as prediction modules. We obtained successful results in some of the KPIs proving that the agent-based method is more suitable than Monte Carlo method. This supports our motivation to proceed with the experiments and to enhance our approach to extend the results on the rest of the KPIs. In our future research we will concentrate on the enhancement of the model.

Acknowledgment This paper was supported by the Ministry of Education, Youth and Sports Czech Republic within the Institutional Support for Long-term Development of a Research Organization in 2015.

References

1. Szarowská, I.: Effects of taxation by economic functions on economic growth in the European Union. In: Jircikova, E., Knapkova, A., Pastuszkova, E. (eds.) Proceedings of the 6th International Scientific Conference: Finance and the performance of Firms in Science, Education and Practice Zlin, pp. 746–758. Tomas Bata University, Czech Republic (2013)
2. Gajdova, K., Cieslarova, G.: Changes in selected characteristics of the EU labour market in times of economic crisis. Conference: In: Proceedings of the19th International-Business-Information-Management-Association Conference. Barcelona, Spain, 2011. Innovation Vision 2020: Sustainable Growth, Entrepreneurship, and Economic Development, pp. 628–635 (2012)
3. Janáková, M.: Challenges of information technologies in the global information society. In: Proceedings of 7th International Conference on Economic Policy in the European Union Member Countries, pp. 90–100. Silesian University in Opava, Karviná, Czech Republic (2009)

4. Suchánek, P.: Business intelligence - the standard tool of a modern company. In: Proceedings of the 6th International Scientific Symposium on Business Administration: Global Economic Crisis and Changes: Restructuring Business System: Strategic Perspectives for Local, National and Global Actors, pp. 123–132. Karviná, Czech Republic (2011)
5. Šperka, R., Spišák, M.: Microeconomic demand functions implementation in java experiments. In: Proceedings Advances in Intelligent Systems and Computing, Volume 296, Agent and Multi-Agent Systems. Technologies and Applications. In: 8th KES International Conference, KES AMSTA 2014, Chania, Greece, pp. 183–192. Springer International Publishing, Berlin (2014)
6. Šperka, R.: Application of a simulation framework for decision support systems. Mitteilungen Klosterneuburg, Hoehere Bundeslehranstalt und Bundesamt fuer Wein- und Obstbau, vol. 64, Issue 1. Klosterneuburg, Austria (2014). ISSN 0007-5922
7. Šperka, R., Vymětal, D., Spišák, M.: Towards the validation of agent-based BPM simulation. In: Proceedings Frontiers in Artificial Intelligence and Applications. 7th International Conference KES-AMSTA`13, Hue city, Vietnam, vol. 252, pp. 276–283. IOS Press BV, Amsterdam (2013). ISBN 978-1-61499-253-0 (print)
8. Šperka, R., Spišák, M., Slaninová, K., Martinovič, J., Dráždilová, P.: Control loop model of virtual company in BPM simulation. In: Proceedings Advances in Intelligent Systems and Computing, vol. 188, pp. 515–524. Springer, Berlin (2013). ISSN 2194-5357
9. Šperka, R., Spišák, M.: Transaction costs influence on the stability of financial market: agent-based simulation. Journal of Business Economics and Management, Taylor & Francis, London, vol. 14, supplement 1, pp. S1–S12 (2013). doi:10.3846/16111699.2012.701227. ISSN 1611-1699
10. Šperka, R., Vymětal, D.: MAREA - an education application for trading company simulation based on REA Principles. In: Proceedings Advances in Education Research. Information, Communication and Education Application, vol. 30, pp. 140–147. USA (2013). ISBN 978-1-61275-056-9
11. Barnett, M.: Modeling and simulation in business process management', Gensym Corporation, pp. 6–7. http://news.bptrends.com/publicationfiles/1103%20WP%20Mod%20Simulation%20of%20BPM%20-%20Barnett-1.pdf (2003). Accessed 16 Jan 2012
12. Tay, N., Lusch, R.: Agent-based modeling of ambidextrous organizations: virtualizing competitive strategy. IEEE Trans. Intell. Syst. **22**(5), 50–57 (2002)
13. Wilkinson, I., Young, L.: On cooperating: firms, relations, networks. J. Bus. Res. **55**, 123–132 (2002)
14. Casti, J.: Would-be Worlds. How Simulation is Changing the World of Science. Wiley, New York (1997)
15. Goldenberg, J., Libai, B., Muller, E.: The chilling effect of network externalities. Int. J. Res. Mark. **27**(1), 4–15 (2010)
16. Heath, B., Hill, R., Ciarallo, F.: A survey of agent-based modeling practices. J. Artif. Soc. Soc. Simul. **12**(4), 5–32 (2009). (January 1998 to July 2008)
17. Rahmandad, H., Sterman, J.: Heterogeneity and network structure in the dynamics of diffusion: comparing agent-based and differential equation models. Manage. Sci. **54**(5), 998–1014 (2008)
18. Shaikh, N., Ragaswamy, A., Balakrishnan A.: Modelling the Diffusion of Innovations Using Small World Networks Working Paper. Penn State University. Philadelphia (2005)
19. Toubia, O., Goldenberg, J., Garcia, R.: A New approach to modeling the adoption of new products: aggregated diffusion models. MSI reports: Working Papers Series, vol. 8, issue 1, pp. 65–76 (2008)
20. Adjali, I., Dias, B., Hurling R.: Agent based modeling of consumer behavior. In: Proceedings of the North American Association for Computational Social and Organizational Science Annual Conference. University of Notre Dame. Notre Dame (2005)
21. Ben, L., Bouron, T., Drogoul, A.: Agent-based interaction analysis of consumer behavior. In: Proceedings of the First International Joint Conference on Autonomous Agents and Multiagent Systems: Part 1. ACM, New York, pp. 184–190 (2002)

22. Marshall, A.: Principle of Economics, 8th edn. MacMillan, London (1920)
23. Mas-Colell, A., Whinston, M., Green, J.: Microeconomic Theory. Oxford University Press, Oxford (1995). ISBN 0-19-507340-1
24. Pollak, R.: Conditional demand functions and consumption theory. Quart. J. Econ. **83**, 60–78 (1969)
25. Varian, H.R.: Microeconomic Analysis, 3rd edn., Chapters 7, 8 and 9. W.W. Norton & Company, New York (1992)

Agent-Based Simulation Model of Sexual Selection Mechanism

Rafał Dreżewski

Abstract Agent-based approach is especially applicable and useful in modeling and simulation of social and biological systems and mechanisms. In this paper a formal agent-based model of sexual selection mechanism is presented. The paper includes results of simulation experiments aimed at showing whether sexual selection can trigger speciation processes and maintain genetic population diversity. We show that in certain conditions sexual selection mechanism and co-evolution of sexes resulting from it, can lead to formation of new species.

Keywords Agent-based modeling and simulation · Speciation · Sexual selection

1 Introduction

In recent years agent-based approach to modeling and simulation becomes more and more popular, especially when social and biological simulations are taken into consideration. Agent-based modeling and simulation (ABMS) approach has been applied in many different areas, such as research on complex biological, social, and economical systems [11, 12, 15, 16] crowd behavior and traffic simulation [19].

ABMS allows for elegant and explicit representation of environment, objects, resources, individuals (agents) and relations between them [15]. It is relatively easy to introduce learning mechanisms, to model dynamic formation of organizations and teams, also with learning capabilities at the organization/team level. Spatial relations are also easy to model because agents and objects are located within the environment with some topography. Usually scientists can develop agent-based model, implement it with the use of some existing tools (review of some such tools can be found in [15, 19]) and experiment with the simulation model in order to observe some emergent phenomena that was not explicitly represented within it.

R. Dreżewski (✉)
AGH University of Science and Technology, Department of Computer Science, Kraków, Poland
e-mail: drezew@agh.edu.pl

G. Jezic et al. (eds.), *Agent and Multi-Agent Systems: Technologies and Applications*, Smart Innovation, Systems and Technologies 38,
DOI 10.1007/978-3-319-19728-9_13

In this paper we will mainly focus on processes of species formation resulting from sexual selection and on agent-based modeling and simulation of such phenomena. Biological models of speciation include [14]:

- *allopatric models* that require geographical separation of sub-populations,
- *parapatric models* that require that zones of organisms only partially overlap and
- *sympatric models* where speciation takes place within one population without physical barriers as a result of, for example, co-evolutionary interactions or sexual selection.

Sexual selection results from the fact that cost of the reproduction for one of the sexes (usually females) is much higher than for the second one and the proportions of both sexes within the population are almost equal [14, 17]. These facts cause that females choose males on the basis of some features. These male's features and female's preferences are inherited by children and reinforced. Such phenomena leads to co-evolution of sexes where one of the sexes evolves in direction of keeping the reproduction rate on optimal level and the second one evolves in a direction that leads to increasing of the reproduction rate.

Model of multi-agent system with biological and social mechanisms (BSMAS) that we use in this paper to describe formally the simulation model of sexual selection mechanism was presented in [6]. BSMAS is reformulated and enhanced version of the model of co-evolutonary multi-agent system (CoEMAS) proposed in [4]. CoEMAS model was based on the idea of combining together multi-agent systems and evolutionary algorithms presented in [3]. CoEMAS approach introduced the possibility of existing of multiple species and sexes within a population and possibility of defining relations between them. BSMAS approach additionally introduced possibilities of defining social relations and structures. CoEMAS approach was used in many different areas, such as multi-modal optimization [5], multi-objective optimization [8, 10] and generating investment strategies [7, 9].

In this paper we apply BSMAS approach to modeling and simulation of sexual selection. We then investigate whether sexual selection can cause speciation and whether it promotes genetic diversity.

2 Agent-Based Model of Sexual Selection Mechanism

The notions *agent* and *multi-agent system* are used in literature, in rather not very formal and strict manner, for naming different kinds of systems and approaches. In this paper we follow the definitions proposed by J. Ferber [13]. In his approach *agent* is considered to be a physical or virtual entity capable of acting within an environment and capable of communicating with other agents. Its activities are driven by individual goals, it possesses some resources, it may observe the environment (but only local part of it), it possesses only partial knowledge about the environment. It has some abilities and may offer some services, it may also be able to reproduce.

Multi-agent system is a system composed of environment, objects (passive elements of the system) and active elements of the system (agents). There can be different kinds of relations between elements of the system. Multi-agent system also includes set of operations that allow agents to observe and interact with other elements of the system, and set of operators, which aim is to represent agent's actions and reactions of the other elements of the system [13].

2.1 Agent-Based Model of Sexual Selection Mechanism

In the model presented in this section sympatric speciation takes place as a result of sexual selection. There are two sexes within the population: female and male (Fig. 1). Each agent needs resources for living so it tries to get them from the environment. During movement agents lose resources and when an agent runs out of resources it dies.

Fig. 1 Multi-Agent System with Sexual Selection

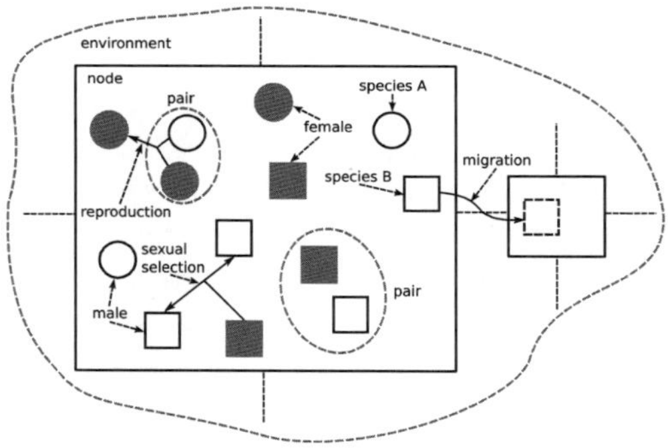

When the level of the resource is above some minimal value, an agent is ready for reproduction and searches for a partner. Female agent chooses a partner from the set of male agents that are ready for reproduction in the given period of time and that are located within the same node of the environment—the decision is made on the basis of genotype similarity (the more the agents are similar the more probable is the decision of choosing the given male agent). Then a pair is formed and agents move together within the environment for some time (Fig. 1)—such behavior helps during reproduction because agents do not have to search for a partner every time when they are ready for reproduction. The offspring is generated with the use of mutation and recombination operators (intermediate recombination [2], and mutation with self-adaptation [1]). The offspring receives some of the resources from parents—more from female agent, so female agents' cost of reproduction is much higher than male agents' cost of reproduction.

The multi-agent system with sexual selection is defined as follows:

$$\begin{aligned} sBSMAS(t) = \langle & EnvT(t) = \{et\}, Env(t) = \{env\}, ElT(t) = \\ & VertT(t) \cup ObjT(t) \cup AgT, ResT(t) = \{rt\}, \\ & InfT(t) = \emptyset, Rel(t), Attr(t) = \{genotype\}, \\ & Act(t) \rangle \end{aligned} \tag{1}$$

where:

- $EnvT(t)$ is the set of environment types in the time t;
- $Env(t)$ is the set of environments of the *sBSMAS* in the time t;
- $ElT(t)$ is the set of types of elements that can exist within the system in time t;
- $VertT(t) = \{vt\}$ is the set of vertice types that can exist within the system in time t;
- $ObjT(t) = \emptyset$ is the set of object (not an object in the sense of object-oriented programming but object as an element of the simulation model) types that may exist within the system in time t;
- $AgT(t) = \{female, male\}$ is the set of agent types that may exist within the system in time t;
- $ResT(t)$ is the set of resource types that exist in the system in time t, the amount of resource of type $rest(t) \in ResT(t)$ will be denoted by $res^{rest}(t)$;
- $InfT(t)$ is the set of information types that exist in the system, the information of type $inft(t) \in InfT(t)$ will be denoted by $inf^{inft}(t)$;
- $Rel(t)$ is the set of relations between sets of agents, objects, and vertices;
- $Attr(t)$ is the of attributes of agents, objects, and vertices;
- $Act(t)$ is the set of actions that can be performed by agents, objects, and vertices.

The set of actions is defined as follows:

$$Act = \{die, reproduce, get\_resource, give\_resource, migrate, choose\} \tag{2}$$

Environment type *et* is defined in the following way:

$$et = \langle EnvT^{et} = \emptyset, VertT^{et} = VertT, ResT^{et} = ResT, InfT^{et} = \emptyset \rangle \tag{3}$$

$EnvT^{et} \subseteq EnvT$ is the set of environment types that may be connected with the *et* environment. $VertT^{et} \subseteq VerT$ is the set of vertice types that may exist within the environment of type *et*. $ResT^{et} \subseteq ResT$ is the set of resource types that may exist within the environment of type *et*. $InfT^{et} \subseteq InfT$ is the set of information types that may exist within the environment of type *et*.

Environment *env* of type *et* is defined as follows:

$$env = \langle gr^{env}, Env^{env} = \emptyset \rangle \tag{4}$$

where gr^{env} is directed graph $gr^{env} = \langle Vert, Arch, cost \rangle$, *Vert* with the *cost* function defined is the set of vertices, *Arch* is the set of arches. The distance between two nodes is defined as the length of the shortest path between them in graph gr^{env}. $Env^{env} \subseteq Env$ is the set of environments of types from *EnvT* connected with the environment *env*.

Vertice type *vt* is defined in the following way:

$$vt = \Big\langle Attr^{vt} = \emptyset, Act^{vt} = \big\{give\_resource\big\}, ResT^{vt} = ResT, \\ InfT^{vt} = \emptyset, VertT^{vt} = VertT, ObjT^{vt} = \emptyset, AgT^{vt} = AgT \Big\rangle \quad (5)$$

where:

- $Attr^{vt} \subseteq Attr$ is the set of attributes of *vt* vertice at the beginning of its existence;
- $Act^{vt} \subseteq Act$ is the set of actions, which *vt* vertice can perform at the beginning of its existence, when asked for it;
- $ResT^{vt} \subseteq ResT$ is the set of resource types, which can exist within *vt* vertice at the beginning of its existence;
- $InfT^{vt} \subseteq InfT$ is the set of information, which can exist within *vt* vertice at the beginning of its existence;
- Vt^{vt} is the set of types of vertices that can be connected with the *vt* vertice at the beginning of its existence;
- $ObjT^{vt} \subseteq ObjT$ is the set of types of objects that can be located within the *vt* vertice at the beginning of its existence;
- $AgT^{vt} \subseteq AgT$ is the set of types of agents that can be located within the *vt* vertice at the beginning of its existence;
- *give_resource* is the action of giving resource to agents.

Element of the structure of system's environment (vertice) $vert \in Vert$ of type $vt \in VertT^{env}$ is given by:

$$vert = \Big\langle Attr^{vert} = \emptyset, Act^{vert} = Act^{vt}, Res^{vert} = \big\{res^{vert}\big\}, Inf^{vert} = \emptyset, \\ Vert^{vert}, Obj^{vert} = \emptyset, Ag^{vert} \Big\rangle \quad (6)$$

where:

- $Attr^{vert} \subseteq Attr$ is the set of attributes of vertice *vert*—it can change during its lifetime;
- $Act^{vert} \subseteq Act$ is the set of actions, which vertice *vert* can perform when asked for it—it can change during its lifetime;
- Res^{vert} is the set of resources of types from *ResT* that exist within the *vert*;
- Inf^{vert} is the set of information of types from *InfT* that exist within the *vert*;
- $Vert^{vert}$ is the set of vertices of types from *VertT* connected with the vertice *vert*;
- Obj^{vert} is the set of objects of types from *ObjT* that are located in the vertice *vert*;
- Ag^{vert} is the set of agents of types from *AgT* that are located in the vertice *vert*.

res^{vert} is the amount of resource of type rt that is possessed by the *vert*. $Vert^{vert}$ is the set of four vertices connected with the vertice *vert* (Fig. 1). Ag^{vert} is the set of agents located within the vertice *vert*.

There are two types of agents in the system: *female* and *male*. *female* agent type is defined in the following way:

$$\begin{aligned} female = \big\langle & Gl^{female} = \{gl_1, gl_2, gl_3\}, Attr^{female} = \{genotype\}, \\ & Act^{female} = \{die, reproduce, choose, get\_resource, migrate\}, \\ & ResT^{female} = ResT, InfT^{female} = \emptyset, ObjT^{female} = \emptyset, \\ & AgT^{female} = \emptyset \big\rangle \end{aligned} \tag{7}$$

where:

- Gl^{female} is the set of goals of *female* agent at the beginning of its existence;
- $Attr^{female} \subseteq Attr$ is the set of attributes of *female* agent at the beginning of its existence;
- $Act^{female} \subseteq Act$ is the set of actions, which *female* agent can perform at the beginning of its existence;
- $ResT^{female} \subseteq ResT$ is the set of resource types, which can be used by *female* agent at the beginning of its existence;
- $InfT^{female} \subseteq InfT$ is the set of information, which can be used by *female* agent at the beginning of its existence;
- $ObjT^{female} \subseteq ObjT$ is the set of types of objects that can be located within the *female* agent at the beginning of its existence;
- $AgT^{female} \subseteq AgT$ is the set of types of agents that can be located within the *female* agent at the beginning of its existence.

gl_1 is the goal "get resource from environment", gl_2 is the goal "reproduce", and gl_3 is the goal "migrate to other vertice". *die* is the action of death—agent dies when it runs out of resources, *reproduce* is the action of reproducing (with the use of recombination and mutation operators), *choose* is the action of choosing partner for reproduction from the set of *male* agents that are located within the same vertice and are ready for reproduction, *get_resource* is the action of getting resource from environment, and *migrate* is the action of migrating to other vertice.

male agent type is defined in the following way:

$$\begin{aligned} male = \big\langle & Gl^{male} = \{gl_1, gl_2, gl_3\}, Attr^{male} = \{genotype\}, \\ & Act^{male} = \{die, reproduce, get\_resource, migrate, \}, \\ & ResT^{male} = ResT, InfT^{male} = \emptyset, ObjT^{male} = \emptyset, AgT^{male} = \emptyset \big\rangle \end{aligned} \tag{8}$$

where gl_1 is the goal "get resource from environment", gl_2 is the goal "reproduce", and gl_3 is the goal "migrate to other vertice". *die* is the action of death—agent dies when it runs out of resources, *reproduce* is the action of reproducing (with the use of recombination and mutation operators), *get_resource* is the action of getting resource from environment, and *migrate* is the action of migrating to other vertice.

Agent ag^{female} (of type *female*) is defined in the following way:

$$\begin{aligned} ag^{female} = \big\langle & Gl^{ag,female} = Gl^{female}, Attr^{ag,female} = Attr^{female}, \\ & Act^{ag,female} = Act^{female}, Res^{ag,female} = \{r^{ag,female}\}, \\ & Inf^{ag,female} = \emptyset, Obj^{ag,female} = \emptyset, Ag^{ag,female} = \emptyset \big\rangle \end{aligned} \tag{9}$$

where:

- Gl^{female} is the set of goals, which agent *female* tries to realize—it can change during its lifetime;
- $Attr^{female} \subseteq Attr$ is the set of attributes of agent *female*—it can change during its lifetime;
- $Act^{female} \subseteq Act$ is the set of actions, which agent *female* can perform in order to realize its goals—it can change during its lifetime;
- Res^{female} is the set of resources of types from $ResT$, which are used by agent *female*;
- Inf^{female} is the set of information of types from $InfT$, which agent *female* can possess and use;
- Obj^{female} is the set of objects of types from $ObjT$ that are located within the agent *female*;
- Ag^{female} is the set of agents of types from AgT that are located within the agent *female*.

Notation $Gl^{ag,female}$ means "the set of goals of agent ag of type *female*". $r^{ag,female}$ is the amount of resource of type rt that is possessed by the agent ag^{female}.

Agent ag^{male} (of type *male*) is defined in the following way:

$$\begin{aligned} ag^{male} = \big\langle & Gl^{ag,male} = Gl^{male}, Attr^{ag,male} = Attr^{male}, \\ & Act^{ag,male} = Act^{male}, Res^{ag,male} = \{r^{ag,male}\}, \\ & Inf^{ag,male} = \emptyset, Obj^{ag,male} = \emptyset, Ag^{ag,male} = \emptyset \big\rangle \end{aligned} \tag{10}$$

Notation $Gl^{ag,male}$ means "the set of goals of agent ag of type *male*". $r^{ag,male}$ is the amount of resource of type rt that is possessed by the agent ag^{male}.

The set of relations is defined as follows:

$$Rel = \left\{ \xrightarrow[\{get\_resource\}]{\{get\_resource\}}, \xrightarrow[\{reproduce\}]{\{choose,reproduce\}} \right\} \tag{11}$$

The relation $\xrightarrow[\{get\_resource\}]{\{get\_resource\}}$ is defined as follows:

$$\xrightarrow[\{get\_resource\}]{\{get\_resource\}} = \left\{ \left\langle Ag^{\{get\_resource\}}, Ag^{\{get\_resource\}} \right\rangle \right\} \tag{12}$$

$Ag^{\{get\_resource\}}$ is the set of agents capable of performing action *get_resource*. This relation represents competition for limited resources between agents.

The relation $\xrightarrow[\{reproduce\}]{\{choose,reproduce\}}$ is defined as follows:

$$\xrightarrow[\{reproduce\}]{\{choose,reproduce\}} = \left\{ \left\langle Ag^{female,\{choose,reproduce\}}, Ag^{male,\{reproduce\}} \right\rangle \right\} \tag{13}$$

$Ag^{female,\{choose,reproduce\}}$ is the set of agents of type *female* capable of performing actions *choose* and *reproduce*. $Ag^{male,\{reproduce\}}$ is the set of agents of type *male* capable of performing action *reproduce*. This relation represents sexual selection mechanism—*female* agents choose partners for reproduction form *male* agents and then reproduction takes place.

Action *choose* chooses male agent from the set of male agents that are ready for reproduction and are located within the same vertice as the given female agent with probability proportional to genetic similarity of both agents.

3 Results of Simulation Experiments

In this section we will present results of simulation experiments with our agent-based model of sexual selection. The main goal of these experiments was to investigate whether sexual selection mechanism is able to trigger speciation within the population and whether it is able to maintain genetic diversity.

In all experiments we used our own multi-agent simulation system based on BSMAS model and implemented in Java.

During experiments we used two fitness landscapes: Rastrigin and Waves (Fig. 2).

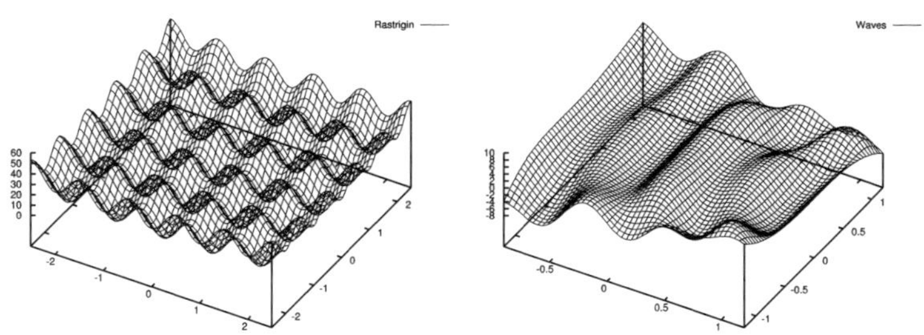

(a) Rastrigin (b) Waves

Fig. 2 Fitness landscapes

Rastrigin multimodal fitness landscape is defined as follows [18]:

$$f_2(\boldsymbol{x}) = 10 * n + \sum_{i=1}^{n} \left(x_i^2 - 10 * \cos(2 * \pi * x_i) \right)$$
$$x_i \in [-2.5; 2.5] \text{ for } i = 1, \ldots, n \tag{14}$$

During experiments $n = 2$ was assumed (Fig. 2a).

Waves fitness landscape is defined as follows [20]:

$$f_4(\boldsymbol{x}) = -\Big(\left(0.3 * x_1\right)^3 - \left(x_2^2 - 4.5 * x_2^2\right) * x_1 * x_2 - 4.7 * \cos\left(3 * x_1 - x_2^2 * \left(2 + x_1\right)\right) * \sin\left(2.5 * \pi * x_1\right)\Big) \tag{15}$$
$$x_1 \in [-0.9; 1.2],\ x_2 \in [-1.2; 1.2]$$

This function has many irregularly placed local minima (Fig. 2b).

Species formation processes during experiments with different fitness landscapes can be observed in Figs. 3 and 4.

In all these cases it can be observed that after initial phase, distinct species were formed. In step 0 of each experiment we can see that population is dispersed over the whole genetic space, and later species are formed—distinct species were depicted

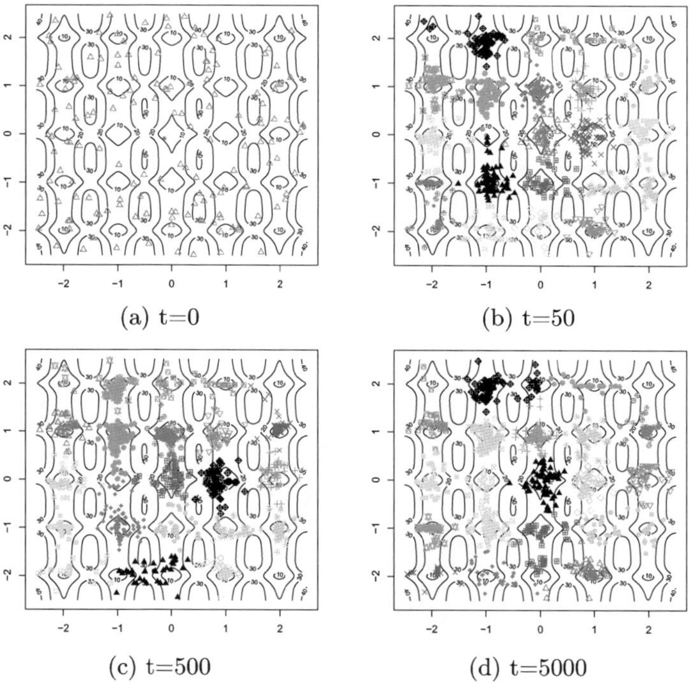

Fig. 3 Species formation processes in sBSMAS with Rastrigin fitness landscape

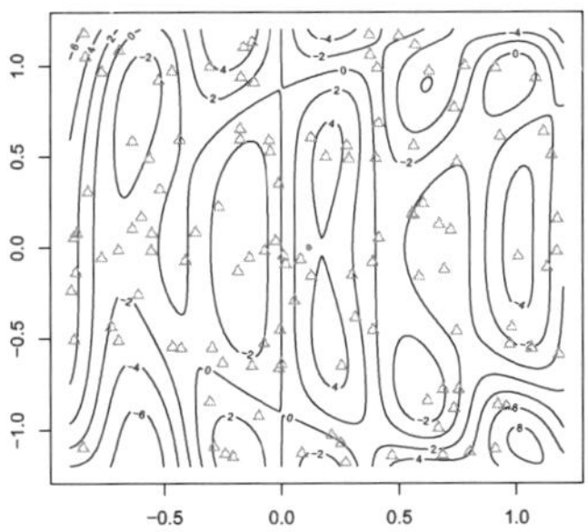

(a) t=0

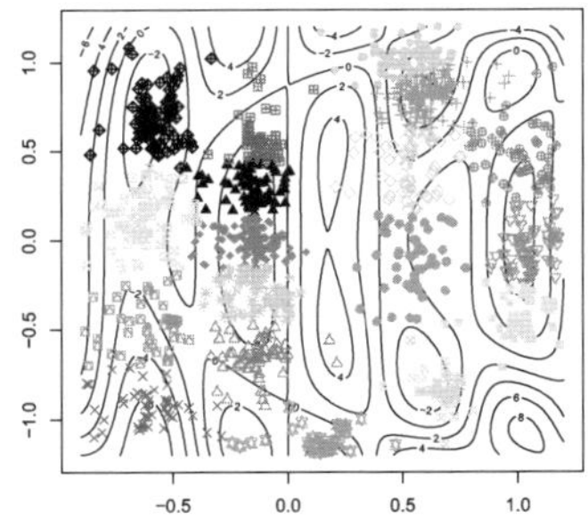

(b) t=50

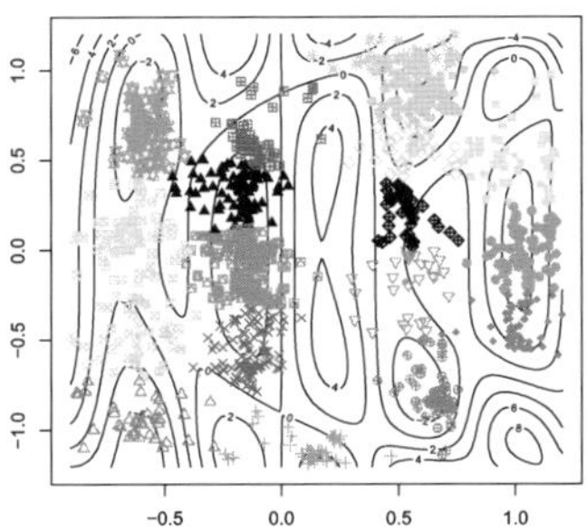

(c) t=500

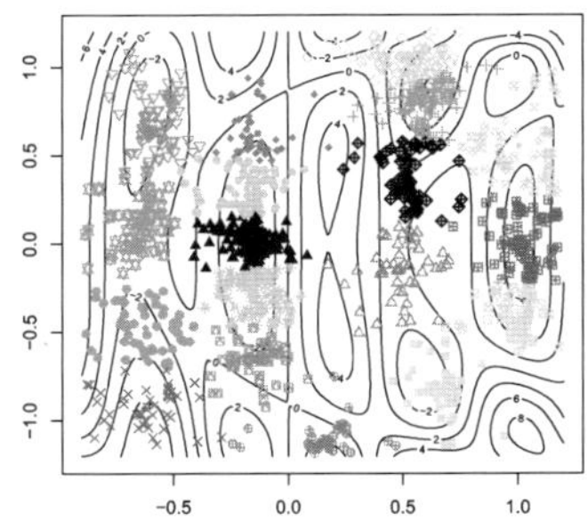

(d) t=5000

Fig. 4 Species formation processes in sBSMAS with Waves fitness landscape

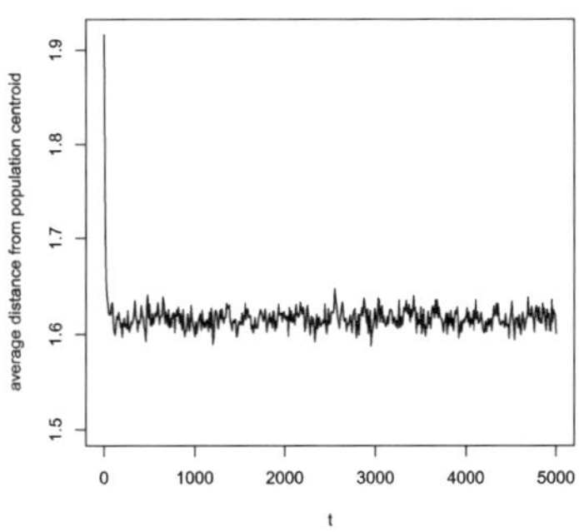

(a) Rastrigin

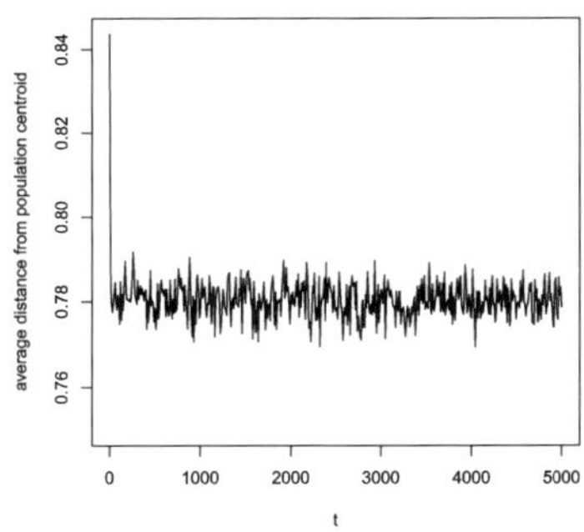

(b) Waves

Fig. 5 Average distance from population centroid

with the use of different shapes and colors. We used k-means clustering algorithm to find clusters (species) within the population of agents occupying different parts of the genetic space. Sexual selection—selecting males on the basis of genetic similarity—causes that agents tend to form species (isolated clusters within the genetic space).

In Fig. 5a, b, results showing population diversity measure are presented. As the measure of genetic population diversity average distance of agents from the centroid of the population was used. It can be observed that sexual selection maintains genetic population diversity. Diversity of course drops after some time, but it is caused by the fact that distinct species are formed and agents are grouping within some regions of the genetic space. But generally in all the cases genetic diversity is maintained and do not diminish.

4 Conclusions

In this paper we presented agent-based model of species formation processes caused by sexual selection. Presented results show that in particular conditions sexual selection can lead to speciation. Sexual selection also helps maintaining genetic population diversity. It has been demonstrated that the agent-based approach to modeling and simulation in general, and proposed BSMAS model in particular, are well suited for constructing models of biological and social systems and mechanisms.

In the future we plan to conduct experiments with more realistic environments, without explicit fitness function. Agents would be placed in such environment and would have to deal with changing conditions and different species co-existing within the same environment.

Acknowledgments This research was partially supported by Polish Ministry of Science and Higher Education under AGH University of Science and Technology Grant (statutory project).

References

1. Bäck, T., Fogel, D.B., Whitley, D., Angeline, P.J.: Mutation. In: Bäck, T., Fogel, D., Michalewicz, Z. (eds.) Handbook of Evolutionary Computation. IOP Publishing and Oxford University Press, Oxford (1997)
2. Booker, L.B., Fogel, D.B., Whitley, D., Angeline, P.J.: Recombination. In: Bäck, T., Fogel, D., Michalewicz, Z. (eds.) Handbook of Evolutionary Computation. IOP Publishing and Oxford University Press, Oxford (1997)
3. Cetnarowicz, K., Kisiel-Dorohinicki, M., Nawarecki, E.: The application of evolution process in multi-agent world to the prediction system. In: Tokoro, M. (ed.) Proceedings of the 2nd International Conference on Multi-Agent Systems (ICMAS 1996). AAAI Press, Menlo Park (1996)
4. Dreżewski, R.: A model of co-evolution in multi-agent system. In: Mařík, V., Müller, J., Pěchouček, M. (eds.) Multi-Agent Systems and Applications III, 3rd International Central and Eastern European Conference on Multi-Agent Systems, CEEMAS 2003, Prague, Czech Republic, June 16–18, 2003, Proceedings. LNCS, vol. 2691, pp. 314–323. Springer, Berlin (2003)
5. Dreżewski, R.: Co-evolutionary multi-agent system with speciation and resource sharing mechanisms. Comput. Inf. **25**(4), 305–331 (2006)
6. Dreżewski, R.: Agent-based modeling and simulation of species formation processes. In: Alkhateeb, F., Al Maghayreh, E., Abu Doush, I. (eds.) Multi-Agent Systems – Modeling, Interactions, Simulations and Case Studies, pp. 3–28. InTech, Rijeka (2011)
7. Dreżewski, R., Sepielak, J.: Evolutionary system for generating investment strategies. In: M. Giacobini, et al. (ed.) Applications of Evolutionary Computing, EvoWorkshops 2008: EvoCOMNET, EvoFIN, EvoHOT, EvoIASP, EvoMUSART, EvoNUM, EvoSTOC, and Evo-

TransLog, Naples, Italy, 26–28 March 2008. Proceedings. LNCS, vol. 4974, pp. 83–92. Springer, Berlin, Heidelberg (2008)

8. Dreżewski, R., Siwik, L.: Multi-objective optimization using co-evolutionary multi-agent system with host-parasite mechanism. In: Alexandrov, V.N., van Albada, G.D., Sloot, P.M.A., Dongarra, J. (eds.) Computational Science—ICCS 2006, 6th International Conference, Reading, UK, May 28–31, 2006, Proceedings, Part III. LNCS, vol. 3993, pp. 871–878. Springer, Berlin (2006)
9. Dreżewski, R., Siwik, L.: Multi-objective optimization technique based on co-evolutionary interactions in multi-agent system. In: Giacobini, M., et al. (eds.) Applications of Evolutinary Computing, EvoWorkshops 2007: EvoCoMnet, EvoFIN, EvoIASP, EvoINTERACTION, EvoMUSART, EvoSTOC and EvoTransLog, Valencia, Spain, April11-13, 2007, Proceedings. LNCS, vol. 4448, pp. 179–188. Springer, Berlin (2007)
10. Dreżewski, R., Siwik, L.: Agent-based co-operative co-evolutionary algorithm for multi-objective optimization. In: Rutkowski, L., Tadeusiewicz, R., Zadeh, L.A., Zurada, J.M. (eds.) Artificial Intelligence and Soft Computing—ICAISC 2008, 9th International Conference, Zakopane, Poland, June 22–26, 2008, Proceedings. LNCS, vol. 5097, pp. 388–397. Springer, Berlin (2008)
11. Epstein, J.M.: Generative Social Science. Studies in Agent-Based Computational Modeling. Princeton University Press, Princeton (2006)
12. Epstein, J.M., Axtell, R.: Growing Artificial Societes. Social Science from Bottom Up. Brookings Institution Press, The MIT Press, Washington, DC (1996)
13. Ferber, J.: Multi-Agent Systems: An Introduction to Distributed Artificial Intelligence. Addison-Wesley, Boston (1999)
14. Gavrilets, S.: Models of speciation: what have we learned in 40 years? Evolution **57**(10), 2197–2215 (2003)
15. Gilbert, N.: Agent-based models. SAGE Publications, London (2008)
16. Gilbert, N., Troitzsch, K.G.: Simulation for the social scientist. Open University Press, Buckingham (2005)
17. Krebs, J., Davies, N.: An Introduction to Behavioural Ecology. Blackwell Science Ltd, Oxford (1993)
18. Potter, M.A.: The design and analysis of a computational model of cooperative coevolution. Ph.D. thesis, George Mason University, Fairfax, Virginia (1997)
19. Uhrmacher, A.M., Weyns, D. (eds.): Multi-agent Systems. Simulation and Applications. CRC Press, Boca Raton (2009)
20. Ursem, R.K.: Multinational evolutionary algorithms. In: Angeline, P.J., Michalewicz, Z., Schoenauer, M., Yao, X., Zalzala, A. (eds.) Proceedings of the 1999 Congress on Evolutionary Computation (CEC-1999), pp. 1633–1640. IEEE Press, Piscataway (1999)

Core of n-Person Transferable Utility Games with Intuitionistic Fuzzy Expectations

Elena Mielcová

Abstract One of the main tasks in the theory of n-person transferable utility games is bound with looking for optimal solution concept. In general, a core is considered to be one of basic used solution concepts. The idea behind core derivation is straightforward; however, in reality uncertainty issues should be taken into consideration. Hence, the main aim of this article is to introduce formalization of the n-person transferable utility games in the case when expected utilities are intuitionistic fuzzy values. Moreover, this article discusses superadditivity issues of intuitionistic fuzzy extensions of cooperative games, and the construction of a core of such games. Calculations are demonstrated on numerical examples and compared with classical game theory results.

1 Introduction

In multiagent systems, agents are assumed to be to some extent autonomous - capable of making decisions in order to satisfy their design objectives [13]. In complex multiagent systems agents are expected to interact with other agents in order to successfully carry out their tasks. One of the standard descriptions of the multi-agent problem of cooperation is given by the theory of cooperative games [12]. Agents – in game theoretical approach players of a model game – are cooperating in order to increase a mutual profit. Considering that the profit can be distributed among players with respect to some coalitional agreement, these games are called also transferable utility (TU) games. Cooperative game theory considers the question of profit distribution, and provides several solution concepts, among them the most simplistic and basic is the concept of core.

E. Mielcová (✉)
Department of Informatics and Mathematics, Silesian University in Opava,
School of Business Administration in Karvina,
Univerzitní Náměstí 1934/3, 733 40 Karviná, Czech Republic
e-mail: mielcova@opf.slu.cz

G. Jezic et al. (eds.), *Agent and Multi-Agent Systems: Technologies and Applications*, Smart Innovation, Systems and Technologies 38,
DOI 10.1007/978-3-319-19728-9_14

The main aim of this article is to discuss the construction core of the TU cooperative game when pay-offs are vague – in this case expressed as intuitionistic (IT) fuzzy numbers. Similar research was done on the sets of TU coalitional games with fuzzy expectations [10], however this concept was not yet discussed on the extension of IT fuzzy sets. In general, the theory of IT fuzzy sets is considered to be an extension of fuzzy set theory, where the degree of non-membership denoting the non-belongingness to a set is explicitly specified along with the degree of membership of belongingness to the set [1]. The IT fuzzy set theory is widely used to describe game theory concepts, as this model realistically describes real-world concepts [6].

The next text is organized as follows; the preliminaries cover basic terms from the theory of TU coalitional games and IT fuzzy sets. The third section introduces IT fuzzy extension of the coalitional games with IT fuzzy pay-off function, and concept of IT fuzzy core. The conclusion followed by references ends the text.

2 Preliminaries

2.1 TU Game, Imputation, Core, Convexity

Let $N=\{1,2,\ldots,n\}$ be a set of n players, let 2^N be the collection of all subsets of N. Any nonempty subset of N is called a coalition. Formally, a TU game (or a cooperative game) in a form of characteristic function is defined as a pair (N,v), where $v\colon 2^N \to R$ is a mapping called characteristic function of the game. The characteristic function connects each coalition $K\subset N$ with a real number $v(K)\subset R$ representing total profit of coalition K; we assume $v(\varnothing)=0$. A cooperative game can by denoted as (N,v), or simply v.

For any pair of disjoint coalitions $K,L\subset N, K\cap L=\varnothing$ a coalition game (N,v) is called

- superadditive if $v(K\cup L)\geq v(K)+v(L)$,
- subadditive if $v(K\cup L)\leq v(K)+v(L)$, and
- additive if $v(K\cup L)=v(K)+v(L)$.

A game (N,v) is additive iff it is superadditive and subadditive. We say that a coalitional game (N,v) is convex if for every pair of coalitions $K,L\subset N$ there hold

$$v(K\cup L)+v(K\cap L)\geq v(K)+v(L); \tag{1}$$

Moreover, for any pair of disjoint coalitions $K,L\subset N$ Eq. (1) changes to superadditivity condition; that means convexity implies superadditivity.

Expected distribution of profit among players is represented by real-valued payoff vectors $\boldsymbol{x}=(x_i)_{i\in N}\in R^n$. A preimputation is an efficient payoff vector; that means any $\boldsymbol{x}$ from the preimputation set $I^*(v)$ of a TU cooperative game (N,v) defined as

$$I^*(v) = \{\boldsymbol{x} \in R^n;\ \sum_{i \in N} x_i = v(N)\}. \tag{2}$$

A payoff vector $\boldsymbol{x} \in R^n$ is individually rational if for each $i \in N$ there is $x_i \geq v(\{i\})$. An individually rational preimputation is called an imputation; that means that imputation is an element from an imputation set $I(v)$ of a TU cooperative game (N, v) defined as

$$I(v) = \{\boldsymbol{x} \in R^n;\ \sum_{i \in N} x_i = v(N), x_i \geq v(\{i\}) \text{ for each } i \in N\}. \tag{3}$$

Let the game (N, v) is superadditive. Then the core of (N, v) is the set of payoff vectors

$$C(v) = \{\boldsymbol{x} \in R^n;\ \sum_{i \in N} x_i = v(N),\ \sum_{i \in L} x_i \geq v(\mathrm{L}) \text{ for each } L \subset N\}. \tag{4}$$

In general, a core can represent stable imputations as a result of bargaining.

2.2 *Intuitionistic Fuzzy Numbers*

Intuitionistic fuzzy sets and related operations and relations on the IT fuzzy sets were introduced in [1]. Formally, let a set $X = \{x_1, x_2, \ldots, x_n\}$ be fixed. Then an intuitionistic fuzzy set is defined as a set of triples $A = \{\langle x_i, t_A(x_i), f_A(x_i)\rangle; x_i \epsilon X\}$. where functions $t_A : X \to L$ and $f_A : X \to L$ for $L = [0, 1]$ define the degree of membership and the degree of non-membership of the element $x_i \in X$ to $A \subset X$, respectively. For an intuitionistic fuzzy set, the condition $0 \leq t_A(x_i) + f_A(x_i) \leq 1$ holds for all $x_i \in X$.

In this analysis, payoffs of players will be expressed as intuitionistic fuzzy numbers, defined in [2, 3, 5]. An intuitionistic fuzzy subset $A = \{\langle x, t_A(x), f_A(x)\rangle; x \in R\}$ of the real line is called an intuitionistic fuzzy number if

(a) A is if-normal (there exist at least two points $x_0, x_1 \in X$ such that $t_A(x_0) = 1$ and $f_A(x_1) = 1$);
(b) A is if-convex (its membership function t is fuzzy convex and its non-membership function is fuzzy concave);
(c) t_A is upper semi-continuous and f_A is lower semi-continuous;
(d) $A = \{x \in X; f_A(\mathrm{x}) < 1\}$ is bounded.

For simplicity, an illustrative example will be given at the special class of intuitionistic fuzzy numbers – triangular intuitionistic fuzzy numbers. In general, triangular intuitionistic fuzzy number (TIFN) A_{TIFN} is a subset of IT fuzzy sets in R with membership function and non-membership functions as follows [7]:

$$t_{A_{TINF}}(x) = \begin{cases} \frac{x-a_1}{a_2-a_1} & \text{for } a_1 \le x \le a_2 \\ \frac{a_3-x}{a_3-a_2} & \text{for } a_2 \le x \le a_3 \\ 0 & \text{otherwise} \end{cases}, \quad f_{A_{TINF}}(x) = \begin{cases} \frac{a_2-x}{a_2-a'_1} & \text{for } a'_1 \le x \le a_2 \\ \frac{x-a_2}{a'_3-a_2} & \text{for } a_2 \le x \le a'_3 \\ 1 & \text{otherwise} \end{cases} \tag{5}$$

Hence, any triangular intuitionistic fuzzy number is fully described by a six-tuple of interval borders $A_{TINF} = (a_1, a_2, a_3, a'_1, a_2, a'_3)$.

Arithmetic operations with triangular and trapezoidal numbers were described in [7, 11]. For a positive number $k\epsilon R$, a triangular intuitionistic fuzzy number $A_{TINF} = (a_1, a_2, a_3, a'_1, a_2, a'_3)$, the membership and non-membership functions of $C_{TINF} = kA_{TINF}$ are:

$$t_{C_{TINF}}(z) = \begin{cases} \frac{z-\mathrm{k}a_1}{ka_2-\mathrm{k}a_1} & \mathrm{k}a_1 \le z \le \mathrm{k}a_2 \\ \frac{ka_3-z}{ka_3-\mathrm{k}a_2} & \mathrm{k}a_2 \le z \le ka_3 \\ 0 & \text{otherwise} \end{cases} \quad f_{C_{TINF}}(z) = \begin{cases} \frac{ka_2-z}{ka_2-ka'_1} & ka'_1 \le z \le ka_2 \\ \frac{z-ka_2}{ka'_3-\mathrm{k}a_2} & ka_2 \le z \le ka'_3 \\ 1 & \text{otherwise} \end{cases} \tag{6}$$

In general, $C_{TINF} = kA_{TINF} = (ka_1, ka_2, ka_3, ka'_1, \mathrm{k}a_2, ka'_3)$. For two IT fuzzy numbers $A_{TINF} = (a_1, a_2, a_3, a'_1, a_2, a'_3)$, and $B_{TINF} = (b_1, b_2, b_3, b'_1, b_2, b'_3)$, the addition operation:

$$C_{TINF} = A_{TINF} \oplus B_{TINF} = \left(a_1 + b_1, a_2 + b_2, a_3 + b_3, a'_1 + b'_1, a_2 + b_2, a'_3 + b'_3\right) \tag{7}$$

The membership and non-membership functions of $C_{TINF} = A_{TINF} \oplus B_{TINF}$ are:

$$t_{C_{TINF}}(z) = \begin{cases} \frac{z-(a_1+b_1)}{(a_2+b_2)-(a_1+b_1)} & \text{for } (a_1+b_1) \le z \le (a_2+b_2) \\ \frac{(a_3+b_3)-z}{(a_3+b_3)-(a_2+b_2)}) & \text{for } (a_2+b_2) \le z \le (a_3+b_3) \\ 0 & \text{otherwise} \end{cases} \tag{8}$$

$$f_{C_{TINF}}(z) = \begin{cases} \frac{(a_2+b_2)-z}{(a_2+b_2)-(a'_1+b'_1)} & \text{for } (a'_1+b'_1) \le z \le (a_2+b_2) \\ \frac{z-(a_2+b_2)}{(a'_3+b'_3)-(a_2+b_2)} & \text{for } (a_2+b_2) \le z \le (a'_3+b'_3) \\ 1 & \text{otherwise} \end{cases} \tag{9}$$

In order to compare two intuitionistic fuzzy numbers, there exist several definitions for the ordering relations over set of intuitionistic fuzzy numbers (for example [2], or [14, 15]). Each of these relations corresponds to specific needs in particular applications. In this analysis, the expected ordering relation is based on the idea of fuzzy ordering relation, described in [8]. The used IT fuzzy ordering relation gives the membership and nonmembership degree of relation between two IT fuzzy numbers. The IT fuzzy ordering relation is derived from the extension principle [4]. More formally, let be the set of all intuitionistic fuzzy numbers. For $a, b \in \mathbb{R}$ we define fuzzy relations $a \succcurlyeq b$ and $a \sim b$ with membership functions $t_{\succcurlyeq}: \mathbb{R} \times \mathbb{R} \to [0, 1]$, $t_{\sim}: \mathbb{R} \times \mathbb{R} \to [0, 1]$ and nonmembership functions $f_{\succcurlyeq}: \mathbb{R} \times \mathbb{R} \to [0, 1]$, $f_{\sim}: \mathbb{R} \times \mathbb{R} \to [0, 1]$ respectively, such that for $x, y \in R$:

$$t_{\succcurlyeq}(a, b) = \sup_{x \geq y}[\min(t_a(x), t_b(y))]; \tag{10}$$

$$f_{\succcurlyeq}(a, b) = \inf_{x \geq y}[\max(f_a(x), f_b(y))]; \tag{11}$$

The equivalence relation $a \sim b$ is defined as:

$$t_{\sim}(a, b) = \sup_{x, y \in R}[\min(t_a(x), t_b(y))]; \tag{12}$$

$$f_{\sim}(a, b) = \inf_{x, y \in R}[\max(f_a(x), f_b(y))]; \tag{13}$$

Logical conjunction of $a \succcurlyeq b$ and $b \succcurlyeq a$ is not $a \sim b$, but in general

$$t_{\sim}(a, b) = t_{\sim}(b, a) \leq \min(t_{\succcurlyeq}(a, b), t_{\succcurlyeq}(b, a)); \tag{14}$$

$$\mathrm{f}_{\sim}(a, b) = f_{\sim}(b, a) \geq \max(f_{\succcurlyeq}(a, b),\ f_{\succcurlyeq}(b, a)); \tag{15}$$

In the case of monotonic score degree, this statement is valid with equality.

3 Fuzzy Extension of Cooperative Games with Vague Payoffs

The main aim of this article is to discuss the cooperative games with vague payoffs, described as intuitionistic fuzzy numbers. In general, the vagueness of payoffs implies also shifts in characteristics of games and in solution concepts of games. Therefore, this part is devoted to the definition of an IT fuzzy extension of the cooperative game and discussion of superadditivity issues, as well as core definition.

Formally, let (N, v) be a cooperative game and let $w(K) = \langle x, t_K(x), f_K(x) \rangle$ be an IT fuzzy number for any $K \subset N$ with the membership function $t_K: R \to [0, 1]$, non-membership function $f_K: R \to [0, 1]$ fulfilling condition $0 \leq t_K(x) + f_K(x) \leq 1$; such that

- $t_K(v(K)) = 1$;
- $t_K(x)$ is non-decreasing for $x < v(K)$ and non-increasing for $x > v(K)$;
- $f_K(x)$ is non-increasing for $x < v(K)$ and non-decreasing for $x > v(K)$;
- $t_{\emptyset}(x) = 1$ iff $x = 0$, and $f_{\emptyset}(x) = 1$ for all $x \neq 0$.

Then w is the intuitionistic fuzzy characteristic function, and the pair (N, w) is called an intuitionistic fuzzy extension of the cooperative game (N, v).

3.1 IT Fuzzy Superadditivity

Let a pair (N, w) be an intuitionistic fuzzy extension of the cooperative game (N, v). Then (N, w) is fuzzy superadditive, if for any two disjoint coalitions K and L, $K, L \subset N, K \cap L = \varnothing$ the next inequality holds:

$$w(K \cup L) \succcurlyeq w(K) \oplus w(L). \tag{16}$$

From (10, 11) we receive:

$$t_{\succcurlyeq}(w(K \cup L), w(K) \oplus w(L)) = \sup_{x \geq y}\left[\min(t_{w(K \cup L)}(x), t_{w(K) \oplus w(L)}(y))\right], \tag{17}$$

$$f_{\succcurlyeq}(w(K \cup L), w(K) \oplus w(L)) = \inf_{x \geq y}\left[\max(f_{w(K \cup L)}(x), f_{w(K) \oplus w(L)}(y))\right]. \tag{18}$$

Then the possibility that the intuitionistic fuzzy coalition game (N, w) is fuzzy superadditive is expressed by a membership function:

$$t_{\text{super}}(N, w) = \min(\{t_{\succcurlyeq}(w(P \cup Q), w(P) \oplus w(Q)); P, Q \subset N, P \cap Q = \varnothing\}), \tag{19}$$

and the impossibility that the intuitionistic fuzzy coalition game (N, w) is fuzzy superadditive is expressed by a nonmembership function:

$$f_{\text{super}}(N, w) = \max(\{f_{\succcurlyeq}(w(P \cup Q), w(P) \oplus w(Q)); P, Q \subset N, P \cap Q = \varnothing\}). \tag{20}$$

The next two remarks valid for the intuitionistic fuzzy extension of the cooperative games are equivalent to statements valid for fuzzy extension of the cooperative games as presented in [9].

Remark 1: Let (N, w) be an intuitionistic fuzzy extension of the cooperative game (N, v). Then for two disjoint coalitions K and L, $K, L \subset N, K \cap L = \varnothing$ inequality $v(K \cup L) \geq v(K) + v(L)$ implies $t_{\succcurlyeq}(w(K \cup L), w(K) \oplus w(L)) = 1$ and $f_{\succcurlyeq}(w(K \cup L), w(K) \oplus w(L)) = 0$.

Proof: This relation is a direct implication of intuitionistic fuzzy extension of the game: from the definition $t_K(v(K)) = 1$, $t_L(v(L)) = 1$, and $t_L(v(K \cup L)) = 1$ which gives $f_K(v(K)) = 0$, $f_L(v(L)) = 0$, and $f_L(v(K \cup L)) = 0$. Consequently the value $\sup_{x \geq y}\left[\min(t_{w(K \cup L)}(x), t_{w(K) \oplus w(L)}(y))\right] = 1$ and after substitution into (17) we receive $t_{\succcurlyeq}(w(K \cup L), w(K) \oplus w(L)) = 1$; $\inf_{x \geq y}\left[\max(f_{w(K \cup L)}(x), f_{w(K) \oplus w(L)}(y))\right] = 0$ and after substitution into (18) we receive $f_{\succcurlyeq}(w(K \cup L), w(K) \oplus w(L)) = 0$.

Remark 2: Let (N, w) be an intuitionistic fuzzy extension of the cooperative game (N, v) and let (N, w) be superadditive. Then $t_{\text{super}}(N, w) = 1$ and $f_{\text{super}}(N, w) = 0$.

Proof: This statement is a direct implication of Remark 1 and Eqs. (19) and (20).

3.2 IT Fuzzy Core

The idea of IT fuzzy core is an extension of the concept of the core, described by Eq. (4) with respect to IT fuzzy extension of the game. That means, that core should be also the IT fuzzy set of payoff vectors. Formally, let (N, w) be an intuitionistic fuzzy extension of the cooperative game (N, v). Then the intuitionistic fuzzy core of (N, w) is an IT fuzzy subset C_{ITF} of R^n with the membership function $t_C: R^n \to [0, 1]$ and non-membership function $f_C: R^n \to [0, 1]$. That means that for any payoff vector $\boldsymbol{x} \in R^n$, there is a number $t_C(\boldsymbol{x})$ evaluating the membership of $\boldsymbol{x}$ in the core, and number $f_C(\boldsymbol{x})$ evaluating the non-membership of $\boldsymbol{x}$ in the core.

For an IT fuzzy superadditive game, condition for membership and non-membership function of the core t_C, and f_C can be derived from the definition of the core (4). Namely, for a membership function, the condition of efficiency is of the form:

$$t_{1C}(\boldsymbol{x}) = t_{\succcurlyeq}\left(w(N), \textstyle\sum_{i \in N} x_i\right) = \sup\left\{t_{\mathrm{N}}(\mathrm{y}); \mathrm{y} \in \mathrm{R}, \mathrm{y} \geq \textstyle\sum_{i \in \mathrm{N}} x_i\right\}. \quad (21)$$

The condition of coalitional rationality $\sum_L x_i \succcurlyeq w(L)$ is described by $t_{\succcurlyeq}\left(\sum_L x_i, w(L)\right)$ and should be valid for all coalitions of N:

$$t_{2C}(\boldsymbol{x}) = \min\left(\left\{t_{\succcurlyeq}\left(\textstyle\sum_{i \in \mathrm{L}} x_i, w(L)\right); \mathrm{L} \subset \mathrm{N}\right\}\right). \quad (22)$$

That means:

$$t_{2C}(\boldsymbol{x}) = \min\left(\sup\left\{t_{\mathrm{L}}(\mathrm{y}); \mathrm{y} \in \mathrm{R}, \mathrm{y} \leq \textstyle\sum_{i \in \mathrm{L}} x_i\right\}\right). \quad (23)$$

Then the membership function of a core should be

$$t_C(\boldsymbol{x}) = \min(t_{1C}(\boldsymbol{x}), t_{2C}(\boldsymbol{x})). \quad (24)$$

As for non-membership function, condition of efficiency is not valid with possibility:

$$f_{1C}(\boldsymbol{x}) = f_{\succcurlyeq}\left(w(N), \textstyle\sum_{i \in N} x_i\right) = \inf\left\{f_{\mathrm{N}}(\mathrm{y}); \mathrm{y} \in \mathrm{R}, \mathrm{y} \geq \textstyle\sum_{i \in \mathrm{N}} x_i\right\}. \quad (25)$$

The condition of coalitional rationality:

$$f_{2C}(\boldsymbol{x}) = \max\left(\left\{f_{\geqslant}\left(\sum_{i\in \mathrm{L}} x_i, w(L)\right); \mathrm{L}\subset \mathrm{N}\right\}\right). \tag{26}$$

That means:

$$f_{2C}(\boldsymbol{x}) = \max\left(\inf\left\{f_{\mathrm{L}}(\mathrm{y}); \mathrm{y}\in \mathrm{R}, \mathrm{y}\leq \sum_{i\in \mathrm{L}} x_i\right\}\right). \tag{27}$$

Then the non-membership function

$$f_C(\boldsymbol{x}) = \max(f_{1C}(\boldsymbol{x}), f_{2C}(\boldsymbol{x})). \tag{28}$$

Then the core of the IT fuzzy extension of a cooperative game is a set of all IT fuzzy numbers on set of vectors $\boldsymbol{x}$, such that levels of their membership and non-membership degrees are different from 0, and 1, respectively:

$$C_{ITF} = \{\langle \boldsymbol{x}, t_C(\boldsymbol{x}), f_c(x)\rangle; \boldsymbol{x}\in R^n \text{ and } (t_C(\boldsymbol{x}) > 0 \text{ or } f_C(\boldsymbol{x}) < 1)\}. \tag{29}$$

Example 1: Let's consider two-players coalitions game (N, v) such that $N = \{1,2\}$, $v\{1\} = 2$, $v\{2\} = 4$, $v\{N\} = 8$. This game is superadditive with a non-empty core $C = \left\{x\in R^2; x_1 + x_2 = 8, x_1 \geq 2, x_2 \geq 4\right\}$ represented by a line segment between points [2, 4, 6].

Let's consider the IT fuzzy extension (N, w) of this game:

$$t_{\{1\}}(x) = \begin{cases} \frac{x}{2} & \text{for } 0\leq x\leq 2 \\ 2-\frac{x}{2} & \text{for } 2\leq x\leq 4 \\ 0 & \text{otherwise} \end{cases}, \quad f_{\{1\}}(x) = \begin{cases} 1-\frac{x}{2} & \text{for } 0\leq x\leq 2 \\ \frac{x}{2}-1 & \text{for } 2\leq x\leq 4 \\ 1 & \text{otherwise} \end{cases}$$

$$t_{\{2\}}(x) = \begin{cases} \mathrm{x}-3 & \text{for } 3\leq x\leq 4 \\ 5-\mathrm{x} & \text{for } 4\leq x\leq 5 \\ 0 & \text{otherwise} \end{cases}, \quad f_{\{2\}}(x) = \begin{cases} 2-\frac{x}{2} & \text{for } 2\leq x\leq 4 \\ \frac{x}{2}-2 & \text{for } 4\leq x\leq 6 \\ 1 & \text{otherwise} \end{cases}$$

$$t_{\{N\}}(x) = \begin{cases} \frac{x}{2}-3 & \text{for } 6\leq x\leq 8 \\ 9-\mathrm{x} & \text{for } 8\leq x\leq 9 \\ 0 & \text{otherwise} \end{cases}, \quad f_{\{N\}}(x) = \begin{cases} 4-\frac{x}{2} & \text{for } 6\leq x\leq 8 \\ \mathrm{x}-8 & \text{for } 8\leq x\leq 9 \\ 1 & \text{otherwise} \end{cases}.$$

There are two coalition structures - $(\{1,2\})$, denoted also $(\{N\})$, and $(\{1\},\{2\})$. For a membership function, for both coalition structures the condition of efficiency is of the form $t_{1C}(\boldsymbol{x}) = t_{\geqslant}\left(w(N), \sum_{i\in \mathrm{N}} x_i\right) = \sup\left\{t_{\mathrm{N}}(\mathrm{y}); \mathrm{y}\in \mathrm{R}, \mathrm{y}\geq \sum_{i\in \mathrm{N}} x_i\right\}$; in this case this part is of a form

$$t_{1C}(\boldsymbol{x}) = \max(0, \min(1, 9-(x_1+x_2)))$$

and can be expressed as:

$$t_{1C}(\boldsymbol{x}) = \begin{cases} 0 & \text{for} \quad 9 \le x_1 + x_2 \\ 9 - (x_1 + x_2) & \text{for} \quad 8 \le x_1 + x_2 \le 9 \\ 1 & \qquad\;\; x_1 + x_2 \le 8 \end{cases}$$

Similarly, the condition for a nonmembership function $f_{1C}(x) = f_{\succcurlyeq}(w(N), \sum_{i \in N} x_i) = \inf\{f_N(y); y \in R, y \ge \sum_{i \in N} x_i\}$ in this case is fulfilled when

$$f_{1C}(\boldsymbol{x}) = \min(1, \max(0, (x_1 + x_2) - 8))$$

and can be expressed as:

$$f_{1C}(\boldsymbol{x}) = \begin{cases} 1 & \text{for} \quad 9 \le x_1 + x_2 \\ (x_1 + x_2) - 8 & \text{for} \quad 8 \le x_1 + x_2 \le 9 \\ 0 & \qquad\;\; x_1 + x_2 \le 8 \end{cases}$$

The condition for coalitional rationality should be valid for both coalitional structures $(\{N\})$, and $(\{1\}, \{2\})$. For the membership function:

$$t_{2C}(\boldsymbol{x}) = \max(0, \min(1, \frac{x_1}{2}, x_2 - 3, \frac{x_1 + x_2}{2} - 3))$$

And for the nonmembership function:

$$f_{2C}(\boldsymbol{x}) = \min(1, \max(0, 1 - \frac{x_1}{2}, 2 - \frac{x_2}{2}, 4 - \frac{x_1 + x_2}{2}))$$

Then the membership function of a core is

$$t_C(\boldsymbol{x}) = \min(t_{1C}, t_{2C}) = \max(0, \min(1, 9 - (x_1 + x_2), \frac{x_1}{2}, x_2 - 3, \frac{x_1 + x_2}{2} - 3)).$$

And the nonmembership function of a core is

$$f_C(\boldsymbol{x}) = \max(f_{1C}, f_{2C}) = \min\left(1, \max\left(0, (x_1 + x_2) - 8, 1 - \frac{x_1}{2}, 2 - \frac{x_2}{2}, 4 - \frac{x_1 + x_2}{2}\right)\right)$$

■

3.3 IT Fuzzy Core – Remarks

In the classical coalitional game, the core is representing set of possible payoff distribution among players such that no coalition has incentive to leave the grand coalition and receive a larger payoff. That means that the core determines achievable and safe imputations.

In the case of games with the IT fuzzy extension, the formula of core assigns to each possible payoff distribution a degree of membership and nonmembership as a measure of possibility and impossibility of such a distribution to be in a core. This uncertainty is bound with superadditivity issues. Even the superadditive IT fuzzy game is to some extent additive. This concept is described in the next example.

Example 2: Expect the same coalitional game with intuitionistic fuzzy extension as in Example 1. In order to evaluate a superadditivity, it is necessary to construct membership function, and non-membership function of the addition $(\{1\}+\{2\})$:

$$t_{\{1\}+\{2\}}(x) = \begin{cases} \frac{x}{3} - 1 & \text{for} \quad 3 \leq x \leq 6 \\ 3 - \frac{x}{3} & \text{for} \quad 6 \leq x \leq 9 \\ 0 & \quad\text{otherwise} \end{cases}, \quad f_{\{1\}+\{2\}}(x) = \begin{cases} \frac{6-x}{4} & \text{for} \quad 2 \leq x \leq 6 \\ \frac{x-6}{4} & \text{for} \quad 6 \leq x \leq 10 \\ 1 & \quad\text{otherwise} \end{cases}$$

Values $t_{\succcurlyeq}(w(\{N\}), w(\{1\}) \oplus w(\{2\})$, and $f_{\succcurlyeq}(w(\{N\}), w(\{1\}) \oplus w(\{2\})$ indicate that the game is superadditive because $w(K \cup L) \succcurlyeq w(K) \oplus w(L)$.

$$t_{\succcurlyeq}(w(\{N\}), w(\{1\}) \oplus w(\{2\}) = 1$$

$$f_{\succcurlyeq}(w(\{N\}), w(\{1\}) \oplus w(\{2\}) = 0$$

However, this game is to some extent additive or subadditive, because membership and nonmembership functions of relation $w(K) \oplus w(L) \succcurlyeq w(K \cup L)$ are:

$$t_{\succcurlyeq}(w(\{1\}) \oplus w(\{2\}), w(\{N\})) = \frac{3}{15}$$

$$f_{\succcurlyeq}(w(\{1\}) \oplus w(\{2\}), w(\{N\})) = \frac{1}{3}$$

■

Directly from the definition from the core we can derive that the deterministic core of the original game is always covered in IT core of the IT extension of original game. Moreover, its membership function is 1, and nonmembership function is 0, as is demonstrated in Example 3. Even if the original deterministic game core is empty, the core of an IT fuzzy extension of this game is not necessarily empty; however membership function of such core will not reach the value 1.

Example 3: Expect the same coalitional game with intuitionistic fuzzy extension as in Example 1. The core of the original game is represented by a segment of a line $x_1 + x_2 = 8$ for $x_1 \in [2, 4]$, $x_2 \in [4, 6]$. The IT fuzzy core membership and nonmembership degrees of this deterministic core are:

$$t_c(\boldsymbol{x}) = \max\left(0, \min\left(1, 9-(x_1+x_2), \frac{x_1}{2}, x_2-3, \frac{x_1+x_2}{2}-3\right)\right)$$

$$t_c(\boldsymbol{x}) = \max\left(0, \min\left(1, 9-8, \frac{2}{2}, 4-3, \frac{8}{2}-3\right)\right) = \max(0,1) = 1$$

$$f_c(\boldsymbol{x}) = \min\left(1, \max\left(0, (x_1+x_2)-8, 1-\frac{x_1}{2}, 2-\frac{x_2}{2}, 4-\frac{x_1+x_2}{2}\right)\right)$$

$$f_c(\boldsymbol{x}) = \min\left(1, \max\left(0, 8-8, 1-\frac{2}{2}, 2-\frac{4}{2}, 4-\frac{8}{2}\right)\right) = \min(1,0) = 0$$

■

4 Conclusion

The main aim of this article was to introduce the intuitionistic fuzzy extension to class of transferable utility function coalitional game when expected payoffs are intuitionistic fuzzy numbers. Intuition behind coalitional game is bound with superadditivity issues; hence this text adds some remarks to the concept of intuitionistic fuzzy superadditivity. Another aim of this article is to discuss a calculation of core, as the concept of core belongs to elementary solution concepts in the deterministic coalition game theory. While the transition from the fuzzy extension of TU coalitional games to IT fuzzy cooperative games is relatively simple in the case of superadditivity, the next step in this research will cover subadditivity and additivity issues.

Acknowledgments This paper was supported by the Ministry of Education, Youth and Sports within the Institutional Support for Long-term Development of a Research Organization in 2015.

References

1. Atanassov, K.T.: Intuitionistic fuzzy sets. Fuzzy Sets Syst. **20**(1), 87–96 (1986)
2. Anzilli, L.L., Facchinetti, G., Mastroleo, G.: Evaluation and ranking of intuitionistic fuzzy quantities. In: Masulli, F., Pasi, G., Yager R. (eds.) WILF 2013, LNAI, vol. 8256, pp. 139–149, Springer International Publishing Switzerland (2013)
3. Burillo, P., Bustince, H., Mohedano, V.: Some definitions of intuitionistic fuzzy number. First properties. In: Proceedings of the 1st Workshop on Fuzzy Based Expert Systems, pp. 53–55 (1994)
4. Coker, D.: An Introduction to intuitionistic fuzzy topological spaces. Fuzzy Sets Syst. **88**(1), 81, 89 (1997)
5. Grzegorzewski, P.: Distances and orderings in a family of intuitionistic fuzzy numbers. In: Wagenknecht, M., Hampel R. (eds.): Proceedings of EUSFLAT Conference 2003, pp. 223–227 (2003)
6. Li, D.-F.: Decision and game theory in management with intuitionistic fuzzy sets. Studies in fuzziness and Soft computing, vol. 308, Springer, Heidelberg (2014)

7. Mahapatra, G.S., Roy, T.K.: Intuitionistic fuzzy number and its arithmetic operation with application on system failure. J. Uncertain. Syst. **7**(2), 92–107 (2013)
8. Mareš, M.: Weak Arithmetics of fuzzy numbers. Fuzzy Sets Syst. **91**, 143–153 (1997)
9. Mareš, M.: Additivities in fuzzy coalition games with side-payments. Kybernetika **35**(2), 149–166 (1999)
10. Mareš, M.: Fuzzy Cooperative Games: Cooperation with Vague Expectations. Phisica-Verlag, Heilderberg (2001)
11. Parvathi, R., Malathi, C.: Arithmetic operations on symmetric trapezoidal intuitionistic fuzzy numbers. Int. J. Soft Comput. Eng. **2**(2), 268–273 (2012)
12. Shoham, Y., Leyton-Brown, K.: Multiagent systems: algorithmic, game-theoretic, and logical foundations. Cambridge University Press, Cambridge (2009)
13. Wooldridge, M.J.: An introduction to multiagent systems. Wiley, New York (2002)
14. Xu, Z.S.: Intuitionistic fuzzy aggregation operators. IEEE Trans. Fuzzy Syst. **15**, 1179–1187 (2007)
15. Xu, Z.S., Xia, M.: Induced generalized intuitionistic fuzzy operators. Knowl.-Based Syst. **24**, 197–209 (2011)

Self-learning Genetic Algorithm for Neural Network Topology Optimization

Radomir Perzina

Abstract The aim of this paper is presentation of encoding for self-adaptation of genetic algorithms which is suitable for neural network topology optimization. Comparing to previous approaches there is designed the encoding for self-adaptation not only one parameter or several ones but for all possible parameters of genetic algorithms at the same time. The proposed self-learning genetic algorithm is compared with a standard genetic algorithm. The main advantage of this approach is that it makes possible to solve wide range of optimization problems without setting parameters for each type of problem in advance.

Keywords Genetic algorithms · Neural networks · Self-adaptation · Global optimization

1 Introduction

Genetic algorithms can be defined as multi-agent systems in which individuals cooperate in population and they try to find the optimal solution by applying basic principles of natural selection and natural genetics. It is well known that efficiency of genetic algorithms strongly depends on their parameters. These parameters are usually set up according to vaguely formulated recommendations of experts or by the so-called two-level genetic algorithm where at the first-level genetic algorithm optimizes parameters of the second-level. A self-adaptation seems to be a promising way of genetic algorithms where the parameters of the genetic algorithm are optimized during the same evolution process as the problem itself. The aim of this paper is to present encoding and genetic operators for self-adaptation of genetic algorithms that is suitable for optimization of neural network topology. Comparing

R. Perzina (✉)
School of Business Administration in Karvina, Silesian University in Opava, University Square 1934/3, Karvina, Czech Republic
e-mail: perzina@opf.slu.cz

G. Jezic et al. (eds.), *Agent and Multi-Agent Systems: Technologies and Applications*, Smart Innovation, Systems and Technologies 38,
DOI 10.1007/978-3-319-19728-9_15

to previous approaches (e.g. [1, 4]) we designed the encoding for self-adaptation not only one parameter but for all or nearly all possible parameters of genetic algorithms at the same evolution process. Moreover, the parameters are encoded separately for each element of a chromosome.

The proposed self-learning genetic algorithm was already successfully used for solving real timetabling problem [7–9]. Here we apply it for solving another optimization problem – Neural Network Topology Optimization. Neural network topology is usually designed by human experts who have knowledge about neural networks and the problem to be solved. There also exist automated approaches which can be divided in constructive algorithms [6], destructive algorithms [12] and evolutionary algorithms [13]. Constructive algorithms start with very complex network and then removing neurons or connections, on the contrary destructive algorithms start with very small network and then adding neurons and connections.

2 Encoding

Encoding is a major element of every genetic algorithm. The structure of the proposed self-learning genetic algorithm's encoding is depicted on Fig. 1. The idea behind the proposed encoding consists in redundancy of information through hierarchical evaluation of individuals.

Fig. 1 The structure of a population

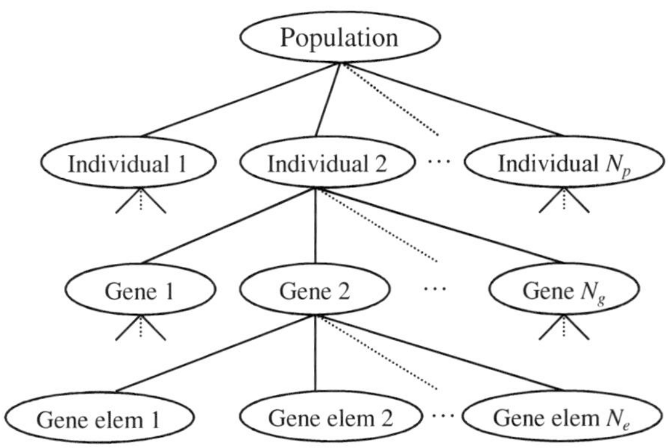

As we can see, in the population each individual is composed of N_g genes where each gene corresponds to exactly one optimized variable. Each gene is composed of N_e gene elements. The number of gene element is different for each gene and it varies through evolution. Each gene element contains low-level parameters, which encode optimized variables and parameters of evolution. All parameters are listed in Table 1.

Table 1 The structure of a gene element

Name	Description	Range
x^E	Optimized variable	<0;1>
q_m^E	Parameter of mutation	<−1;1>
r_m^E	Radius of mutation	<0;0.5>
p_c^E	Probability of crossover	<0;1>
r_c^E	Ratio of crossover	<0;1>
q_d^E	Parameter of deletion	<−0.1;0.1>
q_u^E	Parameter of duplication	<−0.1;0.1>
s_m^E	Identifier of myself for mating	<0;1>
s_w^E	Wanted partner for mating	<0;1>
r_r^E	Ratio of replacement	<0;1>
r_t^E	Ratio of population for selection	<0;1>
r_p^E	Ratio of population for 2nd partner selection	<0;1>
c_d^E	Coefficient of death	<0;1>
N_p^E	Wanted size of population	<0;1>

The upper index “E” denotes, that it is a gene element value of the parameter. As the encoding is hierarchical, there are several levels of the parameters so gene values of parameters are marked by the upper index “G”, individual values by “I” and population values by “P”.

Since genetic operators are applied only to the low level values of parameters (gene element), the upper level values of parameters cannot be updated directly through evolution process, but only indirectly by evaluation mechanism from low level values.

2.1 Mechanism of Gene Evaluation

The proposed encoding is polyploiditial, so each gene is composed of N_e gene elements. The number of gene elements is variable and undergoes evolution. For evaluation of gene values of gene elements we use simple arithmetical average (1).

$$X^G = \frac{1}{N_e} \sum_{i=1}^{N_e} X_i^E \tag{1}$$

where X stands for parameters that must be evaluated, i.e. x, s_m, s_w, r_r, r_t, r_p, c_d, N_p, i_t.

2.2 Mechanism of Individual Evaluation

Parameters concerning the whole individual, such as s^I_m, s^I_w, r^I_r, r^I_t, r^I_p, c^I_d, N^I_p are evaluated as simple arithmetical average (2).

$$X^I = \frac{1}{N_g} \sum_{i=1}^{N_g} X_i^G \tag{2}$$

The number of genes N_g is not variable, because one gene contains exactly one optimized variable.

2.3 Mechanism of Population Evaluation

Parameters concerning the whole population, such as r^P_r, r^P_t, c^P_d, N^P_p are evaluated as weighted average with weights according to their relative fitness w_f, defined as (3).

$$w_f = \frac{N_p^P - i + 1}{\frac{\left(1 + N_p^P\right) N_p^P}{2}} \tag{3}$$

where i is index of ith individual in population sorted by fitness in descending order, i.e. the individual with the highest value of the fitness function has the value of i equal to 1, the individual with the second highest value of the fitness function has the value of i equal to 2 etc.

3 Genetic Operators

As the proposed encoding is specific, the genetic operators must be adjusted to fit the encoding. There are used not only common genetic operators such as selection, crossover or mutation, but also some specific ones, as described in following paragraphs.

3.1 Selection

In genetic algorithms the selection of both mating parents is usually based on their fitness, but this is not true in nature. In nature a winner of a tournament selects his partner according to his individual preferences. Important is that he cannot take into account his genotype, i.e. directly the values of his genes nor his fitness, but only his phenotype, i.e. only expression of the genes to the outside. In a similar way we try to imitate nature by using parameters s_m^I and s_w^I. The parameter s_w^I represents individual preferences for mating and the parameter s_m^I represents individual's phenotype for mating. So the first parent is selected by a tournament selection method with variable ratio of population r_t^P from which the fittest individual is selected. The second parent is selected according to individual's preferences represented by the parameter s_w^I, i.e. the first parent selects an individual with the minimal value of expression $|s_w^I - s_m^I|$, but this selection is made from only limited ratio of population r_p^I.

3.2 Crossover

The crossover operator is applied to every gene element of the first parent with the probability p_c^E. The crossover itself proceeds only between gene elements of mating parents according to formula (4).

$$X_3^E = X_1^E + \left(X_2^E - X_1^E\right) \cdot r_c^E \tag{4}$$

where X stands for all parameters of a gene element (see Table 1), r_c^E is a ratio of crossover of the first parent defined in this gene element, the lower index "$_1$" denotes the gene element of the first parent, the index "$_2$" the second parent and the index "$_3$" denotes the child of both parents. The gene element of the second parent is selected randomly, but it is of the same gene as the gene element of the first parent.

3.3 Mutation

The mutation operator is applied to every gene element with probability $p_m^E = |q_m^E|$. Notice that probability of mutation is calculated as the absolute value of the parameter of mutation $q_m^E \in \langle -1; 1\rangle$, because the mean value of p_m^E should be zero. Moreover, every gene element has its own probability of mutation. The mutation formula is defined as (5).

$$X_{new}^E = X_{old}^E + \left(X_{\max}^E - X_{\min}^E\right) \cdot U\left(-r_m^E, r_m^E\right) \tag{5}$$

where X stands for all parameters of the gene element, $U(a, b)$ is a random variable with uniform probability distribution in the interval $\langle a; b\rangle$, X_{new}^E is the value of the parameter after mutation, X_{old}^E is the original value of the parameter, $X_{\max}^E$ ($X_{\min}^E$) is the maximal (minimal) allowed bit element value of the parameter as defined in Table 1.

3.4 Duplication

The duplication operator is applied to every gene element with probability $p_u^E = |q_u^E|$. The gene element is duplicated (copied) with the same value of all parameters with the only exception that the values of parameter q_u^E of both gene elements are divided by 2, in order to inhibit exponential growth of the number of bit elements.

3.5 Deletion

The deletion operator is applied to every gene element with probability $p_d^E = |q_d^E|$. It means that the gene element is simply removed from the particular gene. By deletion and duplication operators the degree of polyploidity is controlled.

3.6 Replacement of Individuals

For every individual the parameter of a life strength – L is defined. When the individual is created, its life strength L is set to one and in every generation it is multiplied by the coefficient c_L defined as (6).

$$c_L = 1 - c_d^P(1 - w_f) \tag{6}$$

Evidently, through evolution, a less fitter individual causes the greater decrease in L.

In every generation all X^P parameters are evaluated and by using the above listed genetic operators $N_p^P \cdot r_r^P$ new individuals are created. Then a randomly selected individual is killed with probability $(1 - L)$. This process of killing individuals is repeated until only N_p^P individuals survive in the population.

4 Neural Networks

Artificial neural networks are simplification of a human brain. They are widely used in many fields of artificial intelligence, e.g. pattern recognition, clustering, computer vision, robotics or machine learning. There exists wide variety of neural networks, e.g. feed-forward networks, Kohonen's networks, Hopfield networks, etc. Here we focus on feed-forward neural networks.

Functionality of neural networks strongly depends on their topology, i.e. number of neurons, their inner characteristics and how they are connected. Too simple networks may not be able to solve some problems with required precision and on the other hand too complex networks may be hard to learn and suffer from bad generalization. Neural network topology is usually designed by human experts who have knowledge about neural networks and the problem to be solved. There also exist automated approaches which can be divided in constructive algorithms [6], destructive algorithms [12] and evolutionary algorithms [13]. Constructive algorithms start with very complex network and then removing neurons or connections, on the contrary destructive algorithms start with very small network and then adding neurons and connections.

The neural network optimization model can formalized as (7).

$$E = \frac{1}{n}\sum_{i=1}^{n}(y_i - o_i)^2 \rightarrow \min \tag{7}$$

where E is the error of the network, n is number of samples in the training set, y_i is output of the network and o_i is the expected output of the network.

The proposed self-learning genetic algorithm was used for finding the optimal network topology to solve the above problem. Each gene of the chromosome represents one real variable within the interval <0;1>. First gene encodes the number of hidden layers and the remaining genes encode the number of neurons in each hidden layer. When the network is constructed using the chromosome the network is learnt by standard Backpropagation algorithm [2]. The fitness function f for the self-learning genetic algorithm is defined as the negative value of the error of the learnt neural network E defined in Eq. (7), i.e. $f = -E$.

To make the idea behind decoding the chromosome clearer, a simple example is provided. Let's suppose we want to optimize the topology of the neural network with 2 input neurons, single output neuron, maximum number of hidden layers $l_{max} = 5$ and maximum number of neurons in each hidden layer $h_{max} = 10$. Let's assume the gene values in the chromosome are as listed in Table 2.

Table 2 Gene values of the chromosome

Gene name	g_1	g_2	g_3	g_4	g_5	g_6
Gene value	0.24	0.18	0.11	0.45	0.83	0.36

The first gene g_1 encodes the number of hidden layers l which is decoded using the formula (8).

$$l = round(g_1 \cdot (l_{\max} - 1) + 1) \tag{8}$$

where g_1 is the value of the first gene, so the number of hidden layers is $l = 2$. Genes g_2–g_6 encode the number of neurons in each hidden layer, but we use only values of genes g_2 and g_3, because the number of layers l is 2. The number of neurons in i-th hidden layer is decoded using the following formula (9).

$$h_i = round(g_{i+1} \cdot (h_{\max} - 1) + 1) \tag{9}$$

The value of the gene g_2 is 0.18 therefore the first hidden layer will have $h_1 = 3$ neurons and the second hidden layer will have $h_2 = 2$ neurons because the value of the gene g_3 is 0.11. The neural network encoded by the above chromosome is depicted on the Fig. 2.

Fig. 2 Neural network sample

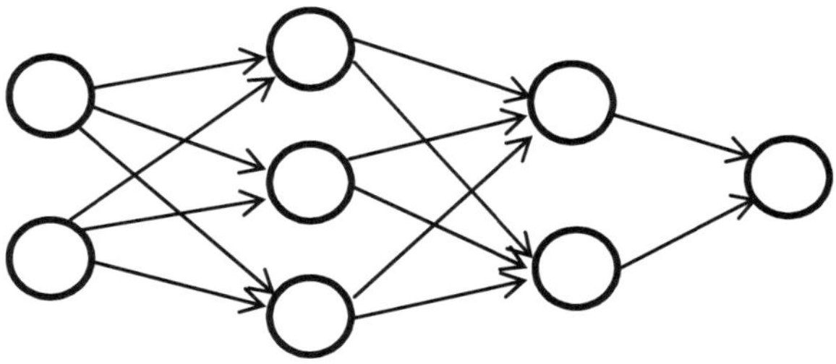

5 Numerical Experiments

The proposed approach was tested on the diabetes data set from the PROBEN1 benchmark datasets collection for neural network testing [11]. The dataset has 8 inputs variables and 1 output variable, therefore the neural network will have 8 input neurons and 1 output neuron. Maximum number of hidden layers $l_{max} = 10$ and maximum number of neurons in each hidden layer $h_{max} = 10$ therefore the chromosome is composed of 11 genes (1st gene is encoding the number of hidden layers and remaining 10 genes are encoding the number of neurons in each hidden layer). The results of the self-learning genetic algorithm (SLGA) were then compared with the simple genetic algorithm (SGA). The simple genetic algorithm used a binary encoding, the size of population was 20 individuals, probability of mutation 0.003 and elitism was used. We ran both algorithms 10 times and measured minimal, maximal and average value of the fitness function. Each algorithm was ended after the fitness function was evaluated $N_f = 1000$ times. The results of both algorithms are shown in Table 3. For better readability the values shown in Table 3 are positive, i.e. they represent total error of the learnt neural network E. As we can see the average error of the learnt neural network is slightly better (by about 10 %) for SLGA than SGA. The minimal error of the neural network is almost the same for both genetic algorithms as well as the maximal neural network error.

Table 3 SLGA and SGA comparison

	SLGA	SGA
Average error	0.050	0.056
Minimal error	0.047	0.049
Maximal error	0.076	0.074

6 Conclusions

In this paper there was proposed a self-learning genetic algorithm with self-adaptation of all its parameters. It was shown that the self-learning genetic algorithm is able to solve not only timetabling problem for which the algorithm was originally designed but also other optimization problems such as the traveling salesman problem or neural network topology optimization.

As a future work there will be investigated possibilities of applying the proposed self-learning genetic algorithm to allow for greater variability of the neural network topology, e.g. different activation function for each neuron, variable degree of connectivity of each neuron, etc.

Acknowledgements This paper was supported by the Ministry of Education, Youth and Sports within the Institutional Support for Long-term Development of a Research Organization in 2015.

References

1. Bäck, T.: Self-adaptation in genetic algorithms. In Proceedings of the First European Conference on Artificial Life. MIT Press, Cambridge (1992)
2. Fausett, L.: Fundamentals of Neural Networks – Architectures, Algorithms and Applications. Prentice Hall, Englewood Cliffs (1994)
3. Garey, M., et al.: Computers and Intractability. W.H. Freeman, San Francisco (1979)
4. Marsili, S.L., Alba, P.A.: Adaptive Mutation in Genetic Algorithms. Soft Computing, vol. 4, no. 2, pp. 76–80. Springer, New York (2000)
5. Michalewicz, Z.: Genetic Algorithms + Data Structures = Evolution Programs. 3rd edn, Springer, New York (1996)
6. Nadal, J.P.: Study of a growth algorithm for a feedforward network. Int. J. Neural Syst. **1**, 55–59 (1989)
7. Perzina, R.: A self-adapting genetic algorithm for solving the university timetabling problem. In: Proceedings of the 8th World Multiconference on Systemics, Cybernetics and Informatics (SCI 2004), pp. 284–288. IIIS, Orlando (2004)
8. Perzina, R.: Solving the university timetabling problem with optimized enrollment of students by a parallel self-adaptive genetic algorithm. In: Proceedings of the 6th International Conference on the Practice and Theory of Automated Timetabling. Masarykova univerzita, Brno (2006)
9. Perzina, R.: Solving multicriteria university timetabling problem by a self-adaptive genetic algorithm with minimal perturbation. In: Proceedings of the 2007 IEEE International Conference on Information Reuse and Integration. IEEE Publishing, Las Vegas (2007)
10. Potvin, J.V.: The traveling salesman problem: a neural network perspective. INFORMS J. Comput. **5**, 328–348 (1993)

11. Prechelt, L.: Proben1 – A set of neural network benchmark problems and benchmarking rules. Fakultat fur Informatik, Univ. Karlsruhe, Karlsruhe. Technical report 21/94 (1994)
12. Reed, R.: Pruning algorithms – A review. IEEE Trans. Neural Netw. **4**, 740–747 (1993)
13. Whitley, D.: Genetic algorithms and neural networks. In: Genetic Algorithms in Engineering and Computer Science. Wiley, Chichester (1995)

Part IV
IS: Business Model Innovation and Disruptive Technologies

Bitcoin: Bubble or Blockchain

Philip Godsiff

Abstract This paper sets out a brief, deliberately non-technical, overview of Bitcoin, a new, but becoming more mainstream, crypto currency, generated and managed by a distributed multi–agent system. Bitcoin was developed in late 2008 by "Satoshi Nakamoto". The nature of Bitcoin as a disruptive currency, payments system and asset, is juxtaposed against the potential for its transactional ledger, the blockchain, to usher in a revolutionary way of recording "digital truth". The main contribution of this paper is to progress the debate around Bitcoin beyond the technical and towards legal and ethical issues and the nature of money and memory itself.

Keywords Bitcoin • Cryptocurrency • Money • Regulation • Disruption • Community

1 Introduction

This paper sets out a conceptual overview of Bitcoin, the crypto-currency invented in 2008 and launched in January 2009 by an entity known as "Satsohi Nakamoto".

Bitcoin is a private crypto-currency and encrypted payment system, relying on a peer to peer network with no centralised authority and hence not intermediated by a trusted (or untrusted) third party. Bitcoins, whose eventual total supply has been deliberately fixed at 21 m, are "mined" by "operator miners" solving increasingly complex problems, requiring increasing amounts of computing resources. This activity of mining also serves to validate transactions which are recorded in the "blockchain", a community validated secure virtual ledger in which transactions are irreversible, and which eliminates the "double-spend" issue associated with non-physical currencies.

P. Godsiff (✉)
University of Exeter Business School, Exeter, UK
e-mail: p.godsiff@exeter.ac.uk

G. Jezic et al. (eds.), *Agent and Multi-Agent Systems: Technologies and Applications*, Smart Innovation, Systems and Technologies 38,
DOI 10.1007/978-3-319-19728-9_16

The table below shows figures for volume, dollar price, and "market capitalisation" annually since inception, clearly demonstrating recent price fluctuations.

As at January	2009	2010	2011	2012	2013	2014	2015
Volume	50	1,629 m	5,027 m	8,008 m	10,617 m	12,204 m	13,675 m
Price (US$)	0	0	0.3	5.2	13.6	746.9	315.7
Capitalisation (US$)	0	0	1.5 m	41.6 m	144 m	9,115 m	4,317 m

Bitcoins may also be purchased from other users through exchanges, and are now sold via some ATM's, are acceptable in many shops and internet mediated services in many cities. It is claimed that Bitcoin fulfils all the requirements of a currency: acting as a unit of account, a medium of exchange and a store of value. In the public conversation surrounding bitcoin, there has been emphasis on its speculative nature, the associated risks, and its alleged use in illegal activities.

Bitcoin is technologically innovative, truly virtual, and potentially disruptive not only to currencies and payment systems, but also to the centralised control, by the state or large corporations, of both. It is also redolent of pre-coinage currencies and means of transacting and recording, based more on collective group memory, and thus less reliant on state underwriting and less susceptible to state interference. The "blockchain", based on community verification and irreversible, offers a potential method of establishing an immutable "digital truth".

In the next section of this paper, aspects of how Bitcoin currently operates as a currency are discussed. Following that the (alleged) more speculative aspects of its ability to behave as or be considered to be an asset (or a liability), are considered, and whether it is a "bubble" or a "fad". Its treatment by various sections of different government authorities is explored. Following that is a discussion on the nature and uses of the blockchain, the nature and uses of money, and the potential for Bitcoin and the blockchain philosophy to support or shape the digital future.

2 Bitcoin: Currency (and Crime)?

2.1 Bitcoin as a Currency

In response to its own rhetorical question: What is Bitcoin? a recent Bank of England report answered "Bitcoin was the first, and remains the largest, functioning digital currency. Several thousand businesses worldwide currently accept bitcoins in payment for anything from pizza to webhosting. Payments can be made at any time and between any two users worldwide" [1]. Bitcoin has been described as a "global phenomenon" with nearly 50,000 "members" in 68 countries and up to 1 m. U.S. digital currency accounts [2]. The first bitcoin ATM opened in Vancouver in October 2013 [3]. Bitcoin fulfils all the requirements of a currency, acting as a unit of account, a medium of exchange and a store of value [3]. But in the public conversation surrounding Bitcoin, there has been emphasis on its speculative nature [4], the associated risk and its alleged use in illegal activities [5].

Bitcoin is truly virtual with each transaction (employing public cryptography, based on well-known standards such as SHA-256 [6]) treated as a "tuple" of buyer, seller and amount [7]. It is non-denominated, which enables splitting; it is infinitely divisible: down to one hundred millionth of a bitcoin, (known as a "satoshi" [8]), and in the absence of intermediaries this facilitates low costs transactions and hence micropayment business models. Bitcoins are stored in "wallets": effectively a cryptographic public key (about 33 digits) hosted on web, pc or phone, or even paper printouts [8]; loss of the "wallet" e.g. by throwing a pc away can be irrecoverable. It is estimated that up to 4 % of created bitcoins have been lost due to errors and theft, but that is just similar to a normal wallet [9].

The need to eliminate the possibility of double spend is a challenge to all virtual (non analog/no physical exchange) currencies. Consensus to changes and transactions is achieved through public key cryptography community verification (by the miners) of proof of work [1], and having a distributed open ledger to which transactions are checked [10]. The community acts as the recording body [11].

Bitcoins, whose supply has been deliberately fixed at 21 m, are created by being mined through "operator miners" solving increasingly complex problems, and hence requiring increasing computing resources [12]. The more bitcoins are brought into existence the harder become the algorithms [1]; some commentators forecast that the last bitcoin will be mined in 2140 [10].

2.2 Exchanges

Bitcoin can be converted to national currencies (leaking out into the "real world" [9]) through exchanges [11]. At any one time there may be as many as 50 exchanges [2]. Less popular exchanges suffer from high closure rates (nearly 50 % over 3 years) with more popular exchanges suffering a higher ratio of security breaches [13], such as malleability attacks [14]. The closure and bankruptcy of Mt. Gox (which had around 600,000 users/customers) was a blow to confidence and price stability, [2] although it should be noted that the system survived. Bitcoin is resilient due to its distributed nature [11].

2.3 Anonymity

Like cash, transactions carried out in bitcoin are claimed to be anonymous. In actuality, the correct aspect is "pseudonymous" [10, 15]; although some claim that these pseudonyms can be traced back to named individuals depending on the processes users have gone through to establish them. It is estimated that 40 % of user profiles could be identified even if the users had followed bitcoin procedural advice [16]. There is a counterintuitive by-product of the protocols in that while *ownership* may be anonymous, the *flows* are globally visible [17]. The impact of this pseudo-anonymity, again like cash (more specifically high value paper), is the use of bitcoin in alleged criminal activity. The nature benefits of anonymity can lead to criminal or secretive activity [18]; which also leads to difficulty in enforcing regulatory anti-money laundering rules where detailed data has to be given to intermediaries [19].

2.4 Crime

Crime may take place *in* bitcoin and *on* bitcoin. Bitcoin offers opportunities for fraud and tax avoidance [11]. It has become the money laundering route of choice for cyber criminals not wishing to deal in cash [20], although some traders in the dark web and "Silk Roads" have expressed concerns over bitcoin instability [21], leading to an alternate view that the concern is over technology rather than the law. Black markets often exist to circumvent discriminatory or discretionary corporate and government practices [22]. There are claims that Bitcoin and its associated exchanges have become a virtual "wild west" with its own breed of bank "heist"; and two of the larger exchanges to close appear to have been robbed or subject to fraud.

But Bitcoin also has legitimate users and these and their aspirations should not be bundled in with its criminal associations [5]; as has been noted, crime also exists in and on analog currencies. However there is concern over whether Bitcoin will work at scale and the length of time taken to validate transactions using existing proof of work protocols [17, 23], and when sufficient network scale and associated ecosystems of payments services and wallet firms can be created [24]. Bitcoin could flourish under circumstances where existing systems are untrusted or expensive [25].

3 Bitcoin: The Bubble?

3.1 Bitcoin: The Asset

As an asset, the price is highly variable though the trend is upwards from its inception. The Bank of England report notes that price rose by 5,000 % in two years. A logarithmic price chart shows a steady rise in value [1]. The value of any commodity is a reflection of supply and demand and because it is perceived to be a source of value [26]. Its reliability as a store of value remains a risk to those who hold it [2].

3.2 Bubbles

Bitcoin trading and price volatility has been likened to "tulip mania in" C17th Holland, where an established but small futures market in spices and tea spread over into tulips and subsequent speculation involving practically the whole nation in the domestic production of a "previously exotic import" led to rapidly rising prices followed by equally rapid collapse; the final settlement value of most contracts being less than 5 % of face value [27]. The failure of Mt Gox led to a halving of Bitcoin value in a week [22].

Others have described Bitcoin as a giant "Ponzi" scheme [8]. There is some evidence of a linkage between google searches and value movements [28, 29]. There is also evidence that such bubbles are being socially created and given the data, now more easily researchable [30]. However the activity levels around bubbles can develop markets and infrastructure as well as raising public awareness, if not interest. A rapidly rising price leads to both an increased number of speculators and entrepreneurs developing new products and services [24]. The volumes and volatility around speculation could also be considered to be a good means of stress testing [10].

3.3 Bitcoin: The Mainstream Investment

Serious institutions are taking Bitcoin seriously, in their investment advice and in their trading. Banks like Goldman Sachs and Merrill Lynch are now covering Bitcoin, with investment companies raising specific funds to invest in Bitcoin [8], as consumer confidence increases [31]. Venture Capitalists are showing both speculative and payments services interest [24]; there has been an estimated $200 m of VC investment in 2014 [2]. Bitcoin has been shown to benefit theoretical portfolios if held in small percentage quantities. Accountants are beginning to run courses on Bitcoin [32]; some are recommending its use in estate planning, where its fluctuating value could bring benefits [31]. The S.E.C. in the U.S. has issued warnings not about its use or existence, but in the potential for fraudsters to target now wealthy individuals holding bitcoins.

Losses on exchanges can be insured against, and a hedge fund is being developed to counter large value fluctuations [24]. Risks from price volatility maybe less than those experienced by investors in e.g. sub-prime mortgages and associated derivatives, a trade that was very destructive of bank, corporate, and personal capital [11].

A limited supply and no central governmental authority (which may be open to temptation to over create fiat money [33]) should mean that the currency bitcoin will not be devalued [10]; possibly quite the converse. The fixed amount promotes "upside potential" [26]. Bitcoin may fall beneficial victim to Gresham's Law: bad money drives out good (from use as a circulating currency), as investors and speculators gravitate towards it when the mining-go-round stops. "Whether Bitcoin is a bubble is too early to tell. It does demonstrate however that anything can be a currency and that emotion can overload even the most mathematical of formulations" [34].

4 Bitcoin: Currency or Commodity?

Virtual currencies like Bitcoin exist in a "gray area" of being able to be considered as both a currency and a commodity [2], so inevitably questions arise as to how Bitcoin should be treated for governance, regulation and taxation purposes.

4.1 Regulation

The views on regulation reveal inter and intra national differences. The US and Canada treat bitcoin as property (commodity/asset) for taxation purposes [3]. However, in spite of this, in June 2014 California removed a ban which prevented the use of currencies other than dollars [3]. It is equally unclear whether transactions are liable to sales tax [11].

The E.U. position is equally confused. Within the European Union, The Payment Services Directive and E-money Directive form the legal framework for consumer protection in mobile payments, but does not cover Bitcoin [35]. Germany treats bitcoin as private money and a financial instrument, while Denmark considers it as neither currency nor an asset [2], the Danish central bank likening it to "glass beads", remarking that it was not a currency because there was no issuer and no utility value, a particularly analog viewpoint. China acted to reduce its use and restrict trade in the currency, as a precursor to an outright ban. France and Korea do not treat it as legitimate [3]. The Japanese central bank suggested that bitcoin was an asset, rather than a currency or financial instrument, and that its use did not constitute banking.

Perversely dealers and exchanges are pushing for regulation as a way of building trust. There is a dichotomy for some users of bitcoin in the issue of regulation. On one hand it can be seen as interference and increasing unnecessary oversight and transaction costs, (e.g. through the need for personal record keeping); on the other it could confer respectability and a degree of legitimacy and support [11]. Whether officially supported or not, it is worth noting that the U.S. government was quite sanguine about potentially benefitting from the sale of 30,000 bitcoin confiscated from the closure of Silk Road 1.

4.2 Central Bank Support

A specific role of the modern central bank is to underwrite its currency and prevent large and erratic fluctuations in its value, and acting as “the ultimate rich uncle” in its support [36]. Bitcoin is not dependent on a central bank or a government standing behind it, but relies on the community to underwrite it. In the absence of centralized support it is claimed that the value fluctuations being experienced will continue and run the risk of the entire system collapsing [26]. However the collapse of Mt Gox whilst affecting value significantly does not appear to have altered fundamental beliefs in its future.

But the issue of central bank involvement runs deeper than underwriting. Central Banks and governments use control over the money supply as a means of managing (for better or worse) inflation, as can be seen with quantitative easing, and promoting growth. The existence of a currency or currencies outside this sphere of influence could lead to such efforts being less successful. National sovereign debt currency crises, like those in Argentina Russia and Thailand Ireland and Greece, leave a lasting political legacy and fiduciary implications [22].

5 Bitcoin: The Blockchain

The Bank of England report considers the impact of the blockchain in some detail; even mooting the possibility of the Bank operating a similar distributed ledger [1], and suggesting the blockchain as the basis of an “internet of finance”.

5.1 The Blockchain

Bitcoin is as much about “on line” as it is about currency; it has myriad uses in and expanding mobile commerce [37]. Bitcoin is about online transactions [5]. It can serve as a “platform” “for financial information” [24]. It “contains the digital blueprints for a number of useful financial and legal services: information can be embedded in the blockchain, such as contracts, bets, and other sensitive or time based material e.g. fines for non performance” [10]. This may be useful for “mutually distrusting parties” engaging in exchange [38]. Because of validation by the community the “fairness” of the contract process can be guaranteed by its protocols [38].

The Bank of England report proposes the possibility, given that the records for most financial assets are now held in electronic form, (albeit centrally in a tiered structure - individual accounts with banks, bank’s reserve accounts with the central bank), of this entire structure of payments services and banking being replaced by distributed systems [1].

Other potential features include (beyond the obvious micropayments) dispute resolution, assurance contracts, and smart property; this could pave the way for enhanced crowd funding applications, translation services and instantaneous processing [10]. Under the Bitcoin protocol legitimacy of transfer and ownership is established beyond challenge [8]. Bitcoin is the "foundation upon which other layers of functionality can be built" [10]. This could easily extend beyond finance providers to providing all the necessary accounting, banking and tax movements on each transaction, either purchase or perhaps on use or consumption in real time. This would give the opportunity for new service providers to disintermediate a range of financial services.

5.2 The Internet of Transactions

In the "internet of things" Bitcoin could be used by two entities to exchange value thus potentially creating a market for sensor data, "exchanging (electronic) data for electronic cash" [39]. The extensive digitisation of data plays a part in this. This is less about how can the data be monetized, a bit of an analogue hangover, but how it can be exchanged in non monetary ways or ways which reduce cost but increase outcome. However "if the web through the internet of things begins to intrude on increasingly more aspects of human and social activity, the issue of security or freedom, the compatibility and how much of each, will need to be addressed" [40].

6 Discussion

Bitcoin acts in one guise as both a currency and a payments system, and in another guise, like any good commodity, as an asset. The blockchain could change the way digital records (of any transaction or asset) are made and kept. Much of current literature is centred on the currency and cryptography, and although it is noted that there has been little research into the economics of Bitcoin [9], there has been even less on the "ethics" of Bitcoin.

6.1 Payments Systems and Disruption

Bitcoin is as much a payments system as a currency. Being virtual, non-denominated and non -intermediated, it is extremely efficient to use in micro or low value payments [37]. But it is potentially an attractive new business model, given that disruption often occurs at the margins [41]. Some speculate that virtual currencies like bitcoin could ultimately replace government forms of fiat money [42], even in the face of government opposition. Opposition may also come from large corporations. Apple withdrew bitcoin apps shortly before announcing their own payments services and Amazon may be considering starting payments services [11, 43]. This poses a potential choice between community intermediation or "big silicon" control. Could Bitcoin or something like it become the "Napster for finance?" [11]. The disruption is not limited to products and processes but, as has been described above, could extend to markets and rules and tax revenues [11]. The Bank of England report remarks that some commentators suggest that bitcoin could become the "internet of money", or even the "internet of finance" [1].

There is a fear that regulation will kill the experiment, for whatever motives [26]. This clash of digital disruption with analog regulation and mind-sets highlights issues of the legitimacy of state regulation itself [11]. But if Bitcoin is a reaction to a lack of liquidity or transparency, some have argued that these underlying causes should be addressed rather than attempting to inhibit Bitcoin's development [22]. Removing the ability of a government to control its currency "can change the economics process and drive deep wedges between various social and political elements" [22]. But there may be effects beyond the potential disruption to financial services and economic structures.

6.2 Law in "Mixed Reality"

The virtual and real worlds increasing mix and confound each other [42]. "If drug or other illegal activity is conducted on the dark web in bitcoin, neither of which is mediated or controlled by government, issues of law enforcement come to the fore" [44]. But what sort of law will regulate these places? There is a lack of current law available to deal with virtualisation technologies where there is "the emergence of a "mixed reality" of virtual and realspace features and geography" [45]. Noting that most virtual world's research is concerned with the impact that "real world" regulations have on online communities, it is important to decide whether, in a world of mixed reality, (real and cyberspace), online or offline law will cover rights of users over property and data [45]. Interestingly, since Bitcoin is essentially a kind of transaction log, where past transactions are public and known to the world, it is of great interest to prosecutors, who have called the coins 'Prosecution Futures' [41].

6.3 What Is Money?

Trust is "essential for virtually all economic activity" [36], be that in the "real/analog" world, (the quality of my cup of coffee, and the paper note with which to pay for it) and the virtual. There needs to be a belief in stated explicit or implicit promises, backed up where necessary by practice and legal frameworks; and there needs to be trust that central bank knows what it is doing in maintaining the value of the currency and of the economy [36]. Certain South Sea islanders use stones as currency, with transactions and ownership being recorded in collective memory [46]. England used wooden sticks, (tallies), as a proxy for money for more than half a millennium and these sticks were very effectively used for taxation purposes [34]. Money is a fairly recent invention, whereas trading debt and credit are not [47].

"Identity is the new money" This quote, suggesting a fundamental change in the nature of currency, is attributed to Sir James Crosby (quondam Chief Executive of HBOS) [46], It indicates the potential for new forms of exchange to emerge as a result of digital disruption. Leading industry commentators [46], see a possible return to those earlier community memory based versions of currencies as stores of value and media of exchange; these would not involve the physical transfer of analogues (e.g. coins or cards). Such developments have the potential to lead to an individual's digital reputation, or access to scarce resources (such as car parking spaces in cities), being of value and tradable directly as a generally accepted means of exchange [46].

Money can be both a series of promises or a fact (like a tally) [34]. If money is about relationships, then the "bitcoin project can best be thought of as a process of financial and communicative experimentation" [10]. Some authors have suggested Bitcoin, a community currency not an individual state mediated one, is a similar concept to Simmel's idea of "perfect money" within a "perfect society", ideals closely aligned to socialist principles [48]. "Perhaps the real problem is money itself" [49]. In existing systems privacy and liberty are allegedly at risk as is the value of the currency itself, due to government or corporate intervention. Bitcoin offers a "practical materialism" reminiscent of previous debates around "privacy, labor and value" [49]. "What is a debt anyway? A debt is just the perversion of a promise. It is a promise corrupted by both math and violence, mediated by state and capitalist institutions" [47].

6.4 Revolution and Motivations

Some commentators [50] have argued that the blockchain system is only, so to speak, being experimentally trialed as a currency, and could be extended beyond the uses already seen, and they envisage the possibility of further innovation in and on the Bitcoin protocol [10]. Bitcoin allows companies and everyone to be peer to peer and distributed not just the currency [1]. It has been claimed that by not relying on trust or asymmetric power relationships, Bitcoin is somehow more "ethical" [51].

What will represent the nature of real and virtual boundaries inside which privacy is shared and outside which anonymity is guaranteed? The Snowden revelations have shown that decisions in this area about trade-offs are hard to make, with privacy too often treated as a "transactional personal good" when it might be considered to be more of a shareable public good [17]. The Bitcoin philosophy may offer an alternative to excessive surveillance.

Bitcoin appears to have an overtly "political" or ideological motivation: the initial "genesis block" in the block chain includes a newspaper headline covering the bail out of the UK banking system [1]. It has attracted strange bedfellows and combinations of technocratic programmers, cypherpunks and crypto-anarchists [2], neo liberals and "crypto-libertarians" [7]. Some "cypherpunks" have intimated that bitcoin is the first step in a much larger project by its founder [50]. This project aims to prevent both private and public fraud by limiting monetary supply and removing the need for a central authority, and having the community authentic transactions and freeze history. These "3 centralities" to Bitcoin could, they argue, be extended to establish a "universally consistent history" in which falsification cannot exist, and separate classes of data emerge [50]. This could for example be an answer to the veracity issue faced for example by big data. Others venture that the "interesting experiment" may form a part of "distributed capitalism" [52], with the recording of "property" and "trust" dis-intermediated and established by the community.

7 In Conclusion

Why use bitcoin when you can use dollars? "Its limited supply and perceived freedom from human interference in a recent era where trust in the traditional monetary system are powerful assets that have transformed an interesting intellectual experiment into a living economy" [34], but how will Bitcoin make it further into the mainstream? This may be achieved by the fuller development of a supportive and innovative ecosystem [24], which will aid the innovation and experiment inherent in the Bitcoin approach to demonstrate ways it could develop [10]. The support, or at least the acquiescence (or absence of outright hostility) of regulatory authorities will be important here [7, 10]. Governments need to support its growth and provide the right legal support; that of course relies on those authorities coming to terms with the potential development of a parallel economy [11], the "internet of finance" [1]. Currencies are the "heart and soul" of society and the governments that rule them [22]; states and "currency" often emerge together, though the recording of debt and obligations precede both [47].

A digital world needs a digital currency but how is living in a non-material world to be managed? Some commentators see the development of Bitcoin as the" beginnings of the struggle" for control over the internet in a new environment but one in which the "age old dilemma" of security against freedom must still be debated [40]. "[T]he idea that any digital currency is to be fully liberated from government or central bank is fascinating; it's like the Fed meets the wild west" [53]. Bitcoin the currency, Bitcoin the blockchain, and Bitcoin the philosophy.

References

1. Ali, R., Barrdear, J., Clews, R., Southgate, J.: Innovations in payment technologies and the emergence of digital currencies. Bank Engl. Q. Bull. **54**, 262–275 (2014)
2. Cofnas, A.B.E.: Bitcoin: currency or commodity? Futures: news. Anal. Strat. Futures Options Deriv. Traders **43**, 10–12 (2014)
3. Blundell-Wignall, A. (2014)
4. Baek, C., Elbeck, M.: Bitcoins as an investment or speculative vehicle? A first look. Applied Economics Letters (2014, in press)
5. Turpin, J.B.: Bitcoin: the economic case for a global, virtual currency operating in an unexplored legal framework. Indiana J. Global Leg. Stud. **21**, 335–368 (2014)
6. Courtois, N.T., Grajek, M., Naik, R.: Optimizing SHA256 in bitcoin mining communications in computer and information science (2014)
7. Van Alstyne, M.: Why bitcoin has value. Commun. ACM **57**, 30–32 (2014)
8. Levin, R.B., O'Brien, A.A., Osterman, S.A.: Dread pirate roberts, Byzantine generals, and federal regulation of bitcoin. J. Taxation Regul. Financ. Institutions **27** (2014)
9. Wu, C.Y., K., P. V.: Breaking News. Timeline (2014)
10. Brito, J., Castillo, A.: Bitcoin: a primer for policymakers. Policy **29**, 3–12 (2013)
11. Smith, A., Weismann, M.F.: Are you ready for digital currency? J. Corp. Account. Finan. **26**, 17–21 (2014)
12. Dev, J.A.: In Canadian conference on electrical and computer engineering (2014)
13. Moore, T., Christin, N.: Beware the middleman: empirical analysis of Bitcoin-exchange risk Lecture Notes in Computer Science (including subseries Lecture Notes in Artificial Intelligence and Lecture Notes in Bioinformatics) (2013)
14. Decker, C., Wattenhofer, R.: Bitcoin transaction malleability and mtgox Lecture Notes in Computer Science (including subseries Lecture Notes in Artificial Intelligence and Lecture Notes in Bioinformatics) (2014)
15. Miers, I., Garman, C., Green, M., Rubin, A.D.: Zerocoin: aonymous distributed e-cash from bitcoin. In: Proceedings IEEE Symposium on Security and Privacy (2013)
16. Androulaki, E., Karame, G.O., Roeschlin, M., Scherer, T., Capkun, S.: Evaluating user privacy in Bitcoin Lecture Notes in Computer Science (including subseries Lecture Notes in Artificial Intelligence and Lecture Notes in Bioinformatics) (2013)
17. Meiklejohn, S., Pomarole, M., Jordan, G., Voelker, G.M., Savage, S.: A fistful of bitcoins: Characterizing payments among men with no names. In: Proceedings of the ACM SIGCOMM Internet Measurement (2013)
18. Peck, M.E.: The cryptoanarchists' answer to cash IEEE Spectrum (2012)
19. Moser, M., Bohme, R., Breuker, D.: An inquiry into money laundering tools in the bitcoin ecosystem. In: eCrime Researchers Summit, eCrime (2013)
20. Shoshitaishvili, Y., Invernizzi, L., Doupe, A., Vigna, G.: Do you feel lucky? A large-scale analysis of risk-rewards trade-offs in cyber security. In: Proceedings of the ACM Symposium on Applied Computing (2014)
21. Van Hout, M.C., Bingham, T.: Responsible vendors, intelligent consumers: silk road, the online revolution in drug trading. Int. J. Drug Policy (2014)
22. Andelman, D.A.: Currency wars. World Policy J. **31**, 115–124 (2014)
23. Singh, P., Chandavarkar, B.R., Arora, S., Agrawal, N.: Performance comparison of executing fast transactions in bitcoin network using verifiable code execution. In: Proceedings 2nd International Conference on Advanced Computing, Networking and Security, ADCONS 2013 (2013)
24. Cusumano, M.A.: The bitcoin ecosystem. Commun. ACM **57**, 22–24 (2014)
25. Surowiecki, J.: Economics: cryptocurrency. Technol. Rev. **114**, 107 (2011)
26. Lemieux, P.: Who is Satoshi Nakamoto? Regulation **36**, 14–15 (2013)
27. Davies, G.: A History of Money (2014)
28. Bhattacharya, J.: Minting pure reason. Economic and Political Weekly (2014)

29. Kristoufek, L.: BitCoin meets Google Trends and Wikipedia: quantifying the relationship between phenomena of the internet era scientific reports (2013)
30. Garcia, D., Tessone, C.J., Mavrodiev, P., Perony, N.: The digital traces of bubbles: feedback cycles between socio-economic signals in the bitcoin economy. J. Roy. Soc. Interface **11**, 0623 (2014)
31. Parthemer, M.R., Klein, S.A.: Bitcoin: change for a dollar? J. Financ. Serv. Professionals **68**, 16–18 (2014)
32. Barry, J.S.: Lawsky makes a bet on New York. CPA J. **84**, 5 (2014)
33. Rogojanu, A., Badea, L.: The issue of competing currencies. Case study-bitcoin. Theor. Appl. Econ. **21**, 103–114 (2014)
34. Swarup, B.: Money Mania (2014)
35. Vandezande, N.: Between bitcoins and mobile payments: will the European Commission's new proposal provide more legal certainty? Int. J. Law Inf. Technol. (2014)
36. Coggan, P.: Trust (not money) makes the world go 'round. OECD Observer, pp. 75–76 (2014)
37. Hurlburt, G.F., Bojanova, I.: Bitcoin: benefit or curse? IT Prof. **16**, 10–15 (2014)
38. Andrychowicz, M., Dziembowski, S., Malinowski, D., Mazurek, Ł.: Modeling bitcoin contracts by timed automata Lecture Notes in Computer Science (including subseries Lecture Notes in Artificial Intelligence and Lecture Notes in Bioinformatics) (2014)
39. Wörner, D., Von Bomhard, T.: When your sensor earns money: exchanging data for cash with bitcoin. In: Proceedings of the 2014 ACM International Joint Conference on Pervasive and Ubiquitous Computing, UbiComp 2014 Adjunct (2014)
40. Doguet, J.J.: The nature of the form: legal and regulatory issues surrounding the bitcoin digital currency system Louisiana Law Review (2013)
41. Ford, P.: Marginally useful. Technol. Rev. **117**, 80–82 (2014)
42. Castronova, E.: Wildcat currency: how the virtual money revolution is transforming the economy (2014)
43. Wiener, H., Zelnik, J., Tarshish, I., Rodgers, M.: Chomping at the bit: U.S. federal income taxation of bitcoin transactions. J. Taxation Financ. Prod. **11,** 35–47 (2013)
44. Barratt, M.J., Lenton, S., Allen, M.: Internet content regulation, public drug websites and the growth in hidden internet services drugs: education, prevention and policy (2013)
45. Michailaki, A.: Mixed reality through the internet of things and bitcoin: how laws affect them. In: Communications in Computer and Information Science, pp. 165–169 (2014)
46. Birch, D.G.W.: Tomorrows transactions: the 2014 Reader. Mastodon Press
47. Graeber, D.: Debt: the first 5000 years (2014)
48. Dodd, N.: Simmel's Perfect Money: Fiction, Socialism and Utopia in The Philosophy of Money Theory, Culture and Society (2012)
49. Maurer, B., Nelms, T.C., Swartz, L.: When Perhaps the Real Problem is Money Itself!: The Practical Materiality of Bitcoin Social Semiotics (2013)
50. Smith, A.: In Sunday Times Magazine. Sunday Times, London (2014)
51. Angel, J.J., McCabe, D.: The ethics of payments: paper, plastic, or bitcoin? J. Bus. Ethics (2014, in press)
52. Kostakis, V., Giotitsas, C.: The (A)political economy of bitcoin. TripleC **12**, 431–440 (2014)
53. Schulaka, C.: Worth a few bitcoins? J. Financ. Plann. **27**, 11 (2014)

A Hybrid On-line Topic Groups Mining Platform

Cheng-Lin Yang and Yun-Heh Chen-Burger

Abstract In recent years, there is a rapid increased use of social networking platforms in the forms of short-text communication. Such communication can be indicative to popular public opinions and may be influential to real-life events. It is worth to identify topic groups from it automatically so it can help the analyst to understand the social network easily. However, due to the short-length of the texts used, the precise meaning and context of such texts are often ambiguous. In this paper, we proposed a hybrid framework, which adapts and extends the text clustering technique that uses Wikipedia as background knowledge. Based on this method, we are able to achieve higher level of precision in identifying the group of messages that has the similar topic.

Keywords Social Network Analysis · Micro-blogging System · Machine Learning

1 Introduction

The information insides on-line communities can be very valuable for public options gathering or commercial marketing. Therefore, mining the user interested topic in large network such as Twitter becomes one of the most important task in social network analytics (SNA).

What is a topic group? Traditionally speaking, a topic group is a group of people who are gathered to embrace the same values or share the same responsibility. Moreover, with the rapid development of communication technology, the Internet has become an indispensable utility in daily life. People exchange information and

C.-L. Yang (✉)
Centre for Intelligent Systems and Their Applications, School of Informatics,
University of Edinburgh, Edinburgh, UK
e-mail: s0969605@inf.ed.ac.uk

Y.-H. Chen-Burger
School of Mathematical and Computer Sciences, Heriot-Watt University, Edinburgh, UK
e-mail: y.j.chenburger@hw.ac.uk

G. Jezic et al. (eds.), *Agent and Multi-Agent Systems: Technologies and Applications*, Smart Innovation, Systems and Technologies 38,
DOI 10.1007/978-3-319-19728-9_17

knowledge on the Internet through various devices, forming a large social network and developing different types of on-line topic group. The user with new communication technology uses the forum or blog system to share his knowledge or experience with multimedia resources. He is able to discuss the topic with users from different countries by Internet. Therefore, a new type of community is formed. In this paper, we call them on-line topic groups.

Members of an on-line topic group are not restricted to the same geographical area unlike the topic group as defined in the traditional sense. An on-line topic group can be defined as a social phenomenon formed by a group of people who communicate with each other through the Internet and share the same interest toward a certain topic. However, due to the short-length of the content used, the precise meaning and context of such texts are often ambiguous. To address these problems, we have devised a new topic mining approach that is an adaptation and extension of text clustering using Wikipedia as background knowledge.

2 Related Work

In this section, we discuss about community mining based on social network analysis approaches and based on other interesting approaches to see the overall perspective of community mining.

2.1 Social Network Analysis Approaches

According to [6], "Social network analysis studies social networks by means of analyzing structural relationships between people". Mining Community using traditional social network analysis approaches usually focuses on structure of the social network that is represented by direct or indirect graph. Each node in the graph represents an instance in the network e.g. person or object whereas links between nodes represent relations between the instances. The relations between the instances in the network can be defined by explicit information such as friends in Facebook or followers in Twitter.

[6] analyzed structure of network to identify communities among Slashdot users. Method of [6] is based on social network analysis approaches with negative weighted edges graph. A study of [1, 3] also detected communities by considering the network structure. [3] took the network structure properties such as loops and edges of the network into the account. [1] considered bi-partite subgraphs to locate communities of websites.

2.2 Wikipedia Concepts Identification and Disambiguation

Recently, Wikipedia is used in many fields that are related to machine learning such as natural language processing, text classification and text clustering. One of difficulties for using Wikipedia is accurately matching between input text and Wikipedia concepts (articles) because each word or each phase in the input text can refer to one or more Wikipedia concepts. For example, a word "apple" can refer to both concepts "Apple (fruit)" and "Apple Inc."

This problem has been interesting for a while and a lot of researchers have been proposed many methods to solve this problem. The review of word sense disambiguation methods can be found at [10].

An alternative approach that performs well in doing word sense disambiguation is using machine learning methods to learn labeled training set and classify ambiguous words. This concept is used in text annotation with Wikipedia links. [4, 5, 8] are several researches regarding to text annotation with Wikipedia links. The first paper published about Wikipedia as a resource for annotation is [4] before significant improvement on this field by [5].

However, most papers did experiment in the context of standard length document. These experiments do not ensure that the approaches used in the papers will perform well in the case of short text document such as tweets, news or search snippets. In 2010, [8] brought the concept of annotating plain-text with Wikipedia links to context of short length document. They used anchor as a resource of identification instead of using only Wikipedia title as [7] because anchors are selected appropriately by people who create the pages. Their approach consists of three main steps: anchor parsing, anchor disambiguation and anchor pruning. Performing those steps extends an ability to deal with short text for the annotation system.

3 Proposed Hybrid Framework

A hybrid system with a three-layered framework: collection, classification and reasoning layers. The architecture of proposed system is shown in Fig. 1.

3.1 Collection Layer

The collection layer contains components that fetch the data from the micro-blogging system, process the raw data into a pre-defined format and convert the data into numerical parameters.

Crawler The crawler is responsible for retrieving the user data from Twitter platform. All fetched tweets will be stored in the storage for further usage. It is designed

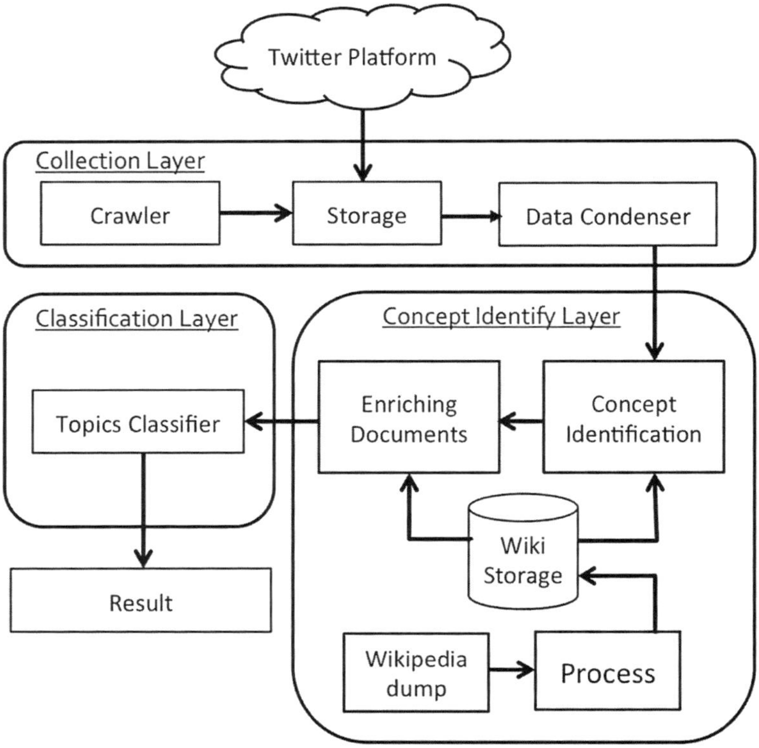

Fig. 1 Overview of proposed hybrid three-layered framework

to be a lightweight but targeted daemon so it can be deployed on multiple machines easily to increase the throughput.

The Storage In order to support fast lookup and flexible schema, the proposed system will take advantage of a distributed key-value database system: MongoDB,[1] which allows us to change the table schema without altering the entire table. Scalability is another concern for any system that handles tremendous amount of data. A distributed database system provides a simple procedure to add new node into the system.

Data Condenser The data condenser reads the raw data from the database. The raw data contains noises like auxiliary words, emoticons or random characters, so, it is the data condenser's responsibility to remove these noises. It is also responsible for converting and normalising the selected fields such as steamed terms among tweets into numerical parameters for the concept identify layer.

[1]MongoDB: http://www.mongodb.org/.

3.2 Concept Indetify Layer

Wikipedia Processing We use 4th July 2012 English Wikipedia article dump which contains 4,012,083 articles and has a size about 8.2 GB compressed. Then, we pre-processed and indexed them into 2 main catalogs to speed up the query.

1. Anchor Dictionary. We extracted all links and their anchors in Wikipedia pages and built them as anchor dictionary. The anchor dictionary is not like English dictionary. It contains only two important information: (1) anchors and (2) their corresponding Wikipedia concepts.
2. Wikipedia Pages. We also indexed Wikipedia pages content, their categories and inlink for speed in querying.

Concept Identification In order to identify Wikipedia concepts related to each tweet, we need to identify all anchors appearing in the tweet. In this sub-process, we use the steps given in Algorithm 1 to find the anchors. $lp(a)$ is link probability that can be calculated by following equation:

$$lp(a) = \frac{link(a)}{freq(a)} \tag{1}$$

where $link(a)$ is number of anchor a used as a link and $freq(a)$ is number of anchor a appearing in all documents in collection.

Algorithm 1 Document Parsing

Require: input document d
  A =ngrams(d, n=6)
  for each $word \in A$ **do**
    if $word \notin dictionary$ **then**
      A=A{$word$}
    end if
  end for
  for $a_1 \in A$ **do**
    for $a_2 \in A$ **do**
      if $a_1 \neq a_2$ and $substring(a_1, a_2)$ and $lp(a_1) < lp(a_2)$ **then**
        A=A{a_1}
      end if
    end for
  end for

Concepts Disambiguation Each anchor in the set of candidate anchors we got from previous sections could refer to several Wikipedia concepts. Therefore, in this step, we disambiguated those concepts and assigned the most appropriate concept to each anchor.

The same anchor may have different meanings and may link to different Wikipedia concepts depending on the context of the document. Therefore, we need the disambiguation process in order to select the appropriate Wikipedia concepts. We use a voting scheme adopted from [8]. The idea behind this voting scheme is that every anchor has to vote for all Wikipedia concepts related to other anchors in the document, except concepts related to itself. Then, the concepts that are given a top e-rank by their voting score are selected as candidate concepts. Finally, we select the most appropriate concept by using a commonness score. The detail of this voting scheme is described as follows:

First, we calculate relatedness between two Wikipedia concepts by using Normalized Google distance $rel(p_a, p_b)$ between the inlink of p_a and p_b where p_b is the Wikipedia concept corresponding to anchor b.

Next, we calculate the voting score of anchor b to concept pa by averaging the relatedness between all corresponding concepts of anchor b to concept pa, with prior probability known as commonness $Pr(p_b|b)$ as shown in the following equation:

$$vote_b(p_a) = \frac{\sum_{p_b \in P_g(b)} rel(p_a, p_b) \cdot Pr(p_b|b)}{\left\| P_g(b) \right\|} \tag{2}$$

After that, the total score assigned to p_a can be calculated by

$$rel_a(p_a) = \sum_{b \in A\{a\}} vote_b(p_a) \tag{3}$$

Using only this score to assign the concepts to the anchors may not be enough because, as mentioned in [5], balancing the score and commonness is the main factor affecting the performance.

Concepts Filtering However, after the disambiguation step is applied, there may be uncorrelated concepts left. Therefore, we have a final concepts filtering step to remove all concepts that are not related to others. We filter out unrelated anchors by using the concept of coherence between selected concepts from the concepts disambiguation step. To calculate the coherence score, we use the average relatedness between selected concepts as follows:

$$coherence(a \rightarrow p_a) = \frac{1}{\|S\| - 1} \sum_{p_b \in S\{p_a\}} rel(p_a, p_b) \tag{4}$$

where S is number of all selected concepts. Next, we filter out unrelated anchors that satisfy the following condition:

$$\frac{coherence(a \rightarrow p_a) + lp(a)}{2} < \epsilon \tag{5}$$

where $\epsilon = 0.2$ and $lp(a)$ is the link probability of an anchor a.

Document Enriching Most terms in the tweets tend to appear only once. With these characteristics of short text document, there are some problems with TF-IDF weighting that we use for modeling the document. In some cases, important words will have lower inverse document frequency compared to those who are not important. This means, in some cases, important words will have lower inverse document frequency compared to those who are not important.

[2] succeed in using 3 different strategies to enrich TF-IDF vector with background knowledge based on WordNet. Three strategies consist of adding corresponding WordNet concepts, replacing terms by WordNet concepts and replacing term vector with concept vector. Also, [7] yielded the good results from enriching the document with semantic related terms based on Wikipedia knowledge. In our proposed method, we adapted the strategies from [2, 7] to enrich Wikipedia knowledge into the tweets.

Strategy 1: Add Wikipedia Concepts We replace and add terms in each tweet with its corresponding Wikipedia concepts. The reason is that one of the problems that reduce the performance of bag-of-word model is that each Wikipedia concept can be mentioned by several different words/phrases. This problem leads to low cosine similarity between these two documents because the word "MS" and "Microsoft" are treated as different words. Therefore, to reduce the error of different representation of the same concept, we replace them with Wikipedia concept which change them into the same representation.

Strategy 2: Add Wikipedia Concepts and Categories We extend the first strategy by adding categories of Wikipedia concepts corresponding to each term in a tweet because another problem of bag-of-word model is that the model cannot capture semantic relationship between two related terms. Adding Wikipedia categories will solve the problem of semantic relationship between the tweets. Adding Wikipedia categories can solve the problem semantic relationship of bag-of-word model. Therefore, in this strategy, we also add Wikipedia concepts in documents besides replacing and adding Wikipedia concepts in Strategy 1.

3.3 Classification Layer

The results from many papers show that affinity propagation is the best among several clustering algorithms in short text clustering task. However, there are many criticism of affinity propagation about a problem with a large dataset. The issue about scalability of affinity propagation was mentioned in [9, 12]. Therefore, due to the size of our dataset, we decided to use Bisecting k-means clustering because of its scalability and efficiency.

4 Evaluation

The aim of this paper is mining groups in Twitter by finding a group of tweets which have the same concepts. Our approaches are based on text clustering and integration of Wikipedia knowledge and vector space model.

4.1 Experiments

In order to evaluate our methods, we did three experiments as follow and applied them to 1,500,000 collected tweets:

- *[Experiment 1]* In order to evaluate our work, we need some approaches to be compared with. Therefore, before experiment on our approaches, we ran experiment on pure clustering algorithm without enriching Wikipedia knowledge.
- *[Experiment 2]* In this method, we did further preprocessing step with our tweets data by adding related Wikipedia concepts as we explained in Strategy 1. After that, we model those enriched tweets with TF-IDF vectors before clustering.
- *[Experiment 3]* This method extended the concepts from Method 2. It does not only add Wikipedia concepts that related to each tweets in the preprocessing step but also add Wikipedia categories of each Wikipedia concepts into the tweets in order to solve the semantic relation problem of bag-of-word model.

4.2 Evaluation Based on Ground-Truth Testset

For evaluation based on ground-truth testset, we manually labelled 400 tweets with appropriate groups. Then, we evaluate all three methods we mentioned earlier by calculating V-Measure score between true labels and labels that were assigned by clustering algorithm.

Evaluation Metric Typically, the basic criteria of a clustering result are homogeneity and completeness. The homogeneity criterion is satisfied, for all clusters, every member of each cluster comes from only one class which is defined as:

$$homogeneity = \begin{cases} 1 & \text{if } H(C,K) = 0 \\ 1 - \frac{H(C|K)}{H(C)} & \text{otherwise} \end{cases} \quad (6)$$

where $H(C|K)$ is the conditional entropy of the classes given assigned clusters and $H(C)$ is the entropy of the class defined as:

$$H(C|K) = -\sum_{k=1}^{|K|}\sum_{c=1}^{|C|} \frac{n_{c,K}}{n} \log \frac{n_{c,k}}{\sum_{c=1}^{|C|} n_{c,k}} \quad (7)$$

$$H(C) = -\sum_{c=1}^{|C|} \frac{\sum_{k=1}^{|K|} n_{c,k}}{n} \log \frac{\sum_{k=1}^{|K|} n_{c,k}}{n} \quad (8)$$

in which n is number of all data points and $n_{c,k}$ is number of data points from class c that clustered into cluster k.

Completeness criterion is quite opposite to homogeneity. It is satisfied if all members of a class are clustered into the same cluster. Mathematically, we can define completeness score as follow:

$$completeness = \begin{cases} 1 & \text{if } H(C,K) = 0 \\ 1 - \frac{H(K|C)}{H(K)} & \text{otherwise} \end{cases} \quad (9)$$

where $H(K|C)$ is the conditional entropy of assigned clusters given the classes and $H(K)$ is the entropy of assigned clusters defined as:

$$H(K|C) = -\sum_{k=1}^{|K|}\sum_{c=1}^{|C|} \frac{n_{c,K}}{n} \log \frac{n_{c,k}}{\sum_{k=1}^{|K|} n_{c,k}} \quad (10)$$

$$H(K) = -\sum_{k=1}^{|K|} \frac{\sum_{c=1}^{|C|} n_{c,k}}{n} \log \frac{\sum_{c=1}^{|C|} n_{c,k}}{n} \quad (11)$$

in which n is number of all data points and $n_{c,k}$ is number of data points from class c that clustered into cluster k.

A good clustering result should satisfy both homogeneity and completeness at the same time. In order to do that, we used V-Measure as a metric for evaluating clustering results. V-Measure which is first introduced by [11] is the harmonic mean of homogeneity and completeness scores bounded in the range of [0, 1]. The closer the value is to 1, the better the quality of a clustering result. It can be defined as the following equation:

$$V_\beta = \frac{(1+\beta) * h * c}{(\beta + h) + c} \quad (12)$$

where β is the weight, if β is set less than 1, homogeneity is weighted more. If β sets to more than 1, completeness is weighted more. In our experiment, we weight them equally by setting $\beta = 1$.

4.3 Results

For each experiment, we ran bisecting k-means multiple times with different setting up of number of clusters k ranging from 64 to 4096. Then, we evaluate the clustering results with our testset using V-Measure as an evaluation metric. Figure 2 shows V-Measure scores of Experiment 1, 2 and 3 with different setting of number of clusters k.

From the figure, it is clear to see that *Method 2 (concepts)* and *Method 3 (concepts+categories)* outperformed *Method 1 (baseline)*. *Method 2* is clearly better than *Method 1* at every setting of number of clusters k. The highest V-Measure that

Method 1 can get is 0.674 at $k = 3800$ whereas the highest V-Measure of *Method 2* is 0.747 at $k = 3400$. The difference between the highest peak of them is 7.3 %.

Furthermore, in the case of *Method 3*, it has clearly higher performance than *Method 1* in the Fig. 2. Comparing with their best performance, *Method 3* gets 14.7 % better with V-Measure 0.821 at $k = 3600$. We can conclude from these results getting from the testset that *Method 2* and *Method 3* have dramatic improvement from *Method 1* with V-Measure 0.747, 0.821 and 0.674. From these results, it confirms that using Wikipedia as a resource for enriching the tweets can improve the performance of topic groups mining.

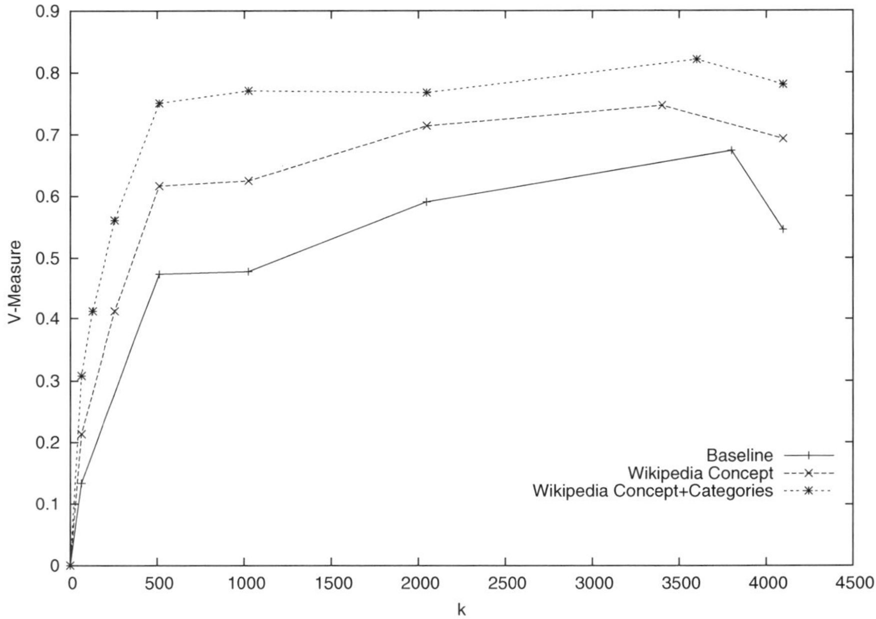

Fig. 2 Comparison among Method 1, 2 and 3

5 Conclusion

In this paper, we proposed a state-of-the-art three layered framework to mine topic groups from large scale short text documents (Tweets) based on Wikipedia knowledge. Mining topic groups based on social network analysis approaches fails to capture hidden concepts in a network because they usually model the network as a graph with explicit relationships that can be found in the network connectivity. The hidden concepts in the network are not taken into account leading these approaches to missing the important concepts among users in the network. In this study, we therefore investigated an alternative approach to identify hidden concepts in Twitter. We used clustering based methods to mine hidden concepts in tweets based on their topics. The optimistic result we have is the method enriched with Wikipedia concept and categories. Based on the evaluation metric in Sect. 4.3, our method has a promising V-measure score up to 0.821 from real Twitter data where baseline method only has 0.674 in V-measure score.

References

1. Kumar, R., Raghavan, P., Rajagopalan, S., Tomkins, A.: Trawling the web for emerging cyber-communities. In: Proceedings of the Eighth International Conference on World Wide Web, WWW '99, pp. 1481–1493, NY, USA (1999)
2. Hotho, A., Hotho, A., Staab, S., Staab, S., Stumme, G., Stumme, G.: Text Clustering Based on Background Knowledge (2003)
3. Newman, M.E.J.: Detecting community structure in networks. Eur. Phys. J. B—Condens. Matter Complex Syst. **38**(2), 321–330 (2004)
4. Mihalcea, R., Csomai, A.: Wikify!: linking documents to encyclopedic knowledge. In In CIKM 07: Proceedings of the Sixteenth ACM Conference on Information and Knowledge Management, pp. 233–242. ACM (2007)
5. Milne, D., Witten, I.H.: Learning to link with wikipedia. In: Proceedings of the 17th ACM Conference on Information and Knowledge Management, CIKM '08, pp. 509–518, NY, USA, 2008. ACM
6. Kunegis, J., Lommatzsch, A., Bauckhage, C.: The slashdot zoo: mining a social network with negative edges. In: Proceedings of the 18th International Conference on World Wide Web, WWW '09, pp. 741–750, NY, USA. ACM (2009)
7. Wang, P., Hu, J., Zeng, H.-J., Chen, Z.: Using wikipedia knowledge to improve text classification. Knowl. Inf. Syst. **19**(3), 265–281 (2009)
8. Ferragina, P., Scaiella, U.: Tagme: on-the-fly annotation of short text fragments (by wikipedia entities). In: Proceedings of the 19th ACM International Conference on Information and Knowledge Management, CIKM '10, pp. 1625–1628, NY, USA. ACM (2010)
9. Fujiwara, Y., Irie, G., Kitahara, T.: Fast algorithm for affinity propagation. In: Proceedings of the Twenty-Second International Joint Conference on Artificial Intelligence, vol. 3, IJCAI'11, pp. 2238–2243. AAAI Press (2011)
10. Navigli, R.: A quick tour of word sense disambiguation, induction and related approaches. In: Proceedings of the 38th International Conference on Current Trends in Theory and Practice of Computer Science, SOFSEM'12, pp. 115–129, Berlin, Heidelberg (2012)
11. Roseberg, A., Hirschberg, J.: V-Measure: A Conditional Entropy-Based External Cluster Evaluation Measure
12. Zhang, X.. Furtlehner, C., Sebag, M.: Distributed and incremental clustering based on weighted affinity propagation. In: Proceedings of the 2008 Conference on STAIRS 2008, pp. 199–210, Amsterdam, The Netherlands. IOS Press (2008)

On Autonomic Platform-as-a-Service: Characterisation and Conceptual Model

Rafael Tolosana-Calasanz, José Ángel Bañares and José-Manuel Colom

Abstract In this position paper, we envision a Platform-as-a-Service conceptual and architectural solution for large-scale and data intensive applications. Our architectural approach is based on autonomic principles, therefore, its ultimate goal is to reduce human intervention, the cost, and the perceived complexity by enabling the autonomic platform to manage such applications itself in accordance with high-level policies. Such policies allow the platform to (i) interpret the application specifications; (ii) to map the specifications onto the target computing infrastructure, so that the applications are executed and their Quality of Service (QoS), as specified in their SLA, enforced; and, most importantly, (iii) to adapt automatically such previously established mappings when unexpected behaviours violate the expected. Such adaptations may involve modifications in the arrangement of the computational infrastructure, i.e. by re-designing a different communication network topology that dictates how computational resources interact, or even the live-migration to a different computational infrastructure. The ultimate goal of these challenges is to (de)provision computational machines, storage and networking links and their required topologies in order to supply for the application the virtualised infrastructure that better meets the SLAs. Generic architectural blueprints and principles have been provided for designing and implementing an autonomic computing system. We revisit them in order to provide a customised and specific view for PaaS platforms and integrate emerging paradigms such as DevOps for automate deployments, Monitoring as a Service for accurate and large-scale monitoring, or well-known formalisms such as Petri Nets for building performance models.

R. Tolosana-Calasanz (✉) · J.Á. Bañares · J.-M. Colom
COSMOS Group—Aragón Institute of Engineering Research (I3A),
Universidad de Zaragoza, Zaragoza, Spain
e-mail: rafaelt@unizar.es

J.Á. Bañares
e-mail: banares@unizar.es

J.-M. Colom
e-mail: jm@unizar.es

G. Jezic et al. (eds.), *Agent and Multi-Agent Systems: Technologies and Applications*, Smart Innovation, Systems and Technologies 38,
DOI 10.1007/978-3-319-19728-9_18

1 Introduction

In the early 60s, at the dawn of the computer era, the conception of computing being organised as a public utility was already envisioned –just as the telephone system or the electrical power networks. John McCarthy, speaking at the MIT Centennial in 1961,[1] imagined that such computing infrastructure would also be integrated with a different and disruptive business model, thereby "*(...) each subscriber needs to pay only for the capacity he actually uses, but he has access to all programming languages characteristic of a very large system (...) Certain subscribers might offer service to other subscribers (...)*". Hence, McCarthy anticipated that the computing utility would become "*the basis of a new important industry*". Half a century later, after significant advances in computing as well as with the required maturity and developments in technology, such a view is becoming a reality with Cloud computing [3, 9].

Nowadays, there is a number of public Cloud providers, such as Amazon[2] or Softlayer,[3] that allow users to rent virtualised resources on demand, namely computers, storage and network links, and to pay per their usage in accordance with a number of Service Level Agreements (SLAs) and enforcement guarantees. Moreover, Cloud computing can also offer additional abstraction levels, by providing not only a virtualised infrastructure (Infrastructure-as-a-Service, IaaS), but also applications (Software-as-a-Service, SaaS) or even actual software development and deployment platforms (Platform-as-a-Service, PaaS). PaaS abstracts the underlying computing infrastructure and provides the user with a language interface so that both the program logic and the SLAs can be expressed. Essentially, such specifications need to be infrastructure-agnostic, that is, without referring to specific details of a particular infrastructure, so that they can subsequently be interpreted, mapped and deployed to a computing infrastructure. Nevertheless, open challenges are emerging with the advent of big data applications, as these applications often have to deal with significants amounts of data that need to be processed *continuously* in (near) real-time, and their data generation rates are often bursty and subject to unpredictability. Such requirements lead to the need for large number of computational distributed resources and the complexity of the management inherently increases, provoked by the appearance of failures, unexpected performance degradation and, in general, unknown behaviours. In such a context, there is some approaches [2, 7] that have studied and considered the introduction a number of policies and mechanisms into PaaS in order to offer some Quality of Service (QoS) guarantees.

In this position paper, we envision a PaaS conceptual and architectural solution for large-scale and data intensive applications. Our architectural approach is based on autonomic principles, therefore, its ultimate goal is to reduce human intervention, the cost, and the perceived complexity by enabling the autonomic platform to

[1] http://www.technologyreview.com/news/425623/the-cloud-imperative/.

[2] http://aws.amazon.com/.

[3] http://www.softlayer.com/.

manage such applications itself in accordance with high-level policies. Such policies allow the platform to (i) interpret the application specifications; (ii) to map the specifications onto the target computing infrastructure, so that the applications are executed and their Quality of Service (QoS), as specified in their SLA, enforced; and (iii) to adapt automatically such previously established mappings when unexpected behaviours violate the expected. Such adaptations may involve modifications in the arrangement of the computational infrastructure, i.e. by re-designing a different communication network topology that dictates how computational resources interact, or even the live-migration to a different computational infrastructure. The ultimate goal of these challenges is to (de)provision computational machines, storage and networking links and their required topologies in order to supply for the application the virtualised infrastructure that better meets the SLAs. Generic architectural blueprints and principles have been provided for designing and implementing an autonomic computing system [4–6, 8, 12]. We revisit them in order to provide a customised and specific view for PaaS platforms and integrate emerging paradigms such as DevOps for automate deployments, Monitoring as a Service for accurate and large-scale monitoring, or well-known formalisms such as Petri Nets for building performance models. The rest of this paper is organised as follows. Section 2 describes the conceptual architectural model for Autonomic Platform-as-a-Service. Section 3 provides a brief related work discussion on PaaS. Conclusions are provided in Sect. 4.

2 Characterisation and Conceptual Model of an Autonomic PaaS

There is a number of well-established and generic architectural blueprints and principles for the design of autonomic systems [4–6, 8, 12]. When designing an autonomic system, a number of challenges emerge: (i) *Autonomic System Specification* – it relates to the appropriate specification and formulation of the autonomic elements, so that the description of what the system needs to do is captured and understood. The specification involves functional and non-functional requirements as well as autonomic behaviours, expressed in terms of high level policies, content and context driven definition, execution and management. (ii) *Autonomic System Design* –which includes the definition of appropriate abstractions, methods and tools for specifying, understanding, controlling, and implementing autonomic behaviours; the provision of models for negotiation among architectural autonomic elements and for detecting, predicting, and correcting potential problems. (iii) *Integration and Consistency of Autonomic Behaviours* –it is related to the autonomic elements constituent of the system, and on how they behave in isolation and in co-operation. It is essential to support the specification of individual and global autonomic behaviours, so that they can be implemented and controlled in a robust and predictable manner. (iv) *Middleware challenges* – the system needs to exploit middleware services for realising autonomic behaviours, including discovery, messaging, security, privacy, trust, etc. Therefore, it will be fundamental to properly identify which functional and non-functional services the middleware layer needs to provide.

In this section, we are characterising and customising such elements for an Autonomic PaaS platform that executes large-scale data-intensive applications. Figure 1 depicts an architectural blueprint for our proposal.

2.1 Application Specification for the Autonomic PaaS

Users need a specification language for expressing their functional and non-functional requirements. Hence, as an input, the PaaS platform receives users' specifications and maps them accordingly to the computing infrastructures. The characteristics of such specifications are, on one hand, application dependent. In the case of data-intensive applications, they may involve, from a functional point of view, compositions of computational processes and data transmission processes. The semantics in how these compositions are accomplished will determine the particular model of computation. Therefore, there is a need for supporting different of models of computation ranging from simple parallelism to data streaming [13]. From a non-functional

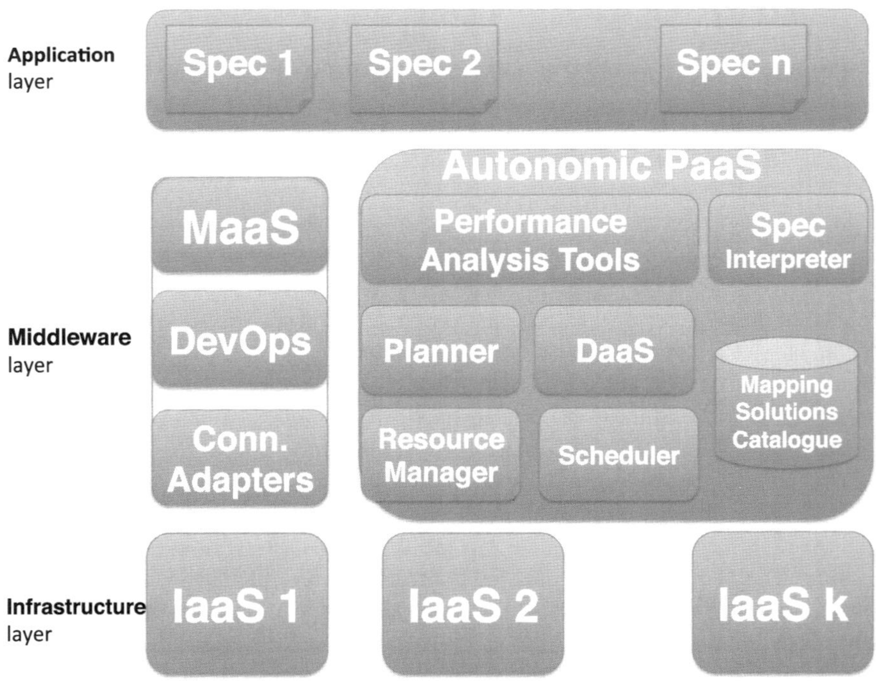

Fig. 1 Conceptual architecture for an autonomic platform as a service

point of view, users can be interested in considering a number of attributes such as throughput/processing time, economic cost (given by the pay-per-usage model of the Cloud), or resilience. Additionally, the desired autonomic behaviour while executing the application can be (partially inferred from the non-functional requirements), or alternatively, high-level policies, which determine the behaviour of the PaaS when executing the application, can also be supported On the other hand, the design of the specification also depends on inherent characteristics of PaaS platforms and their ultimate goal, that is essentially, such specifications need to be infrastructure-agnostic, that is, without referring to specific details of a particular infrastructure, so that they can subsequently be interpreted, mapped and deployed to a computing infrastructure. There is a number of open standards proposed for autonomic computing that can be prescribed for the specification such as OASIS Web Services Distributed Management (WSDM),[4] for the interfaces and their semantics, or Common Information Model – Simplified Policy Language[5] (CIM-SPL) for the specification of policies. In addition to specification languages, and also of key importance, is the specification interpreter of the autonomic PaaS platform, which appears on the right of Fig. 1.

2.2 Autonomic PaaS Design

From the functional and non-functional requirements captured from the user, we envision that a performance engineer needs to conduct an analysis and to generate possible mapping solutions between the application and a computing infrastructure. At such a step, different configurations, i.e. computational machines linked by means of different network topologies, distinct number of computational resources, etc. may be considered. For all of them, the engineer must analyse and identify the boundaries of QoS parameters and how they affect each other. We propose that such a process is accomplished in a two main steps. First, by realising the analysis with generic and abstract resources, studying concurrency from the application and the consumption of resources. Second, by refining and customising the models with the constraints of specific infrastructures. As a result, a *catalogue* of potential application/infrastructure mapping solutions is identified, each with different QoS attribute values and guarantees. Although there is a number of formalisms and generic approaches in the literature, on performance engineering, we developed further that particular idea. We proposed a Petri net-based, model-driven and stepwise refinement methodology for streaming applications over Clouds [14]. In such a case, the complexity of streaming applications arises from the confluence of concurrency, transmission of data and the use of distributed resources. The central role in this methodology is assigned to a set of Petri Net models describing the behaviour of the system including timing and cost information. The goal is to use these models in

[4]https://www.oasis-open.org/committees/wsdm/.

[5]http://www.dmtf.org/documents/policy/cim-simplified-policy-language-cim-spl-100.

an intensive way before the deployment of the application in order to understand its behaviour and to obtain properties of the different solutions adopted. In some cases, the observations may induce or recommend changes into the application with the purpose of modifying parts of the design and assure agreed specifications. The consideration of Petri Nets is based on the natural descriptive power for the concurrency, but also for the availability of analytic tools coming from the domain of Mathematical Programming and Graph Theory. These tools are based on a structural analysis that support the reasoning on properties without the construction of the state space –which for such a class of systems is prohibitive. a methodology for the construction of this kind of applications is proposed based on the intensive use of formal models. Petri Nets are the formalism considered here for capturing the active entities of the system (processes), the flow of data between the processes and the shared resources for which they are competing. For the construction of a model aimed at studying different aspects of the system and for decision-taking design, an abstraction process of the system at different levels of detail is needed. This leads to several system models representing facets from the functional level to the operational level. Petri Net models are used to obtain qualitative information of the streaming application, but their enrichment with time and cost information provides with analysis on performance and economic behaviours under different scenarios. In consequence, a set of performance analysis tools need to be incorporated in the autonomic PaaS architecture.

Moreover, such a catalogue of performance models constitutes an essential element of the system *Knowledge*, and can be subsequently exploited by the autonomic elements of the architecture for their self-management actions. Whenever a violation of the QoS occurs, the platform must decide an action (or combination of actions) that corrects such a deviation. The catalogue is essential for this process, as well as obtaining real-time information from the infrastructure or even from other infrastructures that can be potentially used alternatively. Then, once an action is selected, it has to be planned and accomplished (executed). It may involve a complete reconfiguration of the infrastructure –i.e. a complete different number of machines with different network topologies and storage; or it may require the shift to an alternative computing infrastructure. This raises a number of challenges such as the autonomic deployment of the application (Deployment as a Service, DaaS in Fig. 1) in a new different infrastructure without interruptions in the execution and enforcing the expected QoS.

2.3 Middleware Layer Requirements

Perhaps the monitoring service from the computing infrastructures is the most important one from the middleware layer, but a deployment automation middleware service is also mandatory, so that the required computing infrastructure can be virtualised on-demand. Additionally, connection wrappers and adapters for interacting with different IaaS providers are required. On this regard, the possibility of an Inter-Cloud organisation of the infrastructure may also be considered [1].

Monitoring as a Service Big data analytics and applications often involve a large number of distributed computational resources for their execution and, therefore, monitoring states of resources (i.e. performance, running time, cost, etc.) often requires collecting values of various attributes from a vast amount of nodes. Although there can be recognised a core set of attributes common to most applications, in general terms, the specific requirements of each application determines the monitoring needs, which can even vary in time, i.e. as a result of changing conditions in the application requirements or in the infrastructure. Monitoring as a Service [10, 11] should support not only the conventional state monitoring capabilities, such as instantaneous violation detection, periodical state monitoring, and single tenant monitoring, but also performance-enhanced functionalities that can optimise on monitoring cost, scalability, and the effectiveness of monitoring service consolidation and isolation.

Deployment as a Service Deployment of a mapping solution in the Cloud can be a tedious task for a human being and it is also expected to be fast feedback to new occurring eventualities. Thus, it is a critical competitive advantage to be able to respond quickly. For these reasons, tooling is required to implement end-to-end automation of deployment processes and DevOps [15–17] as an emerging paradigm, which integrates software developers with operational personnel, can be a solution to consider. Automation is the key to efficient collaboration and tight integration between development and operations. The DevOps community is constantly pushing new approaches, tools, and open-source artifacts to implement such automated processes.

2.4 Integration and Consistency of Autonomic Behaviours

In addition to defining the global behaviour of the PaaS platform when executing an application, it is important to identify the autonomic elements forming part of the PaaS architecture as well as third-party autonomic elements, such as autonomic middleware components. Individual policies should dictate their behaviour, but their co-operation and their interactions should also be regulated. Hence, it is essential to support the specification of individual and global autonomic behaviours, so that they can be consistently integrated and the PaaS platform can be controlled in a predictable manner. A number of architectural components in the PaaS platform are likely to be autonomic, such as the planner, the scheduler or the resource manager. On the other hand, some of these internal autonomic components need to interact with third-party autonomic components as well. For instance, due to the complexity of monitoring, the MaaS middleware component may also be designed with autonomic principles.

3 Related Work

PaaS aims at application developing and subsequent deployment. Hence, it typically provides a complete set of tools and programming models and interfaces for processing the logic and automatically deploying and executing them into de underlying infrastructures. According to [2], PaaS providers include Google AppEngine, Microsoft Azure, Bungee Labs, Coghead, Etelos, Google, LongJump, Rollbase, or Salesforce.com, etc. Nevertheless, there is limited support for QoS guarantees in these solutions. The requirements for supporting QoS guarantees in PaaS architectures were analysed in [2]. In this paper, we have revisited these concepts, integrating new emerging paradigms and technologies such as DevOps or MaaS, an paradigms like autonomic computing with well-known formalisms such as Petri nets.

In a previous work [14], a first approach to this architectural model was proposed, but without the autonomic requirement as a final result. There, the goal was to manage the complexity in the construction of applications in the Cloud, but without considering dynamic aspects. Thus, the so called *functional* level there corresponds in here to the *specification* level, and the final result in [14] was a model representing the behaviour of the application without considering a specific infrastructure or third-party applications needed. The operational level, on the other hand, considers a generic infrastructure over which the behaviour of the application is studied. The operational level is described by another model that in combination with the functional level model gives us an overall model for qualitative and quantitative analysis of QoS attributes. Therefore, the approach in [14] represents the required previous step for accomplishing the autonomic PaaS architectural model presented in here.

4 Conclusions

With the advent of large-scale big data applications, which for their execution require complex management of significant number of distributed computational resources. In this position paper, we envision a PaaS conceptual and architectural model. Our architectural approach is based on autonomic principles, aiming at reducing human intervention, the cost, and the perceived complexity by enabling the autonomic platform to self-manage such applications. From the functional and non-functional requirements captured from the user, we envision that a performance engineer needs to conduct a Quality of Service (QoS) analysis (i.e. involving a combination of QoS attributes that includes performance, economic cost or resilience) to explore and assess all the possible mapping solutions of the application to abstract and generic resources. The ultimate goal is to identify the boundaries of QoS parameters and how they affect each other. Subsequent incorporation of the constraints from actual computational infrastructures will refine such models and update the derived QoS boundaries to estimate realistic behaviour of the infrastructure. As a result, a *catalogue* of potential application to infrastructure mapping solutions is constructed, each with different QoS attribute values and guarantees. The autonomic PaaS platform with this knowledge information in combination with accurate monitored real time information of the computational infrastructures can perform autonomic deployments and adaptations of previous ones, so that applications can executed in a flexible and adaptive manner. Our model integrates emerging paradigms such as DevOps for automate deployments, Monitoring as a Service for accurate and large-scale monitoring, or well-known formalisms such as Petri Nets for building performance models.

Acknowledgments This work was supported by the Spanish Ministry of Economy under the program "Programa de I+D+i Estatal de Investigación, Desarrollo e innovación Orientada a los Retos de la Sociedad", project id TIN2013-40809-R.

References

1. Assis, M., Bittencourt, L.F., Tolosana-Calasanz, R.: Cloud federation: characterisation and conceptual model. In: 3rd International Workshop on Clouds and (eScience) Applications Management (CloudAM 2014) (2014)
2. Boniface, M., Nasser, B., Papay, J., Phillips, S., Servin, A., Yang, X., Zlatev, Z., Gogouvitis, S., Katsaros, G., Konstanteli, K., Kousiouris, G., Menychtas, A., Kyriazis, D.: Platform-as-a-service architecture for real-time quality of service management in clouds. In: Fifth International Conference on Internet and Web Applications and Services (ICIW), pp. 155–160 (May 2010)
3. Buyya, R., Yeo, C.S., Venugopal, S., Broberg, J., Brandic, I.: Cloud computing and emerging IT platforms: vision, hype, and reality for delivering computing as the 5th utility. Future Gener. Comput. Syst. **25**(6), 599–616 (2009)
4. Corporation, I.: An Architectural Blueprint for Autonomic Computing. Technical report, IBM (Jun (2005)
5. Hanson, J.E., Whalley, I., Chess, D.M., Kephart, J.O.: An architectural approach to autonomic computing. In: Proceedings of the First International Conference on Autonomic Computing. pp. 2–9. ICAC '04, IEEE Computer Society, Washington, DC, USA (2004)
6. Huebscher, M.C., McCann, J.A.: A survey of autonomic computing -degrees, models, and applications. ACM Comput. Surv. 40(3), 7:1–7:28 (Aug 2008)
7. Keller, E., Rexford, J.: The "platform as a service" model for networking. In: Proceedings of the 2010 Internet Network Management Conference on Research on Enterprise Networking. pp. 4–4. INM/WREN'10, USENIX Association, Berkeley, CA, USA (2010)
8. Kephart, J.O., Chess, D.M.: The vision of autonomic computing. Computer **36**(1), 41–50 (2003)
9. Marinescu, D.C.: Cloud Computing: Theory and Practice. Morgan Kaufmann (2013)
10. Meng, S., Kashyap, S.R., Venkatramani, C., Liu, L.: Resource-aware application state monitoring. IEEE Trans. Parallel Distrib. Syst. **23**(12), 2315–2329 (2012)
11. Meng, S., Liu, L.: Enhanced monitoring-as-a-service for effective cloud management. IEEE Trans. Comput. **62**(9), 1705–1720 (2013)
12. Parashar, M., Hariri, S.: Autonomic computing: An overview. In: Banâtre, J.P., Fradet, P., Giavitto, J.L., Michel, O. (eds.) Unconventional Programming Paradigms. Lecture Notes in Computer Science, vol. 3566, pp. 257–269. Springer, Berlin Heidelberg (2005)
13. Pautasso, C., Alonso, G.: Parallel computing patterns for grid workflows. In: Proceedings of the HPDC2006 Workshop on Workflows in Support of Large-Scale Science (WORKS06), Paris, France 19–23 June 2006
14. Tolosana-Calasanz, R., Bañares, J.Á., Colom, J.M.: Towards petri net-based economical analysis for streaming applications executed over cloud infrastructures. In: Economics of Grids,

Clouds, Systems, and Services—11th International Conference, GECON 2014, Cardiff, UK pp. 189–205, 16–18 September 2014
15. Wettinger, J., Gorlach, K., Leymann, F.: Deployment aggregates—a generic deployment automation approach for applications operated in the cloud. In: IEEE 18th International Enterprise Distributed Object Computing Conference Workshops and Demonstrations (EDOCW), pp. 173–180 (Sept 2014)
16. Wettinger, J., Breitenbücher, U., Leymann, F.: Devopslang—bridging the gap between development and operations. In: Villari, M., Zimmermann, W., Lau, K.K. (eds.) Service-Oriented and Cloud Computing. Lecture Notes in Computer Science, vol. 8745, pp. 108–122. Springer, Berlin Heidelberg (2014)
17. Wettinger, J., Breitenbücher, U., Leymann, F.: Standards-based devops automation and integration using tosca. In: Proceedings of the 7th International Conference on Utility and Cloud Computing (UCC 2014), pp. 59–68. IEEE Computer Society (2014)

Disruptive Innovation: A Dedicated Forecasting Framework

Sanaa Diab, John Kanyaru and Hind Zantout

Abstract This paper describes the design of a forecasting framework to predict disruptive innovations. First, the nature and characteristics of disruptive innovation are presented, as well as the conditions that enable such a phenomenon. Individual factors that feed into disruptive innovations are identified, as well as formulae to allocate quantifiable measurement to these factors. Suitable principles from two existing approaches to forecasting are adopted to put forward a new framework. This will consist of a four-step process that uses both mathematical models and the judgemental method. The findings are based on work that is part of a MSc dissertation [1].

Keywords Disruptive innovation · Forecasting models · Sales drivers modeling · Social media · Technology market

1 Introduction

The latter part of the last millennium has ushered in the era of the knowledge economy. In this new world, the three traditional resources of labor, land and capital were supplemented by a fourth resource, namely knowledge. With the focus on knowledge and learning at its core, it was inevitable that the pace of new developments, or innovations, accelerated.

Innovations can be changes that are introduced to improve the efficiency of the business or the quality of the products and services or, it can be a completely new idea that is targeted to a certain market or one that unintentionally enters a market. Irrespective of the size and area of the innovation, the effect can be limited or have a

S. Diab (✉) · J. Kanyaru · H. Zantout
School of Mathematics and Computer Sciences, Heriot Watt University, Dubai, Uae
e-mail: smd30@hw.ac.uk

J. Kanyaru
e-mail: J.M.Kanyaru@hw.ac.uk

H. Zantout
e-mail: H.Zantout@hw.ac.uk

G. Jezic et al. (eds.), *Agent and Multi-Agent Systems: Technologies and Applications*, Smart Innovation, Systems and Technologies 38,
DOI 10.1007/978-3-319-19728-9_19

far reaching effect that will almost certainly disturb an existing balance in the market, a phenomenon referred to as disruptive innovation.

One way for businesses to foresee such oncoming threat and prepare to mitigate against it is by the use of forecasting models and tools. This paper presents such a model and reports on the findings of the work that is part of a MSc dissertation [1].

2 Disruptive Innovations

The term disruptive innovation was first coined in 1995 at Harvard Business School by Bower and Christensen [2] who investigated the phenomenon that may have a crucial impact on business and affects its ability to survive in the marketplace. They have defined innovations as two types; sustaining and disruptive. A sustaining innovation is the result of the quest for improvements in efficiencies or features of existing products and services and this can be evidenced by a healthy competition in the marketplace. A good example is the mobile phones industry where competition has led to higher quality cameras, longer battery life and a range of other improvements. A disruptive innovation on the other hand is one that will have far-reaching consequences and unexpectedly takes over an established market when the new innovation partially or completely replaces an old established one. Using the mobile phones example, the introduction of smart phones and tablets have created a new market with hundreds of businesses and disturbed the personal computers market.

A number of case studies revealed that some disruptive innovation manifests itself not through incremental enhancement of a product, but often as a product with lower performance or different attributes than the competing product. This often holds true for the technology sector. The initial understanding of the technology market was that existing firms and technologies are only displaced when a superior new firm or technology enters the market. However, Christensen and Bower have challenged this when they suggested that inferior products can also displace existing superior products. They called such innovations "disruptive" [2].

When a certain change creates appeal to a different market or the lower margin of the market, such products ultimately disrupt the market when main market customers eventually find the new product appealing and shift demand, leaving incumbents with great losses. This then may eventually lead to complete business failure [3, 4]. Incumbent firms would typically dismiss the potential products that are not targeted at the main market and don't consider them a threat. The disruptive product then quickly gains higher market share and threatens the status quo [5].

3 Proposed Forecasting Framework

A literature review and an extensive study into forecasting systems have revealed that there are four different approaches to forecasting namely scenarios and simulation, extrapolation and trend analysis, judgmental methods, and models. It was necessary to determine which approach to use for this framework as a first step. The scenarios and simulation approach requires a lot of time and resources, and recent cases of disruption appeared relatively fast, indeed, faster than scenarios and simulations are able to predict and so it was disregarded. Using extrapolation and trend analysis for disruptive innovation forecasting was also disregarded as there is no clear trend in disruption; it often happens unexpectedly. The judgmental methods seemed to be the most commonly used and trusted for cases of high uncertainty. These methods depend on human judgment and extensive analysis of the status quo. However, issues of bias and inexperience of the forecasting team are problematic with this approach and it should be used with care. The last approach to forecasting is the use of mathematical models where a number of factors which are believed to affect the status of something are studied and combined in a mathematical equation to use as a forecasting tool. This approach seemed most feasible for forecasting disruptive innovations.

An analysis of the current forecasting models that exist today have revealed that none have been described as "persistent", and the best ones to date predict a product's emergence with +1/−3 years accuracy [7] Additionally, most models do not employ more than one method of forecasting and are not open to public or voluntary participation. Further studies were done in order to understand better how to design a forecasting framework. Vanston [8] has advised that for any forecasting system to be reliable it should use at least two methods, especially in cases of high uncertainty. For this framework it is not advised to use more than two as speed is key, so the methods agreed on were judgmental and models. Further guidelines from the Committee on Forecasting Future Disruptive Technologies [6] were consulted to design a persistent and dedicated forecasting framework for disruptive technologies.

3.1 The Mathematical Model

Looking at disruption as the change in the market share of a product, it is necessary to first identify the factors that affect its sales. Literature review and historic data analysis concluded four such factors: competitive advantage, business status, marketing and lastly customer reactions. An equation was formulated for each of the four factors, producing four numerical values which are then used to produce a regression test. All data was scaled from 1 (lowest case) to 5 (highest case) on a Likert scale, thus permitting a symmetry preventing any variable from dominating the equation just by having a larger value [9]. The number 0 was not used as it may negatively affect the final results if multiplication was required.

(a) Competitive Advantage

Competitive advantage is a term used to evaluate a certain product or service against those of competitors. It looks at either one of two aspects of competition; the price and the product itself. For this method, both aspects are used as they are considered equally important in affecting a product's sales. Price and product features will be given numerical categories and the simple average is taken of both for the final value. The scale for the price is defined as

1 = Overpriced	4 = Low Price
2 = High Price	5 = Valuable Price
3 = Expected Price	

For the product features, these are evaluated and categorized as follows

1 = Less features/value	4 = Some valuable features
2 = No new features/value	5 = Very valuable/innovative
3 = Almost as the competition	

The equation for competitive advantage is then defined as

$$\text{Competitive Advantage} = \frac{\text{Price} + \text{Value}}{2} \tag{1}$$

(b) Business Status

Business status describes how well-established or well-known a business is at the time of introducing a new product. Finding a numeric value for business status is challenging as the business reputation may vary from country to country. To get a consistent measurement across countries, the net revenue is considered. First, the net revenue of all companies in that industry is listed from lowest to highest and then the company's ranking is considered accordingly. So for example if in an industry there are companies A, B, C, D, E and F, then the net revenues of these companies are found and ranked from lowest to highest such as:

1. Company B (lowest)	4. Company C
2. Company D	5. Company F
3. Company E	6. Company A (highest)

The measurement of company E equals 3/6 and that for company A equals 6/6, so company A has a higher business status. This equation will result in a number between 0 and 1. Thus the general equation for a business status would be:

$$\text{Business Status} = \frac{\text{Company rank}}{\text{Total number of companies in the market}} \tag{2}$$

(c) Marketing

Research into the history of marketing and its effectiveness has shown that it is one of the most important factors of successful sales, especially when it is well designed to suit the target market. With the right message, audience, and reach, marketing can in fact change certain beliefs and even challenge social taboos to increase sales. This is illustrated with an example when Barnay's was able to make women smoke in the 1930s campaign "Torches of Freedom".

The challenge in using marketing as a factor for predicting is that the analysis is done only after the sales figures are available, and pending release of these figures as they are usually private company data. In other words, the measurement can only be done after the disruption has occurred, which is useless for forecasting. For that reason, a new approach had to be found.

The marketing factor was divided based on how it can affect sales, into three parts: cost, reach and effectiveness. Cost is the budget allocated to the marketing campaign, given that generally, the higher the cost of marketing campaign, the higher the expected sales. Reach is the number of media used in marketing, assuming that the more media channels are used, the more exposure and thus the more expected sales. And finally, effectiveness is the measurement of how well the message of marketing is conveyed to the target market, and this is to be evaluated by a marketing analyst. The three parts of marketing were again given a Likert scale from 1 to 5 and considered equally important. However, future data may reveal that a better relationship could be found. In addition, incorporating social media into the marketing model can provide better feedback and could be used to further adjust it. The currently suggested marketing scales are presented in Table 1.

Table 1 Likert scale for cost, effectiveness and reach

Cost ($) [1]	**Value**
500,000 or less	1
500,000 – 5 Million	2
5 Million – 50 Million	3
50 Million – 300 Million	4
300 Million and above	5

Effectiveness	**Value**
Negative reactions (inappropriate for most people, ineffective, ill designed…)	1
Some negative reactions (inappropriate for certain markets/age groups/ religions/ gender… , ineffective, ill designed)	2
Neutral (other competitions, not very attractive, only if person knows the product)	3
Positive reactions (well thought of, attractive, considering culture)	4
Strong influence (correlated with politics, needs, emotions)	5

<table>
<tr><th>Reach</th><th>Value</th></tr>
<tr><td>Print advertising</td><td rowspan="6">For each of the methods of advertising used one point is added. So if print and broadcast and online are used, the value will be 3. The max number is 5</td></tr>
<tr><td>Outdoor advertising (street, booth)</td></tr>
<tr><td>Broadcast (television, radio)</td></tr>
<tr><td>Product Placement (in movies or shows)</td></tr>
<tr><td>Cellphone and Mobile</td></tr>
<tr><td>Online advertising</td></tr>
</table>

[1]The ranges are only a suggestion based on general knowledge of marketing but they are subject to revision according to the type of business, products or market

The equation for marketing would then be:

$$Marketing = \frac{Cost + Reach + Effectiveness}{3} \quad (3)$$

(d) Reactions

The reactions factor gives a numerical value to how good or bad the customers' and reviewers' reactions were to learning about a product or buying and using it. People's reactions may differ based on their beliefs, needs, political agenda or the economy. But also it can differ because of reviews. The abundant availability of online reviews given by experts and users on almost anything sold has made it possible for potential customers to be influenced on future sales by many different sources. Both the reactions of people and the reactions of reviewers must then be considered. In order to validate this relationship, a survey was done online where 52 random people from around the world were asked to read and watch videos about an upcoming product from Google called Google Glass. The product was still in its beta version and the company has tried to create the media hype around it before the selling date. After learning about the product the people were asked whether they would buy it or not. Based on their answer, a review was shown to them to contradict their wishes and they were asked again whether they changed their minds about buying it or not. The survey showed that 100 % of them sought reviews before buying most technological products, 37 % of those who said they would buy it changed their minds after reading a negative review and only 9 % of those who said they wouldn't buy it changed their minds after reading a positive review. This seems to indicate that reviews are in fact an important part of purchases and that negative reviews may have a higher impact on sales than positive reviews. These findings are backed up by another study done on the effect of reviews on sales [10]. Given this, it was possible to determine the elements that go into the reactions equation:

1. Number of unique mentions of a product, those that are done by new authors every day.
2. The sentiment of the customers' mentions, which is the ratio of positive to negative mentions.
3. Customer engagement, this is the number of Facebook likes, Google +1 votes, subscriptions to YouTube channels, Twitter or Instagram followers and other social media channels that may be available.
4. Reviews sentiment which is the ratio of positive to negative reviews given on the product.

These numbers can be taken from automated engines that scan the web and present within certain dates the number of reactions and its sentiment. However, a research into this matter has showed that these engines tend to be less accurate than desirable [11]. Automated engines cannot always understand human language, including sarcasm and emotions. When it comes to languages other than English it becomes even less accurate. For example, the Arabic language includes 30 different dialects in addition to the classical Arabic language, making it difficult to analyze sentiment. A further level of complexity is added when Roman letters, rather than Arabic letters, are used to represent Arabic words on the Internet and using mobile phones. This means that it is not possible to depend on an engine to detect the sentiment. After consulting with experts in the area, it was decided that the sentiment ratio is to be found manually by picking up 500 random mentions of a product and split them into positive and negative.

To construct the final equation, the weighted average was selected as some factors have more impact on the sales than others. The weights were given based on the best knowledge of the researcher. The suggested weights were given as shown in Table 2.

Table 2 Weights given for reactions factor

Value (1-5)	Weight (out of 5)	Reason
Customers' Sentiment	1.0	This is the first indication of how positive is the reaction to a certain product and thus it was given a high weight
Unique Mentions Sentiment	1.5	This number indicates how many good mentions are made by new authors. It is more important than mentions sentiment as that can be purely repetitions of the unique mentions while this indicates further reach to more customers.
Customer Engagement	0.5	While this value is indicative of good or bad engagement it is not very accurate as many social media followers do not tend to follow because they like the brand or product but sometimes they are news agencies, competitors, even people who dislike the company but want to see their news. Thus it was given the less weight.
Reviews' Sentiment	2	From the survey conducted and from research it is found that reviews have a bigger impact on customer reaction thus it was given the highest weight

Therefore the final equation for reactions is:

$$Reactions = \frac{(Customer\ Sent \times 1.0) + (Unique\ Sentiment \times 1.5) + (Engagement \times 0.5) + (Reviews\ Sent \times 2)}{5} \quad (4)$$

(e) The Model

With the four factors described above adopted, the forecasting model can be constructed. In order to create a regression test, one more definitions must be clarified, namely, how to calculate disruptiveness. Since disruptiveness is the actual disturbance in the market when a new product is introduced, the disruptiveness then can be defined as the percentage market share of the product. The forecasting question becomes: what is the expected market share the new product will have in the upcoming weeks or years? Since market share can be calculated based on unit sales or total revenues, the unit sales was considered for this model. This decision was made because revenues can have inaccurate results if a product was sold at a very high price and very few units were sold. Given that, disruptiveness is then calculated by:

$$Disruptiveness(market\ share) = \frac{Total\ Product\ Unit\ Sales}{Total\ Industry\ Unit\ Sales} \times 100\% \quad (5)$$

Using the Eqs. (1) through (5) presented above, data was collected to establish the model. The aim was to collect at least five cases to fit into each one of four categories: successful disruptiveness by a small business, failed products by a small business, successful disruptiveness by established business and failed products by established businesses. Samples for testing included past and forecasting data. There was a challenge finding this data because marketing and sales information is not readily available. For market share figures, it was not clear whether this was based on revenue or sales. For the reactions factor, it is impossible to find at this late stage, also, social media was not widely used before 2005. With these limitations it was still possible to collect and estimate up to 12 cases and run the regression test on them (Fig. 1).

Regression 1: All variables

Regression Statistics	
Multiple R	0.79090386
R Square	0.62552891
Adjusted R Square	0.41154543
Standard Error	0.1466643
Observations	12

ANOVA

	df	*SS*	*MS*	*F*	*Significance F*
Regression	4	0.251521995	0.0628805	2.92325797	0.102486431
Residual	7	0.150572921	0.02151042		
Total	11	0.402094917			

	Coefficients	*Standard Error*	*t Stat*	*P-value*	*Lower 95%*	*Upper 95%*	*Lower 95.0%*	*Upper 95.0%*
Intercept	-0.31257422	0.242508952	-1.28891826	0.23838427	-0.88601677	0.26086833	-0.88601677	0.26086833
Competitive Advantage	0.04811445	0.080298738	0.59919308	0.56792202	-0.14176189	0.23799079	-0.14176189	0.23799079
Business Status	0.06500416	0.146062108	0.44504466	0.66972392	-0.28037784	0.41038617	-0.28037784	0.41038617
Marketing	0.06867625	0.054365597	1.26323002	0.24695274	-0.05987795	0.19723046	-0.05987795	0.19723046
Reactions	0.05520525	0.074268005	0.74332476	0.48148282	-0.12041068	0.23082117	-0.12041068	0.23082117

Fig. 1 Results of the regression test [1]

This regression has resulted in a Significance F of 0.1 but P-Values higher than 0.2 for all the four factors and a sum of errors of 0.07. Thus the model extracted will not be used for forecasting nor disregarded at this stage; it will only be used for demonstration. The model extracted is given as:

$$\begin{aligned} &\textit{Expected Disruption} = -0.31 + 0.048 \times \textit{Competitive Advantage} + \\ &0.065 \times \textit{Business Status} + 0.068 \times \textit{Marketing} + 0.055 \times \textit{Reactions} \end{aligned} \tag{6}$$

3.2 The Judgmental Method

In cases of high uncertainty such as disruptive innovations, there will be instances where human insight is required beyond the available data and numbers, therefore a judgmental process is required to further validate the findings of the mathematical model. After the model has been used and a result is found, five experts from each country where the product will be introduced will be consulted along with a survey to fill within 48 h. The experts will have access to the data, but not the equations nor their results; this is in order to minimize the risk of bias or influence by presumed results. The criterion and conditions for choosing such experts and further details on the process are discussed further in the dissertation paper [1].

3.3 The Complete Framework Process

The complete forecasting process is shown in Fig. 2. It is a merely four step process which starts with the data collection step. A selected team will collect the information needed from online resources with given guidelines [1]. The data is then used in the mathematical model to find a possible disruption percentage in step 2. Simultaneously, the experts' opinions are sought through the survey. The final step brings together the results to the head forecaster for a final validation and consolidation of the report. The complete forecasting process can take a period of two weeks and up to two months depending on the nature of the forecasting request. A long term forecast targets an upcoming product and is done typically on demand

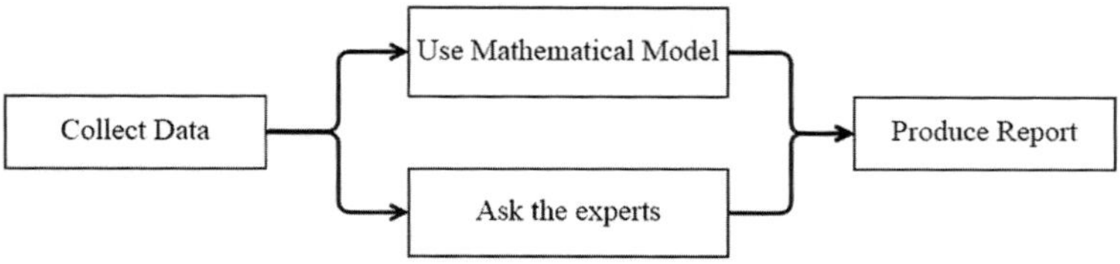

Fig. 2 The forecasting framework

while a complete disruption forecast targets a new technology that may overtake the existing one and is done twice a year. Another version of the forecasting process, the short term forecast, discussed further in [1], is done in 1 day or two where two forecasters are requested to estimate the four factors and use the mathematical model to produce an executive forecast.

4 Conclusion and Discussion

Disruptive innovations create a sudden unexpected disruption in the market causing losses to established businesses. The best way to avoid such losses is by being prepared and that is possible with forecasting. The paper has described work that was the result of a study of current forecasting methods and reported on principles, strengths and most importantly weaknesses of these methods. None was found to specifically forecast disruption. A new framework was suggested using two methods, mathematical models and the judgmental method. The metric that measures disruptiveness to sales in a market is based on four quantifiable factors. It is important to note that the Likert scales given in this paper are only a suggestion based on research of popular products. The forecaster using this model is advised to adjust the scales given for each variable of the equation according to the market/product, while keeping the range between 1 and 5.

The mathematical model is likely to benefit from further validation using "future data" which will indicate whether an adjustment to the calculations provided are needed to improve the accuracy of the results. Such data however, may be challenging to find even in the future as there is no dedicated and trusted body to refer to, there are several resources and perhaps bias involved. This poses a limitation that requires further planning and studying. However, it is our position that the proposed framework covers most of the features the committee [6] has suggested as a persistent forecasting framework. Future work plans also include automation of parts of the process and a continuous refining of the model to achieve highest accuracy possible.

References

1. Diab, S.: Disruptive Innovations and Forecasting. A case study for a dedicated forecasting framework. Unpublished dissertation, Heriot Watt University (2014)
2. Bower, J.L., Christensen, C.M.: Disruptive Technologies: Catching the Wave. In: Press, H.B. S. (ed.) Harvard Business Review on Business Model Innovation, illustrated edn, p. 207. Harvard Business Press, Massachusetts(Boston) (1995)
3. Christensen, C., Overdorf, M.: Meeting the Challenge of Disruptive Change. Harvard Business Review (2000)

4. Vojak, B.A., Chambers, F.A.: Roadmapping disruptive technical threats and opportunities in complex, technology-based subsystems: the SAILS methodology. Technol. Forecast. Soc. Chang. **71**(1–2), 121–139 (2004)
5. Rouse, M.: disruptive technology. http://whatis.techtarget.com/definition/disruptive-technology. Accessed 20 March 2014
6. Committee on Forecasting Future Disruptive Technologies: Persistent Forecasting of Disruptive Technologies Committee on Forecasting Future Disruptive Technologies. The National Academies, Washington (2009)
7. TechCast, N.d.: Accuracy. http://www.techcastglobal.com/accuracy. Accessed 01 March 2014
8. Vanston, J.H.: Better forecasts, better plans, better results. Res. Technol. Manage. **46**(1), 47–58 (2003)
9. The Analysis Factor.: The distribution of independent variables in regression models. http://www.theanalysisfactor.com/the-distribution-of-independent-variables-in-regression-models-2/. Accessed 10 Dec 2014
10. Marketing Land: Survey: 90 % of customers say buying decisions are influenced by online reviews. http://marketingland.com/survey-customers-more-frustrated-by-how-long-it-takes-to-resolve-a-customer-service-issue-than-the-resolution-38756. Accessed 10 Dec 2014
11. GIGAOM: Stanford researchers to open source model they say has nailed sentiment analysis. https://gigaom.com/2013/10/03/stanford-researchers-to-open-source-model-they-ssay-has-nailed-sentiment-analysis/. Accessed 24 Nov 2014

Enabling Data Subjects to Remain Data Owners

Eliza Papadopoulou, Alex Stobart, Nick K. Taylor and M. Howard Williams

Abstract Users have become used to accepting two unfortunate consequences of complying with requests to supply personal data to service providers. Firstly, the personal data that a user supplies becomes the property of the service provider, which means that the data subject loses control over what is subsequently done with their data. Secondly, provision of services is made on an "all or nothing" basis, being dependent upon the user supplying all the personal data requested by a service or forgoing use of that service entirely. We present an approach to personal data management which avoids these two unnecessary disadvantages. Personal Data Stores enable individuals to retain ownership and control of their personal data, granting service providers access to specific items of that data upon request whilst remaining the owners of their data. Trusted third parties will be required to curate the data in order to ensure that it is non-repudiatable. Privacy Policy Negotiation will enable data subjects to negotiate with service providers about how much of their personal data they disclose and how detailed that data is. Different levels of service can be provided depending on what personal data a user is prepared to disclose. In this paper we describe systems and algorithms for Personal Data Stores and Privacy Policy Negotiation which have been implemented and tested separately and show how they can be combined to the benefit of data subjects.

E. Papadopoulou · N.K. Taylor (✉) · M.H. Williams
School of Mathematical and Computer Sciences, Heriot-Watt University, Edinburgh EH14 4AS, UK
e-mail: N.K.Taylor@hw.ac.uk

E. Papadopoulou
e-mail: E.Papadopoulou@hw.ac.uk

M.H. Williams
e-mail: M.H.Williams@hw.ac.uk

A. Stobart
Mydex CIC, Blue Square House, 272 Bath Street, Glasgow G2 4JR, UK
e-mail: alex@mydex.org

G. Jezic et al. (eds.), *Agent and Multi-Agent Systems: Technologies and Applications*, Smart Innovation, Systems and Technologies 38,
DOI 10.1007/978-3-319-19728-9_20

1 Introduction

Users of computer-based services are being subjected to increasing pressures to disclose personal data. These pressures take many forms, from voluntary disclosures on social media to obligatory legal disclosures required by government agencies and required disclosures which users are expected to make in order to avail themselves of proprietary services. Some of the resulting disclosures can have unintended consequences which the typical user cannot be expected to foresee, such as the transfer of their data to third parties and fourth parties, etc. Furthermore, as the number of these disclosures grows and the interoperability of services improves it will become increasingly difficult for users to keep track of, and manage, their disclosures. Laudable developments, such as greater service interoperability, can thus mitigate against an individual–centric approach to service consumption and have the potential to undermine initiatives to improve citizen empowerment such as that advocated in the European Union's Digital Agenda for Europe [1].

How have we arrived at this apparently paradoxical situation, where developments such as interoperability which should be good for service consumers could potentially be perceived as bad and, in the worst case scenario, shunned by them? The traditional service model is one where the user is expected to disclose all the information requested by a service or to forgo use of that service altogether. The choice for the consumer was a simple binary one of all or nothing. If the consumer elected to trust the service and disclose all the information requested then, again historically, the likelihood of that service provider passing their data on to others was minimal and covered by data protection legislation. In a world where most services were stand-alone and the service providers that most consumers interacted with were few in number, this was not an unreasonable model and it proved to be very effective.

However, with the advent of services which made use of other services, delivered by other providers, the traditional model began to unravel. Personal data started to be passed from one service, or provider, to another without the consumer's knowledge. Privacy policies appeared to inform consumers of what might be done with their data but they are typically so long and unfathomable that the majority of consumers do not read them. The one thing we can be sure that privacy policies have achieved is legitimising the passing of personal data from one service, or provider, to another by ostensibly obtaining the service consumer's consent. The myriad uses to which personal data could be put and the inferences that could be drawn from mining it, not least in targeted advertising, made it inevitable that a market in personal data would evolve.

Now it has been recognised that personal data has a real tangible economic value its ownership has come under scrutiny and there is a growing acceptance that the rightful owner of personal data is the subject of that data.

Yahoo's Marissa Mayer said recently that the personalised Internet "is a better Internet," emphasising: "We don't sell your personal data … We don't transfer your personal data to third parties." [1] She said users own their data and need to have control, adding that people give up data to the government for tax assessment, social services and other purposes.

The 2011 World Economic Forum (WEF) report on Personal Data [2] stated that personal data is an economic asset class. It needs to be balanced between the needs and demands of the individual, government and private enterprise. The WEF Global IT Report 2012, "Living in a Hyperconnected World" [3], makes numerous references to the risks inherent in the flow of personal data to individual rights, privacy and cybercrime. WEF, W3C and many others have identified that a personal data ecosystem is emerging in which personal data is becoming a tangible economic asset which rightfully belongs to the subject of that data. The challenge is to devise real workable, convenient, trusted and secure systems that embody these principles.

We argue that a key pillar of individual-centricity and citizen empowerment is returning to users control over their personal data. The tools and infrastructure necessary to achieve this are now available and we present a combination of two such methods which which can readily be deployed across the Internet without the need for any architectural changes to it. In Sect. 2 we describe the first of these tools, the Personal Data Store of Mydex CIC and in Sect. 3 we present a methodology developed at Heriot-Watt University for privacy policy negotiation. In Sect. 4 we discuss some issues relating to data validation and ensuring non-repudiation of data that arise when personal data remains under the control of the consumer. We conclude and present a roadmap for the integration of the two tools in Sect. 5.

2 Personal Data Stores

Mydex Community Interest Company (CIC) works with individuals and organisations to enable control over how personal data is used and shared [5]. This is achieved via a Personal Data Store and a set of tools to manage identity and the consent process for data sharing online as depicted in Fig. 1 [6].

A Personal Data Store is a repository for an individual's data, not unlike the Data Box described in [7]. It is under the complete control of the individual whose date it is and is disclosed under the sole authorisation of that individual. Specific data items can be selectively released to different services. Authorisation can also be given to service providers to share particular data items with other service providers.

Mydex CIC was one of the companies originally selected by the UK Department for Work and Pensions (DWP) for ID Assurance, to provide the service now developed and operated through GDS and known as GOV.UK Verify. DWP recently consulted about new regulations for data sharing in the context of Universal Credit. The consultation seeks to find innovative ways to share personal information between DWP and local support providers, such as local authorities and housing associations. The aim is to enable these organisations to provide the

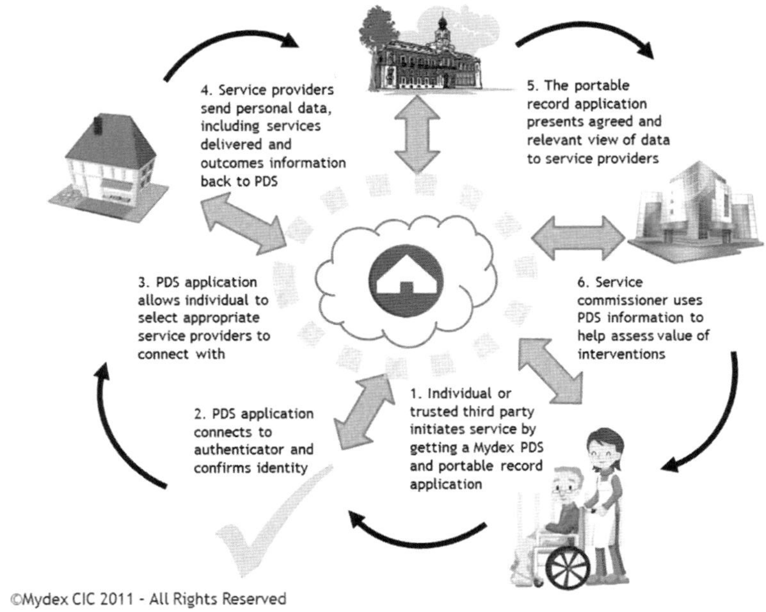

Fig. 1 The Mydex CIC Personal Data Store and its consent process

best support for individuals claiming Universal Credit, the UK Government's new single monthly payment system for those receiving benefits and credits.

DWP is in an influential position to empower UK citizens, and to benefit hugely itself, from starting to implement a more person-centric approach to data sharing. Putting individuals in control of the data sharing process, instead of always having organisations share ever more data about them with each other, offers significant cost, efficiency and regulatory advantages. It will also help individuals acquire and prove trust, learn about, understand, and take a more active part in the digital economy.

There is a further dimension of relevance to DWP's data sharing policy. As well as providing a more elegant solution to the data-sharing challenge, connecting to Personal Data Stores means that DWP clients, notably including those with only a small amount of data about them or "thin file" clients, can then secure the required level of assurance under GOV.UK Verify for access to a wide range of digital public services.

This is much more effectively achieved when "thin file" clients are able to reuse verified evidence provided to them about their existing relationships with government departments and other organisations. This can be achieved with no loss of privacy, and without the need for changes in legislation. The data is delivered directly to the citizen, and placed under their secure control in a verified and protected format. It gives the client what is effectively a digital "proof bank"; of externally verified claims.

The same principles apply to data sharing in many other contexts: health; social care; tax; education and many other non-public services. This is an alternative person-centric approach to the current organisation-centric approach to sharing verified personal information with government agencies and service providers.

Verified data needs to have three distinct properties for it to be useable in any Identity Assurance activity –

- Certainty of issuing authority - verified organisation and issuing endpoint
- Clarity about the process of issuance or generation - this can be any number of things but it has to be understood by the relying parties' processes
- Certainty of accuracy – verification that data has not been modified since being issued

The Mydex CIC Trust Framework and platform make all of these possible with the use of cryptographic solutions to create a seamless chain of trust and a secure end to end process.

3 Privacy Negotiation and Flexible Levels of Service

The Personal Data Store offers a mechanism for service consumers to curate their own personal data. If a data subject only needs to deal with a small number of simple transactions then it is quite feasible for them to remain in direct control of their disclosures, perhaps manually selecting the items they wish to disclose on each and every service usage. However, even with simple transactions, there remains a need for the consumer to attach conditions to their personal data disclosures, such as the duration for which the data might be held by a service provider or to whom that provider may pass the data on. Furthermore, if the transactions require disclosure of significant amounts of data or if the data subject indulges in a large number of transactions with different services then this manual approach to approval is unlikely to suffice, requiring the data subject to approve an unmanageably large number of disclosures or to approve disclosures at each stage in a service composition or other form of interoperation. In order to assist the data subject with these issues we present the dual notions of a privacy policy and a privacy policy negotiation process [8].

A privacy policy is a document that describes how a service collects, stores, uses and disseminates the personal data of its users. More specifically, it states (a) what data are requested; (b) the purpose for which this data is requested; (c) what type of processing will be applied to this data; (d) with whom this data will be shared and (e) for how long this data will be retained. It may also include other statements about the rights of the user as well as itself with regard to the data. With the use of privacy policies, companies and more specifically services, inform their users about what happens to the users' personal data after disclosure.

Privacy Policy Negotiation is the process by which a user negotiates the terms and conditions of the privacy policy with a service. It is a solution to the "take it or leave it" approach that is currently being implemented by millions of services worldwide. Negotiating for privacy means freedom for the user to adjust their privacy as they wish, rather than fitting the preferences of the service provider.

The privacy policy negotiation protocol defines two negotiating entities, a negotiation agent which performs the negotiation on behalf of the service provider and a negotiation client that operates on behalf of the user that wishes to use the service.

As shown in Fig. 2, the negotiation client is tasked to begin the privacy policy negotiation with the service by retrieving anonymously the privacy policy - also termed Request Policy. The service provider defines the terms and conditions for each data item that will be requested from the user, allowing different options to be defined for different data. Each data item or resource (name, age, email, contact book, location etc.) specified in the privacy policy document is defined in a Request Item element in that document. Each Request Item element defines (a) the data item it refers to, (b) the types of actions it will perform on the data item e.g. create, read, update and delete, (c) the conditions for collecting, storing and distributing the data and d) if the data item is optional.

The Privacy Policy Negotiation process involves four steps –

(a) The Negotiation Client retrieves the Request Policy document from the service provider. The Request Policy of the service provider expresses the optimal set of requirements that fit the service provider.
(b) The user is asked to configure the Request Policy with their preferred terms and conditions. A Response Policy document is generated that contains the responses of the user and is sent to the Negotiation Agent for processing.
(c) The Negotiation Agent uses a predefined set of options that indicate what options it can satisfy. It cross matches the original Request Policy with the user's Response Policy and creates a final Response Policy that includes the user's requests that it is able to satisfy and suggests alternatives for the ones that it cannot satisfy. The final Response Policy is returned to the user.
(d) If the final Response Policy contains changes made by the Negotiation Agent, the Negotiation Client presents the changes to the user and allows them to abort the negotiation or continue by accepting the service provider's alternative suggestions. If the user chooses to accept the terms, the Negotiation Client inserts the user's identity in the Response Policy, signs it digitally and returns it to the service provider. If the user chooses to abort, the negotiation halts.

When a privacy negotiation succeeds, the Negotiation Client instructs that access control rules be created reflecting the terms and conditions agreed during the privacy negotiation. The user can make use of the service at this point and the service itself will request the actual data from the user using the Personal Data Store API.

In an environment in which multiple services are combined to offer a composed service, the user would have to negotiate with all of the services separately which would be confusing to the user and would very likely drive users to avoid using such services because of the additional hassle. To avoid this, the component responsible for selecting which services to include in the service composition uses

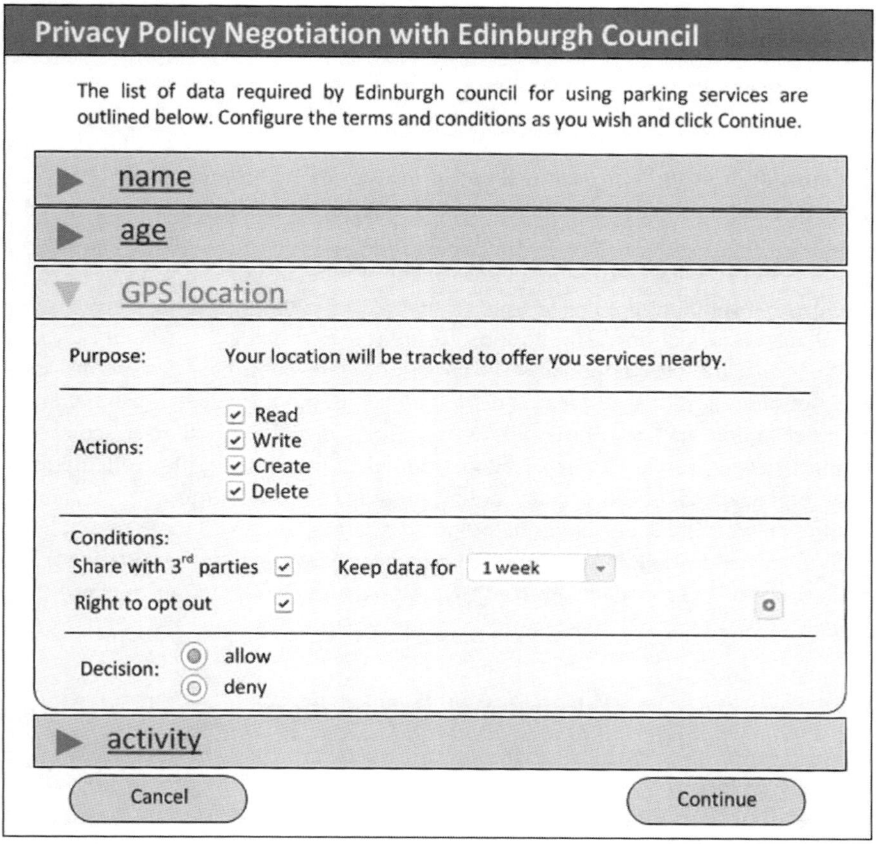

Fig. 2 Example of privacy policy negotiation with Edinburgh Council for using a "parking service". The privacy policy of Edinburgh Council indicates that it needs the name, age, GPS Location and activity of the user

the privacy policies and the predefined set of options of each service as another criterion for service selection. It collects the privacy policies and the predefined set of options from the available services and compares all privacy statements. For each privacy statement, it must find the option that can be satisfied by all services. This produces (a) a single privacy policy that can be used as a starting point to the negotiation with the user and (b) a modified set of options that can be satisfied by the services that will be needed during the privacy policy negotiation with the user. The Negotiation Agent of one of the services is elected to act as a delegate between the Negotiation Agents of the services included in the service composition and the Negotiation Client running on behalf of the user. This means that the delegated Negotiation Agent is authorised to perform a negotiation process without input from the Negotiation Agents of the other services. During the negotiation process, the elected Negotiation Agent has at its disposal the modified set of options that it can use to negotiate with the Negotiation Client. In a successful negotiation, the delegated Negotiation Agent must inform the services in the composition of the result of the negotiation sending them the agreed Response Policy which they will have to adhere to when they receive the user's data.

The option for a service consumer to decide how much information they wish to disclose opens up the possibility of services providing flexible levels of service. For example, a service offering recommendations for nearby restaurants will be able to provide more tailored recommendations if the service consumer is prepared to disclose their location more precisely; disclosing location at the level of a city would result in some recommendations being a long way away from the user but disclosing location at the level of a particular street could provide a more localised set of recommendations. The notion of varying service levels to flexibly accommodate what the user is prepared to provide as input will have a very disruptive effect on the traditional service model which is based on a binary consumer choice of either telling the service everything it requests or not using the service at all.

Furthermore, Privacy Policy Negotiation, when combined with explicitly stated or automatically deduced privacy policy preferences can also resolve the age-old conflict between personalisation of services (which increases as more personal data is divulged) and privacy (which decreases as more personal data is divulged). The privacy preferences of data subjects enable privacy itself to be personalised [9].

4 Provenance, Validation and Repudiation

Data provenance is recognised as an important factor in establishing trust in services [10]. The validity and reliability of all data sources that contribute to decision-making processes undertaken by services has been the subject of the PROV Working Group of W3C [11]. When a data subject is given the primary role in maintaining their own data further issues of provenance arise. Trusted third parties will be required to provide assurance regarding the provenance of the data subject's data. Two key aspects of data provenance which need to be addressed are validating the data and ensuring that it is non-repudiatable.

Data validation might require access to national databases to confirm such things as date of birth, passport and driving licence numbers, etc. It is critical that such data is stamped in some way as officially verified and that the data subject is not able to corrupt that stamp. Services wishing to make use of this data can then be assured that it has not been tampered with.

Data can also change over time and it is important that data provided at one point in time, and which might have formed part of a contract on a given date, is not overwritten by an update which might change the nature of such a contract had it been entered into at a later date. For this reason data items will need to be incontrovertibly time-stamped so that, for instance, somebody getting married cannot repudiate the single status they had at an earlier point in time. Assured versioning will be essential in order to maintain the provenance of data that can change over time.

5 Conclusion

Personal data has now acquired real economic value and it is important that the ownership of that data is returned to the individuals who are the subject of that data. We have both the infrastructure and the tools to achieve this. We have presented one such solution; the Personal Data Store with Privacy Policy Negotiation can be provided via trusted third parties as cloud services and so be available to data subjects at any time.

The personalisation *of* privacy based on privacy preferences that we propose addresses the age-old dichotomy of personalisation *versus* privacy, by enabling data subjects to selectively disclose their personal information to different service providers depending on their context and how much they trust those providers.

In addition, our solution creates the possibility of flexibly varying service levels depending on the information which the data subject is prepared to disclose and this will move service provision on from the traditional binary choice offered to service consumers between providing all the information requested or not using a service at all.

Future work between Mydex CIC and Heriot-Watt University will develop an integrated solution based on these two concepts. Data provenance, validation and non-repudiation remain challenges but the tools to address them exist and we expect further collaborative work to identify efficacious and efficient solutions to these issues.

References

1. de Cockborne, J.-E.: Report on an individual-centric digital agenda for Europe, JEC 130222 (2013)
2. Szalai, G.: Google chairman Eric Schmidt: "The Internet Will Disappear", The holywood reporter. http://www.hollywoodreporter.com/news/google-chairman-eric-schmidt-internet-765989 (2015)
3. World Economic Forum: Personal data: the emergence of a new asset class. World Economic Forum in collaboration with Bain & Company, Inc. (2011)
4. Dutta, S., Bilbao-Osorio, B.: The global information technology report 2012: living in a hyperconnected world. World Economic Forum and INSEAD (2012)
5. Hill, T., Alexander, D.: The third sector and the future of stakeholder engagement in challenging times - a Mydex white paper. Mydex Data Services CIC (2012)
6. Stobart, A.: Scottish government briefing – personal data stores as "Enablers of Reform". Mydex Data Services CIC (2011)
7. Haddadi, H., Howard, H., Chaudhry, A.,Crowcroft, J., Madhavapeddy, A., Mortier, R.: Personal data: thinking inside the box. http://de.arxiv.org/pdf/1501.04737 (2015)
8. Papadopoulou, E., McBurney, S.M., Taylor, N.K., Williams, M.H., Abu Shaaban, Y.: User preferences to support privacy policy handling in pervasive/ubiquitous systems. Int. J. Adv. Secur. **2**(1), 62–71 (2009)

9. Taylor, N.K., Papadopoulou, E., Gallacher, S.M., Williams, M.H.: Is there really a conflict between privacy and personalisation?, Keynote address at ISD 2011. In: Pooley R.J., Coady J., Linger H., Barry C., Lang M., Schneider, C. (eds.) Proceedings of the 20th International Conference on Information Systems Development: ISD 2011, Edinburgh UK. Information Systems Development : Reflections, Challenges and New Directions, pp. 1–9. Springer, Heidelberg, Germany. ISBN 978-1-4614-4951-5 (2013)
10. Townend, P., Webster, D., Venters, C.C., Dimitrova, V., Djemame, K., Lau, L., Xu, J., Fores, S., Viduto, V., Dibsdale, C., Taylor, N., Austin, J., McAvoy, J., Hobson, S.: Personalised provenance reasoning models and risk assessment in business systems: a case study. In: 7th IEEE International Symposium on Service-Oriented System Engineering (2013)
11. W3C: PROV-overview - W3C working group note 30. http://www.w3.org/TR/prov-overview (2013)

A Survey of Business Models in eCommerce

Urszula Doloto and Yun-Heh Chen-Burger

Abstract The last decade of technology developments have permanently changed the way how businesses are operated. Companies are forced to become visible online and stay connected. They recognise adapting to the global dynamic business landscape and responding to customers' demands as key drivers to success. Since technologies have been identified in many studies as one of the most important enabling components for successful innovative business models, it is vital to understand their roles in constructing such models. In particular, we are interested in investigating how business models of eCommerce are enabled by technology innovations. We conduct the survey based on secondary research results. To structure our findings, we developed an evaluation matrix to summarise technologies involved and their contributions to businesses. We also identified gaps in the current research and proposed an extended version of business model classification framework for eCommerce.

Keywords Business model · e-Commerce · Technology innovation · Business model classification

1 Introduction

To understand the relationships between business models and technologies, it is useful to examine how business models may play a role in capturing value (profit creation) for businesses from technology deployment. A business model may be understood as a consistent framework that leverages technologies in their value propositions (product and services offerings), equipping and positioning in the value chain and network, thereby creating profits for the businesses in the long run [4].

U. Doloto (✉) · Y.-H. Chen-Burger
School of Mathematical and Computer Sciences, Heriot Watt University,
Edinburgh, EH14 4AS UK, Europe
e-mail: y.j.chenburger@hw.ac.uk

G. Jezic et al. (eds.), *Agent and Multi-Agent Systems: Technologies and Applications*, Smart Innovation, Systems and Technologies 38,
DOI 10.1007/978-3-319-19728-9_21

Companies seek to adapt the right business models to exploit the full value of technologies by commercializing it [1, 4, 16, 18]. Chesbrough and Rosenbloom pointed out that if an existing business model of a company is not adapted to technologies the company wants to invest in, the company will not be managed effectively. In other words, when a business model is not aligned with new technologies, an organisation will not benefit fully from such innovations.

In the era of rapid technological change and globalisation, it is crucial for organisations to understand the concept of business models and how it contributes to value capture from technological investments [2, 6]. Companies need to identify which business models will suit best the proposed technology. Chesbrough stated that applying the same innovative technology to two different business models will yield two different outcomes [3]. Employing inappropriate business model means that technology will not deliver its estimated expected value. "It is probably true that a mediocre technology pursued within a great business model may be more valuable that a great technology exploited via a mediocre business model" [3].

Built upon this understanding, our next task is to identify existing business models currently enabled by technologies, in particularly, in the business area of eCommerce. Osterwalder [14] provided an earlier example of extensive discussion in this topic, where he used Timmers' eleven electronic business models as a basis to explain the different eCommerce business models [21]. He further provided a survey in business models and a classification framework which enables companies to explicitly describe their business models.

Challenges for capturing all different types of business models in a classification framework have been significantly increased, since the wide spread of Internet eCommerce. The current speed of technology development and deployment has resulted in globalisation, market domination and rapid emerging of even more new business models. ECommerce and eBusiness operations are also becoming more complex and inter-linked, that there may be more than one way to classify and describe these business models. Based on Timmers and Osterwalder confirmed eleven different models, our literature survey has produced a forty-six models classification and this list is not exhausted. In this paper, we report our initial findings in our endeavours, which will create an initial overview of modern business models in the field of eCommerce and eBusiness, with a special focus on those deployed (innovative) technologies.

In order to provide a consistent and comparative analysis of surveyed literature, we have devised an analysis and evaluation framework to generate a summarised report for each selected literature. In this framework, we were particularly interested in the use of new technologies and their contributions to the eCommerce business models. Example papers that reported on deploying innovative technologies in business models are, e.g. Value Alliance Model for supporting cooperation between enterprises, Automation Model for improving order allocation for eCommerce retailers, or negotiation Agent Model using both rule-based and case-based reasoning for eTourism industry. In total, we have included thirty-three papers in our survey. Although the term eCommerce is often used to refer to the business transactions conducted via the Web and eBusiness is referring to the back-end business operations, strategies, organisational goals and processes, etc. In this paper, we do not make such a fine distinction, but will refer to them both in the sense of businesses on the Web.

2 Background

Adapting a right business model is undoubtedly crucial for organisations nowadays. However, there is no one standard model that suits all. For the same business models could be successful for one company, but they will not provide the same results for another. As a result, business models have been a subject of research for both practitioners and academics during the last decade, which had yielded different classifications and also different definitions of what exactly a business model is.

The definition provided by Osterwalder describes a business model as “a conceptual tool that contains a set of elements and their relationships and allows expressing a company’s logic of earning money” [14]. The logic of earning money may be understood as how a company creates and provides value to its customers and business partners in order to generate profit [2]. However, recent technology developments have permanently changed the way how businesses are operated. Internet based operations not only overcome space and time boundaries. It has created totally new forms of companies - from ‘brick and mortar’ to solely virtual forms. When the Internet was opened in 1991 for commercial use, many organisations started trade online, this had a huge impact on business operation processes, particularly on the way how today’s data are collected, stored and used. One prominent evidence on how technologies change business environment and operations is cloud computing, which has become a popular technology as reported by More and Mukhopadhyay [13]. Internet enables companies to exchange information freely and instantly, the volume of data sent electronically grows rapidly not on a yearly basis, but on a daily basis. Based on the Minnesota Internet Traffic Studies, since 1990 the internet traffic of 1 TB/month in the US had grown to 3750 K TB/month in 2011. The growth rate is exponential at 3750 K folds in 21 years (MINTS) [12]. Global Internet showed a similar trend, CISCO reported a 0.002 PB/month for IP and Internet traffic in 1990 to 48,117 PB/month in 2011, a growth rate of 24 million folds (CISCO) [5].

Big organisations are increasingly facing problems with storing, analysing and protecting huge volumes of information. Modern companies seek to optimize information resources by making the most of new technologies. This resulted in cloud computing to be deployed very quickly by businesses and is now widely adopted across different sectors [15]. This raises questions and stimulates research studies in which business models should be adapted by companies in the era of Internet commerce - are existing business models still good enough for new forms of ‘brick and click’ businesses [2]?

Rapid changing business environments have also applied great pressure to organisations to come up with new business models and new strategies. Many of innovative business models have been incorporated and adopted across industries and sectors such as telecommunications, banking and content providers such as newspaper agents [8]. Choosing a right business model means greater business flexibility, efficiency and responding to the demand of customers, but what's more important, it means staying competitive in the era of globalisation. For this reason, it is essential to develop eBusiness models as abstractions of how today's businesses function, i.e. "eBusiness models must clearly describe what information should be shared and when and how to share information" [2]. However, traditionally a business model often assumes that certain customers will pay for products or services offered by a company: "a business model is a conceptual, rather than financial, model of business" [20]. As described by Teece, a business model illustrates the reasoning for organisational and financial structure. However, most of all, correctly adopted business models for new technologies focuses on value propositions.

Based on our research, we observed that technologies enable a range of new eBusiness models. On the other hand, technology in itself provides no guidelines for selecting a suitable model in commercial terms; such guidance to technology development can come from the definition of new models [21]. As a result, we attempt to make a contribution to the computer science community by investigating eBusiness models, thereby giving inspiration for creating innovative technology suitable for commercial development.

3 Our Approach

This research includes secondary electronic resources that are based on academic publications and company reports, as appropriate. An analysis framework was developed to consistently organise and compare the selected literature. Initially, we selected papers from prominent scholarly resources, inc. IEEE, ACM, CiteSeerX, and business management journals. We focused paper discussion in the fields of "business model", "business model innovation" and "digital business model". Subsequently, we selected papers that have the following properties, inc. papers reporting disruptive technologies, innovative business models creation and practices, use of innovative technologies, potential technologies to enable new business models and/or important operational changes, recent or well-cited papers, etc. Based on these selection criteria, we have narrowed down our selection to thirty-three papers. For each of the selected paper, an evaluation and analysis matrix is then applied. This matrix includes the following criteria: paper ID, citation reference, types of the paper, main concepts, business sector, business areas, used/proposed technologies, used/proposed mechanisms, weakness and strength (of the paper and proposed innovation), applicability, case study, evaluation results, whether new technologies and/or mechanism have been proposed and deployed in the real world, whether new business models have been proposed and deployed,

If so, has the proposed business model/technologies/mechanism being successfully deployed, amount of citation to date.

This framework asks one to identify what is the proposed innovation, where new innovation may be (or have been) usefully applied, its strength and weakness, and its relation to business models, so that readers may easier use this work as a guide for their own purposes. The papers were also analysed for case studies - whether the proposition has been based on a company or companies' data. Finally, the framework requires the evaluation of the proposed technology, inc. what were the results, and how it has been tested, whether it has been incorporated in commercial applications. Furthermore, if the proposed innovation has been compared with some other similar work in the field, then what were the results? The framework also attempts to assess how successful the proposition is, in a ranking from 1 to 5, where five is the highest mark, which means successful in linking technology with value creation. The last criterion indicates possible impacts that a paper has/may have by giving the number of citations. We focused more on recent papers published during the last four to ten years. Examples of business areas for investigation were: investment into adoption of e-business across different sectors, (innovative) value creations, business processes changes and strategies.

4 Business Models and Classification

Our first principle is to identify and include representative types of business models on the web - although this is an enormous task and is not possible to be complete when new business models are created continuously. The second principle is to identify business models that leverage technologies, especially where technologies have played a significant role to create and deliver values for businesses. Business' presence on a web page was not simply identified as business model. We looked into the business operations behind the scene, inc. the nature of a business, the position of a business when comparing with their peers, their positions in a value chain, market divisions, their business processes, relevant business partners, target markets, revenue streams, business strategies, trading methods, products and services types offered, etc.

In Osterwalder's thesis, "The business model ontology - a proposition in a design approach", he proposed a relatively comprehensive business model classification framework which enables companies to explicitly describe their business models [14]. This earlier work had attracted a great deal of attention from scholars. It had been cited 1162 times since published eleven years ago. As a result, Osterwalder's above publication was selected for more detailed analysis. We found his review of business models taxonomies of eleven categories by Timmers' along with his own discussion and framework very promising. Nevertheless, this classification has presented a significant gap today. Indeed, advancements in technologies, such as search engine, Web2.0 and the Semantic Web has re-shaped and changed eCommerce and eBusiness significantly. As a result, there are many new emerged business models which have not been captured in existing classifications. Taking into account the focus of this study, we have decided to base on Timmers' classification and Osterwalder's discussion on it as a foundation for expansion; in particular, as this classification is closely related to technology engagements, i.e. some models would only exist, because of innovation technologies.

In Timmers' study 'Business Models for Electronic Markets', he identified eleven different business models, which existed on the Internet. In order to create this classification, Timmers developed a framework which led to the qualitative mapping of these eleven Internet models. His systematic approach was based on two activities: de-construction and re-construction of a company's value chain, which resulted in investigating models along two dimensions: degree of innovation and degree of functional integration.

Following Timmer's classification approach, we have adopted a similar approach to compare with other approaches, e.g. work of Rappa [17] and Tapscott et al. [19], to create an extended classification. Based on web based company presence, we have created a new extended classification that consists of forty-six business model types. For each type of the business model in the classification, we provide an attempted description as well as example businesses that can be found on the web for this type. This list of classification is by no means exhaustive and further work is needed.

For each business model, we also attempt to allocate a Degree of Innovation on a scale of five. The lowest degree of one indicates that a business model merely mimics or supports the original business model exists outside of the web, e.g. its primary purpose is to provide a web based presence. In such cases, the main function of its web counterpart is to communicate and promote a company brand. The highest degree of five indicates a highly innovative eBusiness model that does not previously exist before its web form – and may not exist outside of web. Other lesser degrees of the innovation indicate when an original business model does exist outside of the web form, but the web version of the business is significantly different from the original business model. As similar types of business models may be grouped together, we have also provided initial grouping that does not previously exist in Timmers' work. Below gives a brief description of our business model classification:

4.1 Categories of Business Models on the Web

(1) E-shop: business promotes own brand by selling its goods/services online: www.sony.co.uk; www.ukecigstore.com
(2) E-supermarket: business primarily sells groceries from others, but may also sell own brands: waiyeehong.com; asda.com; tesco.com; ocado.com
(3) E-mall: a collection of e-shops, different brands sell products under one name (the mall): emall.me; emall.karangkraf.com; Amazon.com; johnlewis.com; houseoffraser.co.uk; groupon.com

(4) E-wholesaler: serves B2B market, buyer must have membership to purchase goods: Costco.com; mxwholesale.co.uk; davidssales.co.uk; hktdc.com
(5) E-hubs: brings together a large number of buyers and sellers under one virtual roof to assist information sharing and/or trading, typically serves B2B markets: e-steel.com.sg; PlasticsNet.com; TradeOut.com; W3C.com; o.info
(6) 3rd party marketing and sales channel: a company provides online marketplaces for other businesses: eBay.com; amazon.com; gumtree.com
(7) E-auction: virtual auctions facilitating buyers and sellers regardless of their locations: eBay.com; webstore.com; onlineauction.com;
(8) E-compare: compares services across different vendors, revenue is often generated via web marketing, vendor subscription fees and referral fees: Uswitch.com; GoCompare.com; MoneySupermarket.com; Compare TheMarket.com
(9) E-Supply Chain Management: integrates e-Procurement, e-Billing and/or other e-Business tools to increase the efficiency of logistic, distribution and production: gtnexus.com; quyntess.com; ciltuk.org.uk (see members for more info)
(10) E-Logistics: third party operators that manage the movement of products: dhl.com; logixperience.com; ups.com
(11) E-procurement: connects companies or itself with suppliers while managing interactions between them: owens-minor.com; Department of Defence US; procurement.bristol.gov.uk/supplierselfservice; dwp.bravosolution.co.uk
(12) Value chain integrators: integrate multiple steps of a value chain, with the potential to exploit information flow between those steps as further added value: ictsolutions.co.uk; telkom.co.za; winshuttle.com/solutions-sap/by-function-task/sap-excel
(13) Value chain providers: specialize on a specific function for the value chain, e.g. electronic payment or logistics: paypal.com; www.dpd.co.uk; saildatabank.com;
(14) Information brokerage: provides comprehensive up-to-date news coverage, revenue mostly comes from advertising uk.msn.com; aol.co.uk; news.google.com;
(15) e-Smart Data: catches value from high volumes of data from open networks or from integrated business operations, also called Business Intelligence: nweh.org.uk; alexa.com; urltrends.com; domaintools.com; aiip.org;
(16) E-database: stores and organises information online to make it available to professional and commercial users: saildatabank.com; gmdnagency.org; yellowpages.com
(17) Trust services: certification consultants, electronic notaries and other trusted third party authorities: adviceguide.org.uk; techzone.adviserzone.com
(18) Web host: provides web services to individuals or companies: heartinternet.uk/web-hosting; ntchosting.com/web-hosting
(19) Collaboration platforms: provides a collaboration environment for enterprises: devconnectprogram.com; fuqua.duke.edu/offshoring; collab.net

(20) Collaboration platforms sub-group: file and document sharing: storing and sharing files across members of a group or with public: micorsoft.com (one drive); google.com; dropbox.com; asus.com (web storage)
(21) E-Media: media providers, esp. audio and video based media, revenue comes from advertising, subscribing, renting, and downloading: telegraph.co.uk; lovefilm.com; bbc.co.uk; tvplayer.com; myeasytv.com
(22) E-Government: government information and services online: gov.uk; hmrc.gov.uk; jobsearch.direct.gov.uk
(23) E-Health: healthcare practice supported by electronic processes and online communication, also covers self-monitoring healthcare devices: ehealth.scot.nhs.uk; netdoctor.co.uk; patient.co.uk
(24) E-petition: services to allow public to access, create, forward and sign petitions via the internet: change.org; epetitions.direct.gov.uk; thepetitionsite.com
(25) Virtual Office Environment: business communication and services enabled in virtual settings without a dedicated physical space: www.regus.co.uk/products/virtual-offices/virtual-office-bundles.aspx; www.ereceptionist.co.uk/virtual-office
(26) Virtual Corporate Environment: certain groups of people working online together toward a for-profit goal, with or without having to formally incorporate or form a traditional company: globalhmc.com
(27) Virtual World Organisation: business is operated solely in virtual form: smile.co.uk
(28) E-conferencing: technologies to support professional multi-users communication and 'conferencing' regardless geographical locations: ted.com; adobe.com (adobe connect); bt.com (BT conferencing)
(29) E-telecoms, video conferencing and instant messaging: telecommunication typically provided at a smaller scale, aiming at SME/individual users: skype.com; FaceTime from Apple; adobe.com (adobe connect)

4.2 Virtual Communities (VC)

(30) VC sub-group: topic-specific community: brings together virtual communities that contribute value in an environment that is hosted and managed by a virtual community operator. Revenue comes from: membership fees, advertising, may also be supported by a 3rd party, it can also be found as an add-on to other marketing operations for customer feedback or loyalty building: globalexchange.org; netmums.com; diabetes.org.uk
(31) VC sub-group: person-centric social communities: for creating individual's online profiles, sharing information and communication between users, searching and linking people with similar interests: Facebook.com; Twitter.com

(32) VC sub-group: professional (social) communities: networking sites for business professionals: uk.linkedin.com; efactor.com; odesk.com; ciltuk.org.uk
(33) VC sub-group: community software co-creation: self-selected group of people of suitable technical capabilities work collaboratively to create software and tools for offer often under open source license: mozilla.org; github.com
(34) VC sub-group: online repository: central place where data and/or software is stored and maintained: sourceforge.net; xp-dev.com; lboro.ac.uk (repository)
35) VC sub-group: media sharing: sharing images and videos through social network: flickr.com; youtube.com; Instagram.com
(36) VC sub-group: information sharing: knowledge and information sharing: big.uk.com; semanticweb.org/wiki/GoodRelations; linkeddata.org; cips.org
(37) VC sub-group: tool sharing: tools are shared across members: madiproject.co.uk;

4.3 E-Tool

(38) E-tool: online off-the-shelf tools provided to end users: prezi.com; studyblue.com; padlet.com
(39) E-tool sub-group: on-line calendar: for organising schedules and meetings, to share within a company, between business partners or private individual users: doodle.com; google.com/calendar; calendar.live.com
(40) E-tool sub-group: e-organiser: organising events and selling tickets online: Eventbrite.com; eorganiser.com.au; omnigroup.com/omnifocus

4.4 Mobile Commerce (MC)

(41) Mobile money: brings the web in the form of mobile applications to the end user, T-mobile (mobile money) via Apple iTune; myringgo.co.uk (park)
(42) MC sub-group: Mobile App as a broker: customers order services online/via mobile devices, often the service providers do not have own business websites, so need to advertise/offer their services via a 3rd party: uber.com; just-eat.co.uk/apps;
(43) MC sub-group: Mobile App for existing systems: extend enterprise software that has been already implemented (CRM applications, HR, etc.) onto mobile devices: e.g. Aeroprise as acquired by BMC Software (bmc.com)
(44) MC sub-group: Mobile App as a subscription: content available to subscribers only: wired.com; pumpone.com
(45) MC sub-group: Mobile App as extension of a web business: operations of an existing online business which can be accessed on internet-enabled mobile devices switchfly.com; dpd.com/nl_en/home/about_dpd/mobile_website
(46) MC sub-group: Mobile Business Intelligence: an extension of business intelligence (BI) from desktops and laptops based to mobile devices

including Blackberry, iPhone, and iPad: oracle.com/us/solutions/business-analytics/business-intelligence/mobile/overview/index.html; sap.com/pc/analytics/business-intelligence/software/overview/mobile-bi.html

For other related work, Rappa provided a more recent taxonomy of Internet business models that, at its highest level, it distinguished nine e-business model categories: Brokerage, Advertising, Infomediary, Merchant, Manufacturer (Direct), Affiliate, Community, Subscription and Utility [17]. For each category, he identified its source of revenue streams and business strategies and has backed up these categories with example businesses from the web. Rappa's proposed taxonomy is just one of the examples on how business models may be defined and categorized. There are also other classifications, e.g. Tapscott et al. [19] and Linder and Cantrell [10]. The differences between these taxonomies lie in their definitions of a business model. Since these categories are created based on different business model definitions, the created categories give different perspectives. Separately, Hayes [9] provided a comparison between the different classifications.

5 Conclusions

We investigated existing e-Business models based on secondary research results using a structured analytical framework. We were particularly interested in business models that employed (innovative) technologies and business practices. Based on such studies, we then assessed as to what extent technologies may have affected the process of realising business models. The aim of our research is an attempt to improve our understanding in the poorly-understood relationships between technologies and their potential influences in business models, and from a more technical focused perspective. Based on such research, we hope to inspire technical communities to create innovative technologies suitable for commercial development. In the long run, we therefore hope to motivate and bring together the business and computer science communities alike to create new disruptive technologies and innovative business models, where possible, and eventually lead to the creation of common wealth and well-being for the society.

We have identified substantial gaps in terms of understanding existing business model on the web. We have developed a new classification of forty-six categories in business models – an improvement from the eleven model types as originally identified in 1998 by Timmers. This is due to the rapid advancements in technologies and wide spread use of Internet technologies during the last two decades. We have also backed up each of our identified categories with exemplified real businesses on the web. Our work is by no means exhaustive, as new business models emerge continuously at a global scale, while at the same time becoming more complicated and inter-linked. As a result, further and continuous work is needed to provide a more comprehensive and in-depth understanding of this very important piece of puzzle that moves forward today's Digital Economy rapidly.

Acknowledgments The work reported here is jointly funded by the Research Councils UK (RCUK)'s Digital Economy NEMODE research programme and the Heriot-Watt University, UK.

References

1. Alt, R., Zimmermann, H.-D.: Preface: introduction to special section-business models. Electronic Markets **11**(1), 3–9 (2001)
2. Chen, Y.-T., He, Y.-S.: Study on Value Alliance Model: A new E-Business Model for Enterprise. Ternopil, IEEE, pp. 388–392 (2009)
3. Chesbrough, H.: Business Model Innovation: Opportunities and Barriers. Long Range Planning, pp. 354–363 (2010)
4. Chesbrough, H., Rosenbloom, R.S.: The Role of the business model in capturing value from innovation: evidences from Xerox Corporation's technology spin-off companies. Ind. Corp. Change **11**(3), 529–555 (2002)
5. CISCO, Visual Networking Index. Cisco Systems
6. Dominici, G.: E-Business Model: a content based taxonomy of literature. Int. J. Manag. Adm. Sci. **1**(7), 10–20 (2012)
7. Euripidis, L., Pazalos, K., Salagara, A.: Transforming e-services evaluation data into business analytics using value models. Electron. Commer. Res. Appl. (2012)
8. Gunzel, F., Holm, A.B.: One size does not fit all - understanding the front-end and back-end of business model innovation. Int. J. Innov. Manag. (2013)
9. Hayes, J., Finnegan, P.: Assessing the of potential of e-business models: towards a framework for assisting decision-makers. Eur. J. Oper. Res. (2005)
10. Linder, J., Cantrell, S.: Changing Business Models: Surveying the Landscape. The Accenture Institute for Strategic Change (2000)
11. Lu, J., et al.: A web-based personalized business partner recommendation system using fuzzy semantic techniques. Comput. Intell. Int. J. **29**(1) (2013)
12. MINTS: Minnesota Internet Traffic Studies. University of Minnesota
13. More, A., Vij, S., Mukhopadhyay, D.: Agent Based Negotiation Using Cloud-an Approach in e-Commerce. Springer, pp. 489–496 (2014)
14. Osterwalder, A.: The Business Model Ontology a Proposition in a Design Science Approach. Univeriste de Lausanne (2004)
15. Paredes-Moreno, A., Martinez-Lopez, F.J., Schwartz, D.: A methodology for the semi-automatic creation of data-driven detailed business ontologies. Inf. Syst. (2010)
16. Petrovic, O., Kittl, C., Teksten, R.D.: Developing Business Models for eBusiness. JAI Press (2001)
17. Rappa, M.: Business Models on the Web: Digital Enterprise (2010)
18. Rosenbloom, R., Spencer, W.J.: Engines of Innovation: U.S. Industrial Research at the End of an Era. Harvard Business School Press, Boston (1996)
19. Tapscott, D., Lowy, A., Ticoll, D.: Digital Capital: Harnessing the Power of Business Models. Harvard Business School Press, Boston (2000)
20. Teece, D.J.: Business models, business strategy and innovation. Long Range Plan. **43**, 172–194 (2010)
21. Timmers, P.: Business models for electronic markets. Electron. Markets **8**(2) (1998)

Part V
IS: Anthropic-Oriented Computing

Human-Related Factors in Knowledge Transfer: A Case Study

Sergey V. Zykov

Abstract Knowledge transfer in IT is challenging due to a number of factors, including cultural issues, organization maturity level, and mentality. However, the root cause of these key factors is human nature. The paper describes early steps and findings of the knowledge transfer from world-renowned Carnegie Mellon University to a brand new Russian startup, Innopolis. The Innopolis City will integrate experience of academicians, researchers and practitioners. Innopolis will be a self-sufficient ecosystem with a kindergarten, STEM school, IT university and IT park. The university is going to be a unique project, since it there are no pure IT universities in Russia. Successful knowledge transfer for the new university is challenging due to a number of human-related factors. The focus of the paper is to detect and manage the key factors, which may help or hinder knowledge transfer. The paper contains author's findings, which may differ from the official Innopolis University position.

Keywords Human factor · Knowledge transfer · IT education · Software engineering · Software development · Innovative skills

1 Introduction

The aim of the paper is a review of the human-related factors impact on the knowledge transfer from School of Computer Science of the CMU to the Innopolis, the new Russian IT ecosystem. The Innopolis idea is to create a unique IT city for software development and recreation based on education, research and practice

S.V. Zykov (✉)
National Research University Higher School of Economics, Moscow, Russia
e-mail: szykov@hse.ru

S.V. Zykov
Innopolis University, Kazan, Russia

G. Jezic et al. (eds.), *Agent and Multi-Agent Systems: Technologies and Applications*, Smart Innovation, Systems and Technologies 38,
DOI 10.1007/978-3-319-19728-9_22

integration. Currently, the project is in its early stage of intensive construction and knowledge transfer.

Russian government plans to accelerate development of the major industries [14]. However, the country IT sector is facing dramatic challenges including inadequate technological infrastructure, weak IP laws, poor regulatory environment, insufficient professionals number and problematic recruitment. Among these, insufficient quality of university training deserves special attention.

To assist solving the above problems, a brand new Innopolis city will be built, sometimes referred to as Russian Silicon Valley. The location is in the central part of Russia, next to Kazan', the capital of Tatarstan Republic. Russian and Tatarstan governments and a number of private investors sponsor the project. The university, Innopolis brainpower, will be a major education and research center; it will assist in boosting hi-tech implementations, and growth of innovative GDP [13].

The planned capacity of Innopolis City is around 100,000 people, a fair amount of which are the university students and staff. To support the lifecycle, Innopolis will also include a kindergarten, a STEM training school, and an IT park.

The paper focus is to identify key success factors of knowledge transfer for the master curricula in software engineering (SE). Direct curricula cloning would probably make little benefit because of a number of human-related factors, including mentality, cross-cultural issues, and a few others. So, the paper is to develop possible ways of software engineering curricula customization for the Innopolis. The CMU SE master's program is a balanced synthesis of academic research and project practice. Human-related factors might impede direct curricula duplication. Let us discuss the influence of these factors on the knowledge/skills transfer.

2 Critical Human-Related Factors for Knowledge Transfer

The result of CMU training was detection of the key factors for the knowledge transfer between CMU and Innopolis. The terms used further follow the "seven principles" concepts [2], which "inform" the knowledge transfer processes. The critical factors to consider are prior knowledge, knowledge organization, mastery, feedback and practice, course climate, motivation, and a few others. Let us analyze the nature and impact of the major factors listed over the knowledge transfer between CMU and Innopolis.

The role of prior knowledge in the software engineering education is not that straightforward as in a number of other domains. Oftentimes, particularly in the Software Architecture courses, the students relate what they learn to what they have got to know previously. Thus, the acquired information is often interpreted using prior knowledge, beliefs, and assumptions [15]. Though many researchers agree that students usually connect the prior knowledge to the new knowledge in order to learn [18], the knowledge generated by students within the IT domain may often be incomplete or inappropriate [1]. A positive way to approach the prior knowledge challenge is to self-reflect and to use case-based reasoning. Research suggests that mentoring the students through a process of analogical reasoning helps them to focus on deeper problem domain relationships rather than on superficial similarities, which are quite tricky [9]. After engagement in analysis of multiple business cases, the students tend to build more effective knowledgebase and learn more efficiently [19].

Another human-related factor, which can substantially influence the knowledge transfer processes, is self-motivation [3]. However, a distinction should be made between the subjective value, and the expectations for successful attainment of the goal [21]. In addition, the students with learning goals, as opposed to performance goals, often use complex strategies for deeper mastering the curricula and conquering challenges [5]. To develop mastery in software engineering, the students must acquire component-based skills (such as architectural diagramming and quality measurements), practice to integrate them (for example, analyze software architecture trade-offs), and consider the applicability limitations (such as customer and stakeholder priorities, business and technical constraints).

Mastery factor changes from novice to expert level, and includes issues of competence and consciousness [20]. While acquiring mastery in software engineering, the students seldom succeed on their own in the application of skills to a different context, however, thoughtful mentor prompts, which may seem subtle at first, usually help a lot [8, 10].

Practice and feedback are the closely interwoven factors, and they are critical for achieving a balance of theoretical knowledge and practical skills. To facilitate knowledge transfer, the both factors should focus on the same learning objectives [2]. The essential qualities of feedback are frequency and timeliness. These suggest a trade-off with feedback size, however, even in case the size is sacrificed, a better knowledge transfer still can be achieved [11]. Feedback, together with one-to-one mentoring sessions, often helps to set the appropriate level of assignment difficulty [4].

Self-directed learning is another critical factor. To achieve it, the students should efficiently implement their own metacognitive cycle (Fig. 1): evaluate their own knowledge and skills, plan and monitor progress, and adjust strategies [17]. For successful knowledge transfer, students need to apply metacognitive skills [22]. The key factor to improve the metacognitive skills is self-reflection, which helps to continuously monitor and adjust problem-solving progress. Other techniques to improve software engineering-specific metacognitive skills include team-based concept mapping and brainstorming [16].

3 Building Source-to-Target Interface for Knowledge Transfer

The concept of the Innopolis project originated in 2010. Russia has a long academic tradition; however, the project is a challenge in itself. This is so because prior to Innopolis, none of the universities built was an entirely IT school. In addition, in contrast to classical Russian universities focused on basic research, the Innopolis will specifically aim at industry-focused IT applications. Innopolis will include a kindergarten, a STEM school, a university, and an IT park. Facilities will be within a close 15 min walking distance, and the living standards are going to be world competitive.

Innopolis location is really unique: Kazan' city, Tatar capital, is located just between Europe and Asia with average flight time to major European and Asian destinations nearly 1.5 h shorter than that from Moscow. The location is international: Kazan' city has over 1200-year history [12], it is world famous for academic and research. The dominating sectors are oil-and-gas, automobile industry, construction, transportation and communication. The city has been recently renovated; its infrastructure supported a number of major international events. Large IT multinationals, such as Fujitsu-Siemens, are located there. The above allows building state-of-the-art infrastructure for Innopolis. The Innovation City is going to host around 100,000 inhabitants by 2030; the first stage of the project will be completed in 2015, and it is to include the main university building, residential halls and the IT park offices.

Since the university will be Innopolis brainpower, the knowledge transfer issues are mission-critical.

The CMU MSIT-SE training schedule (see Table 1) is very busy, i.e. one calendar year is has three semesters rather than two as usual. These are fall, spring, and summer semesters. Each of the three semesters is nearly four months long; the midterm breaks are nearly one week short.

Table 1 CMU MSIT-SE required courses list

Course code	Course name
17-602	Introduction to personal software process
17-651	Models of software systems
17-652	Methods: deciding what to design
17-653	Managing software development
17-654	Analysis of software artifacts
17-655	Architectures for software systems
17-656	Communication for software engineers I
17-657	Communication for software engineers II
17-677	MSIT Project I
17-678	MSIT Project II

The CMU MSIT/SE core courses are tightly interwoven, and integrated with the practically oriented software development projects. The Architectures for Software Systems and Personal Software Process are discussed later as the knowledge transfer instances.

4 Human-Related Factors Application to Knowledge Transfer

The key human-related factors identified in Sect. 1 are critically important for knowledge transfer. Let us link these factors to the recent training observations and findings.

One of the human-related factors is the prior knowledge, which may either help or hinder the knowledge transfer. In case of actual faculty training, the amount of prior knowledge on Architectures for Software Systems course was large enough. However, a number of the prior knowledge concepts was inappropriate and/or irrelevant, which hindered the transfer significantly. One example is excessive focusing on real-life implications of the economic and business constraints, which are predominant in Russia rather than merely justifying technologies as prescribed by the course objectives. As a result, the knowledge transfer was incomplete, and the certification has resulted in Guest Lecturer level rather than Primary Instructor position. The Guest Lecturer is usually responsible for not more than 20 % of the course delivery. However, since Architectures for Software Systems course is generally out of the primary domain of the trainee expertise, the assignment of the Guest Lecturer status was perhaps the best option possible.

Other human-related factors are knowledge organization and mastery. Concerning knowledge organization, experts oftentimes omit evident reasoning steps, or "make shortcuts", to use the terms of training side. This "shortcuts" policy was accepted as a default, due to a certain level of trainee mastery.

The other course taken was Personal Software Process, which heavily focuses on process definition and software development discipline. Due to the nature of the course, quite a number of such "shortcuts" was hardly acceptable for the training side. Thus, the first attempt of taking the Personal Software Process did not result in a satisfactory level of knowledge transfer, and a valid instructor certification. However, after self-reflection (evidenced by the Final Report) and adjustment to the training side requirements (such as data presentation standards and formats, levels of detail etc.), the knowledge transfer has been successful, and the trainee had been certified in the course as the Primary Instructor. Another critical factor of success was following the process rigorously. This was a "must-have" from the standpoint of the training side. However, due to cultural diversity, the trainee's initial understanding set up a different "default level" of process compliance degree, which resulted in certain certification issues for the initial course take. During the retake, however, there were certain process adjustments. These adjustments were based both on self-reflection of the trainee, and on explicit feedback from the Carnegie Mellon University. These process adjustments resulted in eventual successful Primary Instructor certification.

The Guest Lecturer is aware of the critical relations between knowledge items (concepts); however, his/her network of the course-related concepts is very different from that of the Primary Instructor. This may inhibit the initial path of the teacher-to-learner knowledge transfer as originally established, so that the knowledge transfer may become somewhat inaccurate, and/or inappropriate. In case of Architectures for Software Systems, identification of mission-critical software quality attributes (such as availability for a robot-controlled warehouse) would be a success. Evaluation of students' assessments was also adequate. However, delicate probing and hinting students on creating their own context-specific application of such a mission-critical concept (e.g. suggesting a well-justified architectural framework) appeared to be quite problematic. Thus, the far knowledge transfer (in terms of applicability of the knowledge base acquired to multiple/cross-context problem domains) was somewhat inhibited. The rich prior knowledge (perhaps excessive in certain aspects such as designing for fault tolerance, data consistency etc.) and complex organization of knowledge, however, promoted a number of the problem domain knowledge areas, such as enterprise and mission-critical systems.

A lesson learned was that the educational framework and terminology, core readings, influenced the transfer largely. As an example, the other trainee, a USA University Professor specializing in Software Architectures, was physically present at 30 % less classes. However, his prior knowledge and knowledge organization structure were more relevant to these of the knowledge transmitting side. He had been delivering courses on Software Architectures in the USA for a number of years. His courses were based on the same core reading, a famous book by three prominent Software Engineering Institute experts [6]. The Software Engineering Institute is located next to the Carnegie Mellon University, and the research activities of the two organizations are tightly interwoven. Therefore, the CMU courseware, and that the other trainee were using were very similar prior knowledge bases which were based on the same knowledge sources. Moreover, concerning practical application of the concepts and principles of designing software intensive systems, both the Carnegie Mellon University, and the other trainee were building the knowledge bases for their courses on similar real-world projects, which they were using for their business cases.

The knowledge transfer can be divided into near and far. The near transfer means knowledge applicability in adjacent problem domains only, while the far transfer usually allows a multi-domain application in different contexts, some of which are substantially far from the original one. The big question is how far the CMU-Innopolis knowledge transfer is possible. By far, there it was hard to find an evidence of the far transfer. Our experience shows that the near transfer is quite possible based on 100 % cloning of the curriculum with certain micro adjustments. However, the far transfer generally requires at least a two-pass (and/or retake) strategy, with non-trivial reflection, multi-context real-world applications and adjustments in between.

One more mission-critical human resource-centered factor for knowledge transfer is the relevant practice based on timely feedback. In theory, these two, practice and feedback, should be aligned. Concerning the training side delivery, the feedback was not always ideally aligned with the assignment submission schedule. The primary reason for that was intensive mobility (a number of business trips) and non-synchronized holidays of both sides. Another challenge was the feedback, which sometimes seemed too general/vague for the trainee to generate a certain solution. For the other course, Personal Software Process, the feedback was sometimes based on an automatically generated template for student submissions. A batch-processing tool assisted in creating the feedback. Thus, a certain portion of the feedback seemed not quite targeted to meet personal trainee's issues, and link him to prior assessments. Therefore, the knowledge transfer was somewhat inhibited.

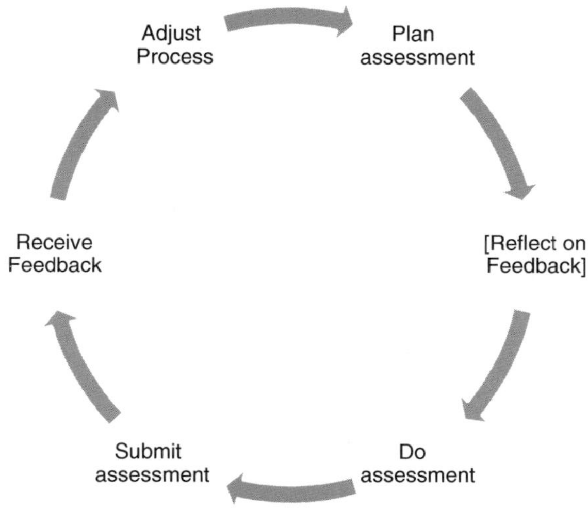

Fig. 1 Faculty training lifecycle diagram

Because of delayed start, the most intensive part of the training was postponed until the summer semester, and was performed distantly. However, due to summertime, both teaching and learning sides were traveling intensively, so, the season and distance factors had hindered timely and detailed feedback. After the initial certification attempts for both Architectures for Software Systems and the Personal Software Process courses were not quite successful, a step-by-step assessment submission strategy was selected. This followed explicit recommendation of the training side as the direct feedback message in reply to the self-reflective course report produced by the trainee. Eventually, both retakes resulted in successful certification, which confirmed that the knowledge transfer did eventually occur. The practical assignments for both course retakes were accompanied by a more detailed feedback as compared to the initial training attempts. Even though both courses were taken remotely, the cognitive load had been essentially decreased due to: (i) less submissions done "on-the-go" (while traveling etc.); (ii) a step-by-step training plan has been created and agreed by both sides (which included time estimation, deliverables etc.); (iii) a well-planned and justified amount of time was invested into the course retake; (iv) detailed and timely feedback followed every assessment submitted; (v) every next submission considered the feedback on the previous one(s), and has been adjusted accordingly. The training lifecycle diagram with reduced cognitive load is presented in Fig. 1.

To further reduce the cognitive load, the Innopolis University seized a number of concurrent learning activities and focused on doing "one thing at a time" as much as it was possible. As opposed to the previous attempt, finalizing the Architectures for Software Systems course by preparing the Final Report was separated from any other knowledge transfer-related activities. A fair amount of time (around 100 h in just two weeks) had been invested into the Final Report on Architectures for Software Systems. After the report had been submitted strictly following the deadline set by CMU, the Personal Software Process course retake started. The planning allowed certain gaps to revise materials, install, configure and test the new development environment, and to do some other preparatory activities. The Final Report on Personal Software Process (though it took approximately an order of magnitude less time) was produced according to similar scenario than that on Architectures. Again, this reduced the cognitive load to minimum, and allowed concentrating on course-specific issues, such as process planning and tracking, self-evaluating and adjusting – see Fig. 2.

Fig. 2 Personal software process training diagram

The feedback on Architectures for Software Systems course included a number of interaction activities. These were discussions on reading questions, grading individual and group assignments, and writing the final report. In a way, incomplete knowledge transfer, and partial certification, could result from insufficient feedback. As for the intermediate assignments (both reading questions and individual/group practical ones), the feedback received was probably not always as prompt and goal-directed as desired by the trainee. Another reason was a heavy cognitive load for the dedicated instructor, which again was a human factor. The dedicated instructor had a bulk of administrative load, and he thus had to arrange intensive online meetings and take frequent business trips. Additionally, an eight-hour time shift between Moscow/Kazan' and Pittsburgh, and different national holiday schedules inhibited feedback efficiency. Thus, a number of meetings with course instructors and some of feedback emails were delayed. Probably, this could be a reason for hindering the Architectures for Software Systems knowledge transfer.

Three final report revisions for the Architectures for Software Systems were submitted. An exhaustive feedback, with clear guidelines, followed the first one. Based on the initial feedback, two more revisions of the Final Report were submitted, however, the feedback was not that detailed as before. Thus, the primary human factor that influenced the process was probably the feedback.

As for the other instructor certification course, Personal Software Process, the outcome was quite different. Again, the primary human factor was probably the feedback. For the initial attempt, the feedback for earlier submissions was delayed for several days. This resulted in trainee's misunderstanding of the ground rule that any further assignment was dependent on the previous one(s). However, the feedback that followed clarified that nearly none of the assignments met that ground rule, and thus were subject to resubmission.

To scaffold the further learning process and to promote knowledge transfer, CMU suggested a context change, which included both programming paradigm and programming language. At first, this could seem a misleading over-complication. However, the multi-context training resulted in better knowledge organization scheme in terms of matching curriculum-specific objectives. In addition, the multi-context approach allowed scaffolding and ordering prior knowledge conceptual scheme so that the representations of the knowledge transmitting and receiving sides were better aligned.

However, the Personal Software Process context change required certain preparatory efforts, such as installing integrated development environment and revising language concepts. However, the result was successful due to the following human-related factors: (i) better meta-knowledge transfer, (ii) improved feedback quality and (iii) adjusted meta-cognitive cycle. The meta-knowledge aspects included training plan, certification requirements, submission ground rules, and a few other aspects. Feedback interaction became faster, it was one assignment at a time, focused on submission benefits; it allowed a timeframe to analyze, self-reflect and adjust. Therefore, knowledge transfer appeared to be critically dependent on feedback. The general requirements for the instructor feedback are frequency, timeliness, focus, and goal-orientation. Late feedback inhibits the knowledge transfer, and it may result in non-certification.

The multi-context approach was also applied to the Architectures for Software Systems. The initial context was traditional MSIT-SE classes; the second context was LG company professional training. The second context had a realistic hardware-and-software system as a goal to implement, so that subtle theoretical concepts were articulated more accurately, and received a multi-dimensional coverage. Thus, multi-context application of the course key concepts had adjusted the knowledge organization scheme of the trainee so that course specific issues and challenges were identified, highlighted and scaffolded. Concerning the other course, Personal Software Process, the same effect had been achieved by using structural and object-oriented development paradigms, Free Pascal and NetBeans environments, and Pascal and Java programming languages.

Another key human-related factor of the knowledge transfer is achieving certain level of mastery, i.e. a set of domain-specific knowledge, skills and attitudes. Concerning software engineering, building and transferring mastery involves a number of non-technical soft skills, such as business communication, teamwork, and time management [13]. Though these skills are non-core to the discipline, they are critical for successful knowledge transfer. However, the Russian software engineering academy generally underestimates the value of the soft skills, whereas CMU explicitly focuses on them. For cross-cultural contexts, these skills are mission-critical. The aim of software engineering is product-based development and maintenance of large-scale and complex software systems, which is impossible without efficient client-to-developer communication. A software engineering product is a result of teamwork. Design, development, testing, maintenance team(s) often differ; however, they should collaborate efficiently to build and utilize a competitive product. Thus, teamwork skills are critically important for the knowledge transfer in software engineering curriculum.

One more human-related factor hindering knowledge and mastery transfer is maturity level, which is more ad hoc at the Innopolis, and more formal at CMU [7].

Additional human-related factors, which affected the knowledge transfer, are related to motivation and course climate. Though they are not critically important for faculty training, they still are influential. Initial late training start resulted in skipping at least 25 % of the classes at the beginning of semester. Since these classes included trainee orientation and course key concepts, not only climate and self-motivation, but also knowledge organization and mastery clearly suffered. Moreover, the late start resulted in uncertainty of the expectancies of both the trainee and trainers, and their organizations. The trainee's initial idea was that some of the learning goals were so flexible that they could be adjusted "on the fly". However, the training side attitude appeared to be clearly different, and all the critical training processes were supposed to be prearranged and approved as training plans, otherwise certification goals would likely be missed. Of course, the training side was probably as flexible as possible. However, this resulted in a more complex training environment with context switches, and increased the cognitive load for both sides. Eventually, the training objectives were met, and the trainees were certified. This confirms that the application of the key human factors was wise, in spite of cross-cultural issues and certain differences in expectancies and maturity levels.

5 Conclusion

The paper discusses the knowledge transfer from the CMU masters' program in software engineering to the Innopolis University. This knowledge transfer requires special faculty training.

The key factors were identified, which usually hinder knowledge transfer. These include differences in culture, geography and process maturity.

The key factors were identified, which usually the help knowledge transfer. These include multiple contexts and scaffolding; reducing cognitive load; frequent, timely, directed and goal-oriented feedback; communication, time management, and teamwork soft skills; self-directed learning.

The lessons learned at CMU MSIT/SE faculty training program include the following. Prior knowledge, knowledge organization, feedback, mastery, and practice are essential human-related factors for successful knowledge transfer, while first three factors are mission-critical. An adequate meta-cognitive cycle organization is very important for learning quality. The learning objectives should address entire meta-cognitive cycle including reflection, analysis, justification, adjustment, and practical application of the theory and principles learned. Therewith, the students should act as software engineering professionals and use their own expert-level judgment. The multi-context education should include not only theoretical software development, but also hands-on practice. Teamwork, communication, and time management are essential "soft" skills. These multi-disciplinary skills are mission-critical in the rapidly changing, multi-context environments of practical software engineering. Even a project failure is a possible starting point to reflect, adjust and become agile. The author is going to continue his studies of knowledge transfer from CMU to Innopolis.

References

1. Alvermann D.E., Smith L.C., Readence J.E.: Prior knowledge activation and the comprehension of compatible and incompatible text. Read. Res. Q. **20**(4), 420–436 (1985) (John Wiley & Sons)
2. Ambrose S.A., Bridges M.W., DiPietro M., Lovett M.C., Norman M.K.: How Learning Works. Seven Research-Based Principles for Smart Teaching. John Wiley & Sons (2010)
3. Ames, C.: Motivation: What teachers need to know. Teachers Coll. Rec. **91**, 409–472 (1990)
4. Anderson, J.R., Corbett, A.T., Koedinger, K.R., Pelletier, R.: Cognitive tutors: lessons learned. J. Learn. Sci. **4**, 167–207 (1995)
5. Barron, K., Harackiewicz, J.: Achievement goals and optimal motivation: testing multiple goal models. J. Pers. Soc. Psychol. **80**, 706–722 (2001)
6. Bass L., Clements P., Kazman R.: Software Architecture in Practice, 3rd ed. SEI Series in Software Engineering. Addison-Wesley, Boston, MA, USA, p. 640 (2012)
7. CMMI product team CMMI for development. http://www.sei.cmu.edu/pub/documents/06.reports/pdf/06tr008.pdf (2006)
8. Cognition and Technology Group at Vanderbilt: From visual word problems to learning communities: changing conceptions of cognitive research. In McGilly, K. (ed.) Classroom Lessons: Integrating Cognitive Theory and Classroom Practice, pp. 157–200. MIT Press/Bradford Books, Cambridge, MA (1994)
9. Gentner, D., Loewenstein, J., Thompson, L.: Learning and transfer: a general role for analogical encoding. J. Educ. Psychol. **95**, 393–405 (2003)
10. Gick, M.L., Holyoak, K.J.: Analogical problem solving. Cogn. Psychol. **12**, 306–355 (1980)
11. Hattie, J., Timperley, H.: The power of feedback. Rev. Educ. Res. **77**, 81–112 (2007)

12. Innopolis University presentation. http://innopolis.ru/files/docs/uni/innopolis_university.pdf (2014)
13. Kondratiev, D., Tormasov, A., Stanko, T., Jones, R., Taran, G.: Innopolis University – a new IT resource for Russia. In: Proceedings of the International Conference on Interactive Collaborative Learning (ICL). Kazan, Russia (2013)
14. Kuchins, A.C., Beavin, A., Bryndza, A.: Russia's 2020 Strategic Economic Goals and the Role of International Integration. Center for Strategic and International Studies, Washington (2008)
15. National Research Council: How people learn: brain, mind, experience, and school. National Academy Press, Washington (2000)
16. Novak, J.: Learning, Creating, and Using Knowledge: Concept Maps as Facilitative Tools in Schools and Corporations. Erlbaum, Mahwah (1998)
17. Pascarella E., Terenzini P.: How College Affects Students: A Third Decade of Research. Jossey – Bass, San Francisco (2005)
18. Resnick, L.B.: Mathematics and science learning. Science **220**, 477–478 (1983)
19. Schwartz, D.L., Bransford, J.D.: A time for telling. Cogn. Instr. **16**, 475–522 (1998)
20. Sprague, J., Stuart, D.: The Speaker's Handbook. Harcourt College Publishers, Fort Worth (2000)
21. Wigfield, A., Eccles, J.: Expectancy-value theory of achievement motivation. Contemp. Educ. Psychol. **25**, 68–81 (2000)
22. Zimmerman, B.J.: Theories of self-regulated learning and academic achievement: an overview and analysis. In: Zimmerman, B.J., Schunk, D.H. (eds.) Self-Regulated Learning and Academic Achievement, 2nd edn, pp. 1–3. Erlbaum, Hillsdale (2001)

Enterprise Applications as Anthropic-Oriented Systems: Patterns and Instances

Sergey V. Zykov

Abstract Managing development of large and complex software systems is a key problem in software engineering discipline. So, the issues related to software-and-hardware system development, often referred to as systems of systems, tend to become even more essential and critical in the enterprise context. The paper discusses general architecture patterns for such complex systems, and provides instances for particular industries, such as oil-and-gas and nuclear power production. Anthropic-oriented issues of the mission-critical system components are discussed.

Keywords Anthropic-oriented system · Enterprise application · System of systems · Software development · Software architecture

1 Introduction

Research and development of the software systems gives rise to the problems, which are similar to these of enterprise systems development. Most complex and mission-critical enterprise-level software systems usually include both "higher" and "lower" level software systems. The former ones assist in business processes management, while the latter ones are intended for interfacing the supporting hardware. Therewith, the architectural design under the above conditions generally reaches out of the software engineering scope, and is related to system engineering to a certain extent. Moreover, the enterprise software components are essentially anthropic-oriented, since they serve information requests of quite different groups of users, such as top and middle management, analysts, designers etc.

S.V. Zykov (✉)
National Research University Higher School of Economics, Moscow, Russia
e-mail: szykov@hse.ru

S.V. Zykov
Innopolis University, Kazan, Russia

G. Jezic et al. (eds.), *Agent and Multi-Agent Systems: Technologies and Applications*, Smart Innovation, Systems and Technologies 38,
DOI 10.1007/978-3-319-19728-9_23

It can be demonstrated that general principles of describing the architecture, i.e. key components and relationships of the software system (for instance, in terms of components and connectors), commonly used in software engineering [7], are generally applicable for enterprise system engineering. Therewith, top level of architectural design is critical since it determines key points and interfaces for software and hardware subsystems of large-scale and complex objects of system engineering. Within this level, a theoretically adequate and practically applicable is the approach based on high-level patterns, which describe the interfaces for the software-and-hardware system components. The foundations the approach have been laid by E. Gamma [1], and applied later on to enterprise software systems in [2, 4, 5], where the models for pattern representation and means of pattern management are given sufficient coverage. However, each instance of the patters is essentially dependent upon the nature of the end users, and on the problem domain of the enterprise application.

The paper objective is "embedding" of the pattern-based approach to enterprise software systems representation into system engineering with subsequent generalization for hardware-and-software systems. Therewith, due to problem domain nature and scale, a rigorous proof of the approach adequacy is beyond the scope of the paper. Instead, the research focus is both the general "pattern", which describes architecture of the software-and-hardware system, and a set of practical examples, which demonstrate applicability of the approach for a number of certain cases. Once the high-level architectural pattern is identified and described, it becomes possible to instantiate it with domain-specific implementations. Specifically, software-and-hardware systems for oil-and-gas industry and nuclear power production domains, including anthropic-oriented aspects, is discussed. Surprisingly, the domain-specific examples clearly indicate that their architectural foundations have very much in common. Thus, the high-level architectural pattern theoretically allows a more unified and better approach for building system-of-systems. Practically speaking, the pattern-based development allows more efficient high-level cloning for system-of-systems in terms of time and cost.

2 The General Architectural Outline for an Enterprise Software-and-Hardware System

Let us present the general outline of the high-level "pattern", which describes enterprise hardware-and-software system conceptual scheme (Fig. 1 shows the diagram). Let us see the structure of the "layers" of the "pattern" in a more detailed way.

The top layer (or Layer 5 in Fig. 1) of the enterprise software-and-hardware system is represented by the software toolkit for integrate representation of strategic data "slices" for enterprise management. It is the "dashboard", which allows enterprise top management to track dynamics of key performance indicators. The

"Dashboard" of key business indicators (portal)						5 Strategy
Supply Chain Management	Channel Management (Fax, email)	CRM	Campaigns	Marketing Plans		4 Relations
Accounts Payable	Accounts Receivable	General Ledger	Manufacturing	HR	PLM	3 Resource Management
Warehouse		Payroll		Inventory		2 Accounting
SCADA			CAD	CAM	CAE	1 "Drivers"
Hardware			Goods / Parts			0 "Hardware"
Databases / Data warehouses						-1 Data

Fig. 1 Enterprise hardware-and-software system architecture

data is aggregated from software systems for planning enterprise resources, such as human resources, financials, time and the like. An enterprise internet portal or a similar tool is used to integrate and visualize high-level reports, and to provide flexible, reliable and secure online access for the management by means of "dashboard"-like interface.

The next layer (or Layer 4 in Fig. 1) purpose is to inform the end-user employees of urgent updates of standard business processes, such as document approval, communication with clients and suppliers, and target email messaging. Naturally, each two adjacent layers are interrelated. Figure 1 omits the relationships for the sake of space; however, it is clear that the Layer 5 aggregates and consolidates the data objects mentioned in Layer 4 to provide a more strategic view of the key business indicators. In essence, the architectural Layer 4 layer of the enterprise software-and-hardware system is represented by the software system for interaction of the key organizational units and responsible enterprise employees with the respective clients and partners. Thus, the architectural Layer 4 of the enterprise software-and-hardware system is functionally similar to customer relations management (CRM) system. Among the anthropic-oriented functional features of the second architectural layer of the enterprise software-and-hardware system are:

1. Managing the channels of interaction with the clients (such as email, fax, and IP telephony);
2. Event planning and management (such as special offers, sales, and distribution channels);
3. Partner/client data management (such as custom clients, marketing campaigns, and preferred communication methods).

The next architectural layer (Layer 3 in Fig. 1) of the enterprise software-and-hardware system is linked to the adjacent ones, and it represents the tier of Enterprise Resource Planning (ERP) software systems. Again, Layer 4 consolidates the Layer 3 data to get a more strategic representation of key business indicators. The Layer 3 includes the ERP components, which assist end-users in management and planning of:

1. Fixed assets;
2. Payables (supplies);
3. Receivables (orders);
4. Production and services;
5. Human resources;
6. Financials;
7. Documentation.

The next architectural layer (Layer 2 in Fig. 1) of the enterprise software-and-hardware system contains more "tactical", lower level software systems as compared to the previous ones (Layers 3–5). Therewith, the previous level (Layer 3 in Fig. 1) is considered a more "analytical" one, as it allows end-users in forecasting dynamics of key production indexes (such as revenues, profits, overheads, personnel defections, expenditures etc.). Naturally, the architectural Layer 2 of the enterprise software-and-hardware system is linked to the two adjacent layers (i.e. Layer 3 and Layer 4); it includes software systems for accounting, warehousing, inventory management and the like.

Generally speaking, abstraction level, data aggregation degree and abilities for strategic analysis, justification and decision-making are growing bottom-up across the system layers hierarchy, i.e. Layer 5 is the most strategic, while Layer 1 is the most operational.

The next architectural layer (Layer 1 in Fig. 1) of the enterprise software-and-hardware system contains the software systems, incorporates the "drivers", i.e. the interfaces between software and hardware components. This layer contains the systems for the end-users who develop and manage design documentation (CAD/CAM/CAE), and for the end-users who interact with field-based devices and sensors, which perform plant operations, such as assembly-line production. Naturally, similar to the previous cases, this layer communicates with the neighboring Layer 0 and Layer 2.

The data level (Layer 0 in Fig. 1) is a dedicated layer, which could be drawn either to the side of the above five (Layer 1–Layer 5), representing their interactive "penetration", or below them, limiting the interaction to Layer 1 only. The data level (Layer 0) is represented both by databases and by data warehouses. Naturally, the family of software systems for this level include DBMS with data mining plug-ins, analytical and online transaction processing (OPAL/OLTP), middleware, and enterprise anthropic-oriented content management tools. In case of the enterprise software-and-hardware system, essential features of the data level are: (i) big data size, (ii) high availability, and (iii) data base/warehouse heterogeneity. Therewith, the heterogeneity can be subdivided into architectural (such as non-normalized data, legacy systems data), and structural (such as weak structured flows of audio- and video data, and scanned documentation).

The data level is marked as Layer 0 because it contains system level software rather than applications.

Below the data level, one more tier (Layer -1 in Fig. 1) can be identified, which is the hardware level. It includes such devices as programming logic controllers, sensors and the like. The hardware layer operates in terms of analog data, which is aggregated at the Layer 0, where it is transformed into digital form, stored, and used for enterprise applications.

3 Software-and-Hardware System for Oil-and-Gas Enterprise

Let us illustrate the above general pattern of the architectural layers of the enterprise software-and-hardware systems by an example of the systems set used in the production and distribution businesses of an oil-and-gas enterprise. Therewith, let us focus on the anthropic-oriented activities of a vertically integrated enterprise, which performs exploration, production, processing, and, possibly, transportation of the oil-and-gas products.

In this case, the upper level (Layer 5) of the enterprise software-and-hardware systems architecture is represented by the software toolkit for integrate representation of strategic data "slices" for the enterprise management. In this case, an enterprise internet portal represents the "dashboard", which allows enterprise top managers to track dynamics of key activity indicators. The portal aggregates data and visually represents it in terms of high-level reports, and provides data access for various anthropic-oriented activities (typically, for top and middle management business-oriented queries). The portal predominantly aggregates data for the software enterprise resource planning systems, including general ones: HR management, financials, time, and domain-specific: gas balances, deposit assets, oil-and-gas upstream and downstream, seismic exploration data etc.

The purpose of the next layer (Layer 4) of the enterprise software-and-hardware systems architecture is to support interaction of the key departments and employees of the oil-and-gas enterprise with clients and partners. The basic functional features of the layer include informing the employees of urgent updates of standard business processes, such as anthropic-oriented activities document approval (e.g. oil-and-gas shipment contracts), communication with clients and suppliers (e.g. oil-and-gas distributors), suppliers (e.g. gas pipeline producers), and target email messaging, such as product articles, price lists for produced and processed items. In this case, the architectural layer of the enterprise software-and-hardware system is functionally similar to customer relations management (CRM) system. Among the functional features of this architectural layer of the enterprise software-and-hardware system are: management of the employee interaction channels by gas communication, email, fax, intelligent IP telephony etc., distribution networks planning and management, and partner/client data management (including regular customers, VIP clients etc.).

The next architectural layer (Layer 3) of the oil-and-gas enterprise software-and-hardware system is linked to the adjacent ones, and it represents the tier of the ERP – resource planning and management – software systems. In this case, the third level includes the ERP software modules and subsystems, which assist in management and planning of oil-and-gas products including: fixed assets, facility supplies for deposit construction, and oil/gas processing, human and financial resources, and documentation management.

The next architectural layer (Layer 2) of the oil-and-gas enterprise software-and-hardware system contains the "tactical", lower level software systems. This layer contains the software systems for such anthropic-oriented activities as payroll, product supply management, which includes track oil-and-gas transportation by pipelines, railways and ships, and the like.

The next architectural layer (Layer 1) of the oil-and-gas enterprise software-and-hardware system contains the software systems for developers and managers of the design documentation including exploration and seismic data maps, and the systems interacting with the devices and sensors, which perform plant operations, such as drilling exploration wells, and oil-and-gas production.

The data level (Layer 0) for an oil-and-gas enterprise is represented both by databases and by data warehouses. Naturally, the family of software systems for this level include DBMS with data mining plug-ins, anthropic-oriented activities analytical and online transaction processing, and content management tools. In case of the oil-and-gas enterprise software-and-hardware system, reliable and fault-tolerant Oracle DBMS-based solutions are often utilized, which are integrated with both domain-specific ERP applications for upstream and downstream management, and with online data visualization tools. Other reasons for Oracle-based implementations are big data size, and high availability requirements. Heterogeneity, as usual, is both architectural and structural [2].

The hardware level (Layer -1) is located below the data level. It includes such devices as programming logic controllers and sensors used for exploratory drilling, oil-and-gas production. The hardware level generally operates with the raw analog data.

4 Software-and-Hardware System for Nuclear Power Industry

Let us illustrate the general pattern of the architectural layers of the enterprise software-and-hardware systems by an example of the systems set used in the production and distribution businesses of a nuclear power enterprise. Therewith, let us focus on the anthropic-oriented activities of design and construction of a reaction units for a nuclear power plant.

In this case, the upper level (Layer 5) of the enterprise software-and-hardware systems architecture is represented by the software toolkit for integrate representation of strategic data for nuclear power production enterprise management. In this case, an enterprise-level internet portal represents the anthropic-oriented "dashboard", which allows the top managers of a nuclear holding company to track dynamics of key activity indicators. The portal aggregates data and visually represents it in terms of anthropic-oriented reports, and provides data access for various management levels, typically, top and middle management. The portal predominantly aggregates data for the software enterprise resource planning systems, including general ones: human resource management, financials, time, and domain-specific: assembly maps, technical documentation, nuclear power plant (NPP) monitoring data, detailed down to separate reaction units and their parts, production and distribution and of the electricity processed, nuclear fuel supplies, and waste utilization.

The purpose of the next layer (Layer 4) of the enterprise software-and-hardware systems architecture is to support anthropic-oriented interaction of the key departments and employees of the oil-and-gas enterprise with clients and partners. The basic functional features of the layer include informing the employees of urgent updates of standard business processes, such as document approval (e.g. design documentation), communication with clients and suppliers (e.g. reaction unit customers), suppliers (e.g. gas pipeline producers), target email messaging (product articles, price lists for produced and processed NPP units etc.). In this case, this architectural layer of the enterprise software-and-hardware system is functionally similar to customer relations management (CRM) system. Among the functional features of this architectural layer of the enterprise software-and-hardware system are: management of the employee interaction channels by NPP-specific communication, email, fax, VoIP telephony etc. Distribution networks planning and management is less essential in terms of anthropic-oriented approach as compared to the oil-and-gas enterprise.

The next architectural layer (Layer 3) of the NPP enterprise software-and-hardware system is linked to the adjacent ones, and it represents the tier of the ERP and PLM – anthropic-oriented resource planning and production lifecycle management – software systems. In this case, the third level includes the ERP software modules and subsystems, which assist the managers to monitor and plan the NPP reaction units and their components production including: fixed assets, facility supplies for NPP unit construction, and electricity processing, human and financial resources, and documentation – predominantly, design documents. Possible instances of such systems for the NPP industry include Siemens and Catie software products [3].

The next architectural layer (Layer 2) of the oil-and-gas enterprise software-and-hardware system contains the "tactical", lower level of anthropic-oriented software systems. This layer contains the software systems for product supply management, which track NPP payroll, human resource management, reaction unit construction, shipment, and assembly, monitoring NPP unit assembly maps, and technical conditions, and the like.

The next architectural layer (Layer 1) of the NPP enterprise software-and-hardware system contains the software systems for developers and managers of the design documentation, such as NPP unit assembly maps, technical conditions, and the anthropic-oriented systems interacting with devices and sensors, which perform plant operations such as heat generation and reaction unit temperature/pressure control.

The data level (Layer 0) for an NPP enterprise is represented by databases and data warehouses. Therewith, the family of the software systems at this level is DBMS with add-ons for data mining and OLAP, and anthropic-oriented enterprise content management tools. In case of the NPP enterprise software-and-hardware system, reliable and fault-tolerant Oracle DBMS-based solutions are utilized, which are custom-integrated with both domain-specific PLM and ERP applications for NPP design and production lifecycle management, electricity production and distribution, and with online 6D modeling and data visualization tools. The 6D models include 3D visualization of the units to be designed, and domain-specific models for resources required to design and construct NPP, such as time, human resources and financials. Other reasons for Oracle-based implementations choice are discussed in the previous section.

The hardware level (Layer -1) is located below the data level. It includes such devices as NPP programming logic controllers and sensors, and operates with the raw analog data.

5 Conclusion

The paper suggests an approach for representing anthropic-oriented enterprise software systems based on high-level design patterns. The development problems for such complex software-and-hardware systems are discussed. A general functional-based architectural pattern for such systems of systems is suggested, which includes five application levels and two data levels. The application levels range from strategic decision-making and interfacing key components down to "drivers", and the data levels include the processed digital and the raw analog data. The architectural pattern is instantiated by functional outlines for systems of systems in oil-and-gas and nuclear power industries. Anthropic-oriented parts of the mission-critical components are identified and discussed.

References

1. Gamma, E., Helm, R., Johnson, R., Vlissides, J.: Design Patterns CD: Elements of Reusable Object-Oriented Software. Addison-Wesley (1998)
2. Zykov, S.: Designing patterns to support heterogeneous enterprise systems lifecycle. In: Proceedings of the Software Engineering Conference in Russia (CEE-SECR), 2009 5th Central and Eastern European (2009)
3. Zykov, S.: Pattern development technology for heterogeneous enterprise software systems. J. Commun. Comput. **7**(4), 56–61 (2010)
4. Fowler, M.: Patterns of Enterprise Application Architecture. Addison-Wesley (2002)

5. Freeman, E., Bates, B., Sierra, K., Robson, E.: Head First Design Patterns. O'Reilly (2004)
6. Hohpe, G., Woolf, B.: Enterprise Integration Patterns: Designing, Building, and Deploying Messaging Solutions. Addison-Wesley (2004)
7. Lattanze, A.: Architecting Software Intensive Systems: A Practitioner's Guide. Auerbach (2008)

Fault Detection in WSNs - An Energy Efficiency Perspective Towards Human-Centric WSNs

Charalampos Orfanidis, Yue Zhang and Nicola Dragoni

Abstract Energy efficiency is a key factor to prolong the lifetime of wireless sensor networks (WSNs). This is particularly true in the design of *human-centric wireless sensor networks* (HCWSN) where sensors are more and more embedded and they have to work in resource-constraint settings. Resource limitation has a significant impact on the design of a WSN and the adopted fault detection method. This paper investigates a number of fault detection approaches and proposes a fault detection framework based on an energy efficiency perspective. The analysis and design guidelines given in this paper aims at representing a first step towards the design of energy-efficient detection approaches in resource-constraint WSN, like HCWSNs.

1 Introduction

Energy efficiency represents a key research issue in resource-constraint wireless sensor networks (WSNs). In particular, energy efficiency is a crucial aspect of the emerging concept of *Human-Centric WSNs (HCWSNs)*, where sensor nodes are more and more embedded in the environment and even in the human body. These *human-based sensors* must be able to communicate with each other in a resource-constraint, open and dynamic setting. Resource limitations in this kind of networks make fault detection greatly different from traditional, static WSNs. As a result, energy efficiency (which is mainly related to the amount of communication exchanges) becomes a key design aspect to build robust HCWSNs.

C. Orfanidis
DTU COMPUTE, Technical University of Denmark, Lyngby, Denmark

Y. Zhang
Shanghai Key Lab for Trustworthy Computing, East China Normal University, Shanghai, China

N. Dragoni (✉)
Centre for Applied Autonomous Sensor Systems, Örebro University, Örebro, Sweden
e-mail: ndra@dtu.dk

G. Jezic et al. (eds.), *Agent and Multi-Agent Systems: Technologies and Applications*, Smart Innovation, Systems and Technologies 38,
DOI 10.1007/978-3-319-19728-9_24

While current literature discusses detection approaches in different ways, it is hard to find one explicitly discussing the position of message exchanging during the process of fault detection, and how this message exchanging impacts the energy efficiency of a fault detection approach. Yu et al. [39] investigate the three-phase fault management process, i.e., fault detection, diagnosis and recovery. They discuss explicit and implicit detection, centralized and distributed approaches, neighbour coordination, clustering and distributed detection techniques. Paradis and Han [30] also give a survey to fault management in WSNs. They describe fault prevention, detection, isolation, identification, and recovery techniques separately. Mahapatro and Khilar [22] adopt a fault type model from [1] and provide their own taxonomy of fault detection techniques. They discuss both centralized and distributed fault diagnosis approaches. Particularly, they classify distributed approaches into several categories, including Hierarchical Detection, Node Self-Detection, and Clustering-Based Approaches from architectural viewpoint; Test-Based Approaches, Neighbor Coordination Approaches, Soft-Computing Approaches, Watchdog Approaches, and Probabilistic Approaches with their focuses on how to make decision; and also Diagnosis in Event Detection Domain. What is worth mentioning is that the neighbor coordination in [22] concerns majority voting and wighted majority voting, instead of focusing only on coordination between neighbors discussed in [39]. Jurdak et al. [10] present a model including different types of WSN anomalies. They illustrate a set of anomaly detection strategies and divide them according to centralized, distributed and hybrid architectures. They also provide some design guidelines for anomaly detection strategies. Rodrigues et al. [32] evaluate fault diagnosis tools in WSNs in a comparative way. The comparison framework consists of architectural, functional, and dynamic aspects as different dimensions.

Contribution of the Paper. This paper extends current literature by adding a perspective on energy efficiency, as this represents a key aspect to design and build the emerging concept of HCWSNs. In the rest of the paper, we first illustrates our fault detection framework (Sect. 2). Then, in Sect. 3, we present and use a set of evaluation criteria to compare fault detection approaches, with emphasis on energy efficiency. In Sect. 4 we sum up some guidelines for energy-efficient fault detection. Sect. 5 concludes the paper.

2 Fault Detection Framework

The process of fault detection is mainly about making a judgement based on related information. Most of the information is collected within the whole or part of the WSN by message exchanging, which has great impacts on energy efficiency. Here we identify *information collection* and *decision making* as two major components of fault detection framework and describe several design considerations of these two components in the following.

2.1 Information Collection

We focus on three characteristics that can have an impact on message exchanging, namely *Message Exchange Pattern*, *Message Design*, and *Communication Range*. Table 1 lists these characteristics and possible options.

Table 1 Design considerations of information collection

Characteristics		Options	
Message exchange pattern		Active probing	Passive observing
Message desing	Content	Status indication	Sensor readings
	Size	Binary bit	User defined
Communication range		Global	Local

Message Exchange Pattern (MEP) is the way the nodes exchange messages inside the network. Two typical patterns may be used during message exchanging, two-way request-reply and one-way broadcasting. The first one uses pair-wise query-based messages, mostly in hierarchical topologies. In this paper we call it *Active-Probing (AP)*. The second one is called *Passive-Observing (PO)*, which is more common on flat topologies, with messages sent without requested.

Message Design (MD) mainly concerns about the content and the size of the message during the information collection step. The content of the message may be an environmental measurement such as the temperature or a network metric. The content of message is greatly related to the type of fault that the fault detection approach is looking for. For instance, if we have be a periodic "IAmAlive" message, indicating the health status of the node, most probably the fault detection approach is dealing with functional faults. The size of the message is also an attribute that can affect the performance and the energy efficiency. To this end, it is very important to have a tradeoff between the message size and comprehensive meaning.

Communication Range (CR) can be defined by how many sensors are involved during the information collection step. In centralized fault detection most of the times the messages are exchanged among the central node and the nodes in the network. For the case of distributed fault detection approaches the CR may include the one hop neighbours or a set of nodes in a cluster or only one sensor.

2.2 Decision Making

In order to decide whether there is a fault or not, sensor nodes need an input that can be obtained from the exchanged messages. The context information is always application-depended and it is hard to have comprehensive view. We describe the characteristics of the context information as a list of *Assumptions*. The *Calculation Method* and the *Output Range* of calculation are the other critical parts of the decision making phase.

Assumptions (ASMPs). The characteristics of the context of a fault detection approach might have several dimensions. Some of them may be too application-specific to describe. We focus on those which are general enough and organize them according to the components of fault detection in WSN. Except functional, informational and communicational components of WSNs, faults themselves are another fundamental component in fault detection. In Table 2 we illustrate a summary with an indicating name *ASMP_X_i*: *ASMP* stands for the assumption, *X* stands for the component category, it can be *FU* for functional, *IN* for informational, *CO* for communicational components, *FA* for fault itself and *i* stands for the number of the assumption.

Calculation Method (CM). Each approach uses a different calculation method for detecting a fault. A fault may be detected by a threshold test, or by complex inferences based on a specific probability model with temporal and spatial correlation considered. Message exchanging may also happen during inferencing. Some calculations based on inference are carried out in an iterated way, which means some information may be collected again and again until the calculation converges. The information collection usually occurs within temporal or/and spatial correlated nodes.

Table 2 Assumptions in fault detection approaches for WSNs

Label	Description
ASMP-FU-1	The computation for decision making is fault-free
ASMP-FU-2	The sensor nodes are mobile
ASMP-FU-3	The sensor nodes are heterogeneous
ASMP-IN-1	There is a correlation between sensor readings
ASMP-CO-1	The communication channels are fault-free
ASMP-CO-2	The network has a specific topology
ASMP-CO-3	The network needs a certain degree of nodes
ASMP-FA-1	The fault is static
ASMP-FA-2	There is a correlation between faults

Output Range (OR). This states the fault status of the fault detection method. The content, format and size are always application-specific, but the range of the output is related to the network structure. For example, in flat networks without hierarchy, the output is usually about the node itself. On the contrary, in hierarchical networks, like a tree-based, the fault status may concern the children or the parents of the node.

3 Evaluation of Fault Detection Approaches

In this section we introduce a set of evaluation criteria which will be used to evaluate the selected fault detection approaches. Next we present the tables with the obtained data of the selected approaches. Finally we analyze the energy-efficiency and the performance of the fault detection approaches, by using the data from the tables and the evaluation criteria.

3.1 Evaluation Criteria

A fault detection approach can be evaluated as an algorithm, from its computation complexity, correctness, robustness and etc. Mahapatro et al. [22] analyze several terminologies, including correctness, completeness, consistency, latency, computational complexity, communication complexity, diagnosability, detection accuracy, false alarm rate. In this paper, we adopt the following application-independent criteria, which we consider the most relevant ones:

- *Detection Accuracy (DA):* the ratio of the number of faulty nodes detected to the actual number of the actual number of faulty nodes in the network.
- *False Alarm Rate (FAR)*: the ratio of the number of fault-free nodes detected to the actual number of of fault-free nodes in the network.
- *Communication Complexity (COMM)*: the number of messages exchanged in a given network structure in WSN used for detecting faults.

Besides application-independent criteria, there are several application-dependent criteria. Such criteria are the Fault Type (FTYPE), which is what types of fault the approach is able to detect. In WSNs, faults are categorized into different types according to different viewpoints. Ni et al. [27] classify faults with *data-centric* and *system-centric* views. Mahapatro et al. [22] discuss fault types according to the view of *fault-tolerant distributed systems* (Crash, Omission, Timing, Incorrect Computation, Fail-Stop, Authenticated Byzantine, Byzantine faults) and *duration* (Transient, Intermittent, Permanent faults). In this paper, we classify fault types with a more general view according to the *components* of WSNs. Here we mainly focus on three major parts: software and hardware of sensor nodes as functional components, sensor readings as informational components, and networking part as communicational components. Accordingly, there are *Functional Faults (F)*: every hardware or software malfunction which prevents the sensor node to deliver the requested services. *Informational Faults (I)*: sensor readings that are correctly sent from a sensor node, but deviates from the true value of the monitored phenomenon. *Communicational Faults (C)* caused by the network component of the WSN. We also consider some other application-dependent criteria, namely Message Exchange Pattern (MEP) and Communication Range (CR) (Sect. 2.1), and Assumptions (ASMPs), Calculation Method (CM), and Output Range (OR) (Sect. 2.2).

3.2 Evaluation Data

We organize the data extracted from the selected papers in Table 3. The first row list the application-dependent and application-independent evaluation criteria mentioned in Sect. 3.1. Table 4 illustrates the notation for the COMM criterion and Table 5 lists the assumptions of each approach.

3.3 Analysis

To evaluate the energy efficiency of each approach we focus on the relationship between those design considerations listed as different columns in Table 3 and the COMM criterion.

COMM vs. Topology. Some fault detection approaches work with a specific topology (ASMP-CO-2). We focus on *cluster-based* and *tree-based* topology. The tree-based topology requires less messages to complete a fault detection.

COMM vs. MEP. In most cases, the approaches which use active probing as MEP consume more energy. The reason is also obvious, because they require more messages to complete a fault detection and consequently, more energy.

COMM vs. CM. The CM may also affect the energy-efficiency of the process. Here we evaluate three CMs *Bayesian Network-based*, *Message Coordination Protocol*, and *Threshold Test*. The Bayesian network-based CMs use basic principles from the Bayesian network model. The Message Coordination Protocol are based on message exchanging e.g. periodic test with "Hello-IAmAlive" messages. The last category of CMs is based on threshold tests to detect a fault. Regarding the CMs based on Bayesian networks, they appear to be the most energy efficient. Many of them are based purely on a mathematical model and the result is calculated locally. The fact that there is no need of extra messages makes these CMs energy-efficient. The threshold-test CMs are consuming more energy than the previous category. The reason for the increased energy consumption is that the threshold tests are disseminated after being calculated and need extra information to be calculated. The message coordination protocol CMs consume more energy than the previous two categories because it functions with messages which increase in great degree the energy consumption and makes them the least energy efficient between the three categories of CMs.

Performance Analysis. To evaluate the performance of each approach, we investigate the relationships between DA and FAR and some other key criteria.

DA and FAR vs. Topology. The topologies we consider are *cluster-based* and *tree-based*. Figure 1 depicts the mean values regarding the performance of these two topologies. As we can see the approaches using tree-based topology seem to present slightly higher DA but also slightly higher FAR.

DA and FAR vs. Message Exchange Pattern. We calculate the mean value of DA

Table 3 Evaluation data extracted from fault detection approaches

Paper	FTYPE	MEP	MD	CR	CM	OR	DA	FAR	COMM
[35]	I	N/A	Sufficient statistics	Leader node-collaborators	Markov linear state	Cluster	0.88	N/A	$N(H + reading)$
[4]	I	AP	Hello, location, readings	Neighbourhood	Weighted voting	itself	>0.8	<0.2	$N(H) + Nn(H) + Nn(H + double)Nn(H + bool)$
[34]	I	PO	Readings	Node-Sink	Expectation maximization algorithm	Itself	0.95	0.05	$N(H + reading)$
[29]	I	PO	Binary decisions	Fusion center	Threshold test	Itself	N/A	<0.05	$ND(H + reading)$
[6]	I	AP	Readings	Neighbours pairwise	Group voting	Itself	>0.926	<0.014	$Nn(H + reading) + Nn(H + bool)$
[14]	I	PO	Readings (vectors)	Pair-wise	Posterior probability	Itself	>0.73	N/A	$N(H + reading) + N(H + array)$
[5]	I	PO	Readings	Central server	Hierarchical Bayesian space time modelling	Itself	>0.7	0.04 – 0.07	$N(H + reading)$
[25]	I	PO	Readings	Node-Fusion center	Bayesian Networks, Neyman Person	N/A	0.7	0.11	$N(H + reading)$
[26]	I	PO	Readings	Fusion center	Hierarchical Bayesian space time modelling	Itself	0.686 > 0.965	0.142 < 0.023	$N(H + reading)$

(continued)

Table 3 (continued)

Paper	FTYPE	MEP	MD	CR	CM	OR	DA	FAR	COMM
[18, 20]	I	N/A	Readings	Pair-wise	ARX model	Itself	>0.9	N/A	$N(H + reading)$
[19]	I	PO	Readings	Arbitrary group	Group testing, Kalman filtering	Itself	>0.8	<0.02	N/A
[24]	I	AP	Readings	Neighbours	Group voting, time series analysis	Itself	0.8 – 0.92	0.05 – 0.26	$Nn(H + reading) + N(H + reading)$
[38]	I	PO	Readings	Pair-wise	HMM, threshold test	Itself	>0.96	N/A	$N(H + reading)$
[8]	I	PO	Readings, fault reports	Neighbours voting	Neighbour	Itself	>0.8	<0.38	$Nn(H + reading) + fn(H)$
[11]	I	AP	Readings, requests, data	Neighbours	Verifier node mechanism	One hop neighbours	0.88 – 0.98	0.0052	$2(H) + 3N(H) + N(H + reading) + 4N(H + double)$
[7]	I, F	PO	Readings	Neighbours	MRF, correlation with neighbours	Itself	>0.96	<0.01	$Nn(H + reading)$
[33]	I,F	PO	Readings	BS-Nodes	Decision tree J48	Itself	1	0.074	$N(H + reading)$
[3]	F	PO	Readings, test results	Neighbours	Neighbour readings comparison	Itself	>0.97	<0.0025	$Nn(H + reading) + 3Nn(H + double)$
[37]	F	PO	Hello, location, energy, ID	Cluster	Threshold test	Itself	N/A	N/A	$Nn(H+double)+M(H+double)$

(continued)

Table 3 (continued)

Paper	FTYPE	MEP	MD	CR	CM	OR	DA	FAR	COMM
[16]	F	PO	Fault status, useful data	Neighbours	Threshold test decision dissemination	Itself	>0.91	<0.1	$Nn(H + boolean)$
[9]	F	PO	Detection status	Neighbourhood	Based on [3]	Itself	>0.94	N/A	$Nn(H + reading)$
[36]	F	AP	Test packet	Cluster	Comparing test results	Itself	N/A	N/A	$2 * D * LN * (H)$
[12]	F	PO	Notify packet	Cluster	Notify packct declares if the node is alive	Itself	N/A	N/A	$2N(H) + fn(H)$
[15]	F	PO	End-to-End delay, readings	Node-Sink	Centralized naive Bayes Detector	Network, itself	>0.6	<0.05	$N(H + reading)$
[23]	F,C	PO	22 metrics, readings	BS-nodes	Temporal/Spatial correlation in system metrics	BS range	N/A	N/A	$Nn(H + reading)$
[28]	F,C	PO	Readings	BS-Nodes	Failure knowledge library	Itself	>0.9[1] 0.75[2]	>0.35[1] 0.3[2]	$N(H + reading)$
[2]	F,I	AP	Hello packet	Within CHs	Request-reply message mechanism	Itself	0.9	N/A	$2CH(H)$
[31]	F,C	AP	Scenario data input/output	Observer-BS, BS-nodes	Time constraints check	Itself	N/A	N/A	$2OB(H + double) + 2N(H + double)$

(continued)

Table 3 (continued)

Paper	FTYPE	MEP	MD	CR	CM	OR	DA	FAR	COMM
[17]	F,C	PO	Readings, useful data	One hop	Fault detection based on FSM	Itself	>0.9	<0.1	$N(H + double)$
[13]	F,C	AP	Readings, test results	Neighbours	Threshold test, evaluation from neighbours	Itself	N/A	N/A	$N(H + double)$
[40]	F,I,C	PO	Uniform distribution	Neighbours	Threshold test	Itself	1	N/A	$Nn(H + double)$
[21]	F,C	AP	Beacons, local evidence	Parent children	Naive Bayesian classifier, evidence fusion	Parent, children	>0.86	<0.16	$fn(n(2H) + (LN(2H) + m(H + char) + M(h) + (H)))$

Table 4 Notation for evaluating the COMM criterion

H	Header
M	Number of the parents
m	Number of the children
N	Number of the nodes in the WSN
n	Number of the nodes in the neighbourhood
CH	Number of the Cluster Heads
reading	The sensor measurement
ff	Number of fault free nodes
fn	Number of faulty nodes [2, 8]
D	Depth of the tree [36]
LN	Leaf nodes [36]
OB	Number of observer nodes [31]
int	integer variable
array	array variable
bool	boolean variable
double	double variable
char	char variable

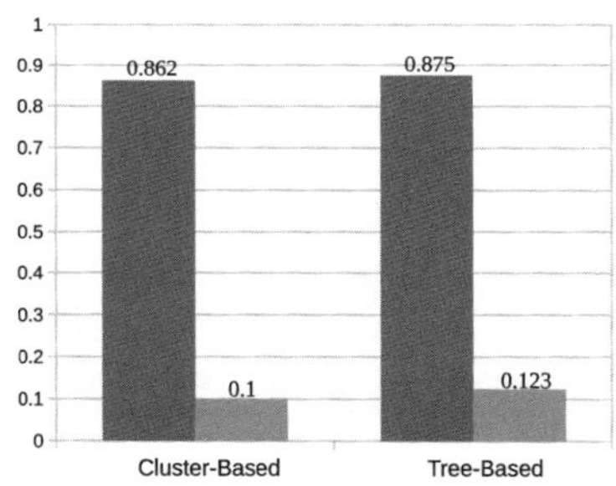

Fig. 1 DA/FAR vs topology

and FAR of the approaches which use PO and AP (Fig. 2). Those using PO have slightly lower DA and higher FAR.

DA and FAR vs. Calculation Method (CM). The CMs we consider are the same as those in the previous part. We can see in Fig. 3 that the threshold test CMs have the highest accuracy and the CMs based on Bayesian network have the lowest DA. Regarding the FAR the Bayesian network CMs have the lowest and the CMs based on message coordination protocols have the highest.

DA and FAR vs. Correlation Assumption. In Fig. 4, we examine how the correlation of the sensor readings (ASMP-IN-1) can affect the performance of an approach. When we adopt ASMP-IN-1, we can achieve higher results in detection accuracy, although the false alarm rate is slightly increased also.

Table 5 Assumptions (ASMPs) of Fault Detection Approaches

Paper	FU_1	FU_2	FU_3	IN_1	CO_1	CO_2	CO_3	FA_1	FA_2
[4]	Y	N	N	Y	Y	Y	N	Y	N
[9]	Y	N	N	Y	Y	N	Y	Y	N
[18, 20]	Y	N	N	Y	Y	N	N	Y	N
[23]	Y	N	N	Y	Y	Y	N	Y	N
[14]	Y	N	N	Y	N/A	Y	N	Y	N
[5]	Y	N	N	Y	N/A	N	N	Y	N
[25]	Y	N	N	Y	N/A	N	N	Y	N
[26]	Y	N	N	Y	Y	N	N	Y	N
[29]	Y	N	N	N	Y	N	N	Y	N
[15]	Y	N	N	N	N	N	N	Y	N
[34]	Y	N	N	N	N	Y	N	Y	N
[28]	Y	N	N	N	N/A	N	N	Y	N
[33]	Y	N	N	Y	N	N	N	Y	N
[37]	Y	N	N	Y	Y	Y	N	Y	N
[36]	Y	N	N	N	N	Y	N	Y	N
[12]	Y	Y	N	N	N/A	Y	N	Y	N
[2]	Y	N	N	N	N	Y	Y	Y	N
[31]	Y	N	N	N	N	N	N	Y	N
[17]	Y	N	N	N	N	N	N	Y	N
[16]	Y	N	N	Y	N/A	N	Y	Y	N
[19]	Y	N	N	Y	Y	N	N	Y	N
[3]	Y	N	N	Y	Y	N	Y	Y	N
[13]	Y	N	N	Y	N	Y	N	Y	N
[6]	Y	N	N	Y	Y	Y	N	Y	N
[40]	Y	N	N	Y	N	N	N	Y	N
[24]	Y	N	N	Y	N	N	N	Y	N
[21]	Y	N	N	N	N	Y	N	Y	N
[8]	Y	N	N	Y	Y	N	N	Y	N
[7]	Y	N	N	Y	N/A	N	N	Y	N
[35]	Y	N	N	N	N/A	Y	N	Y	N
[38]	Y	N	N	Y	N/A	N	N	Y	N
[11]	Y	N	N	Y	N	Y	Y	Y	N

Fig. 2 DA/FAR vs. MEP

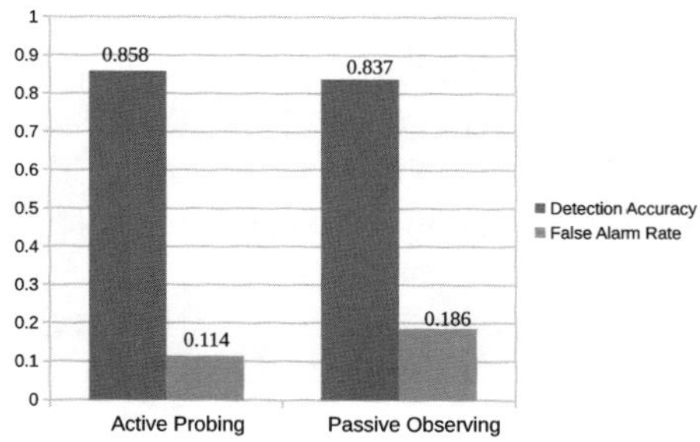

Fig. 3 DA/FAR vs. CM

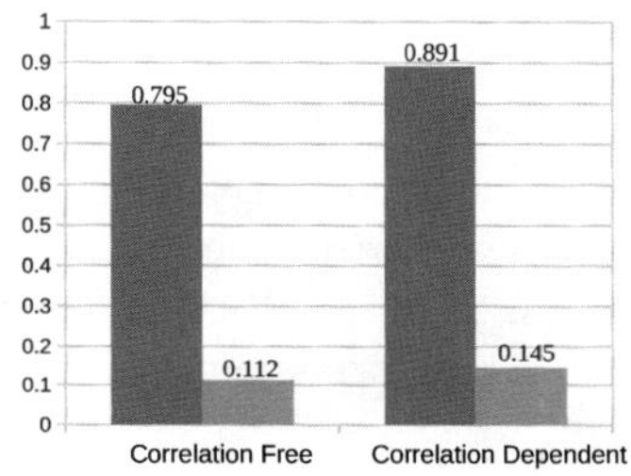

Fig. 4 DA/FAR vs. ASMP-IN-1

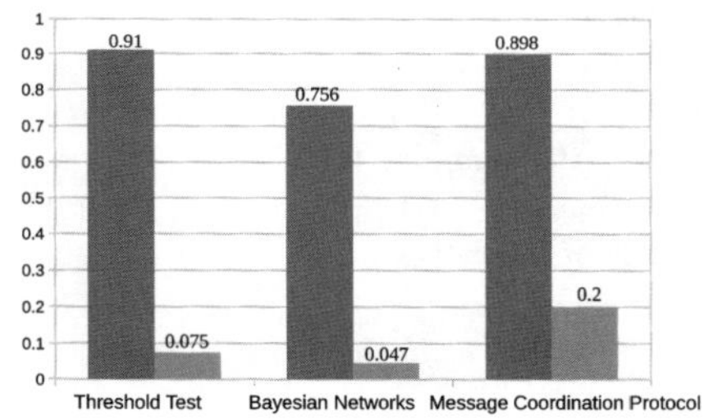

4 Design Guidelines

Designing an appropriate fault detection method for a WSN is a not easy task. Since WSN applications are dependent on the requirements and on the deployment environment, each fault detection method should be designed regarding application specific criteria.

Over the selected approaches the topologies we examined are the cluster-based and the tree-based. An advice regarding the topology is that the tree-based may consume less energy than a cluster-based. If a designer has the option to choose between the two MEPs, the PO is the more energy efficient one. The CMs we distinguish over the selected approaches are the *threshold-test, Bayesian Network-based* and *message coordination protocol*. The Bayesian Network-based CM resulted to be more energy efficient than the others.

Regarding the performance, the topology in fault detection approaches cannot offer tremendous changes, but between the cluster-based and tree-based topologies, the former may have slightly lower FAR and the latter little more DA. Regarding the option of the Message Exchange Pattern, by using the Passive Observing we may have lower FAR but for having slightly higher DA we have to use the Active Probing. According to the selected fault detection approaches, the Calculation Method which offer the higher DA is the threshold test and the one which offer lower FAR is the Bayesian networks. We have to mention that the CMs we consider are the same as the previous section.

Finally, if the design is based on the correlation of the sensor readings, it will have higher DA but slightly higher FAR.

5 Conclusion

This paper complements current literature on fault detection methods for WSNs by adding a perspective on energy efficiency, as this represents a key aspect of future Human-Centric WSNs. In particular, we have proposed a two-phase fault detection process, information collection and decision making, with emphasis on figuring out where and when message exchanging occurs. After defining application-independent and application-dependent evaluation criteria, we have investigated the relationships between some major design factors and performance parameters and also summed up some design guidelines.

Acknowledgments This paper has been supported by the China Scholarship Council during Yue Zhang's visit at DTU in the context of the IDEA4CPS project and National Natural Science Foundation of China (No. 61361136002).

References

1. Barborak, M., Dahbura, A., Malek, M.: The consensus problem in fault-tolerant computing. ACM Comput. Surv. **25**(2), 171–220 (1993)
2. Bhatti, S., Xu, J., Memon, M.: Energy-aware fault-tolerant clustering scheme for target tracking wireless sensor networks. In: ISWCS Conference (2010)
3. Chen, J., Kher, S., Somani, A.: Distributed fault detection of wireless sensor networks. In: DIWANS Workshop, p. 65 (2006)
4. De, D.: A distributed algorithm for localization error detection-correction, use in in-network faulty reading detection: applicability in long-thin wireless sensor networks. In: IEEE WCNC Conference, pp. 1–6 (2009)
5. Dereszynski, E.W., Dietterich, T.G.: Spatiotemporal Models for Data-Anomaly Detection in Dynamic Environmental Monitoring Campaigns. ACM TOSN (2011)
6. Dobson, S., Hughes, D.: An Error-free Data Collection Method Exploiting Hierarchical Physical Models of Wireless Sensor Networks
7. Farruggia, A., Vitabile, S.: A novel approach for faulty sensor detection and data correction in wireless sensor network. In: Conference on BWCCA (2013)
8. Gao, J., Wang, J., Zhang, X.: Hmrf-based distributed fault detection for wireless sensor networks. In: IEEE GLOBECOM Conference, pp. 640–644 (2012)
9. Jiang, P.: A new method for node fault detection in wireless sensor networks. Sensors **9**(2), 1282–1294 (2009) (Basel, Switzerland)
10. Jurdak, R., Wang, X.R., Obst, O., Valencia, P.: Chapter 12 Wireless Sensor Network Anomalies: Diagnosis and Detection Strategies, pp. 309–325 (2011)

11. Kamal, A.R.M., Bleakley, C., Dobson, S.: Packet-level attestation (pla): a framework for in-network sensor data reliability. ACM Trans. Sen. Netw. **9**(2) (2013)
12. Karim, L., Nasser, N.: Energy efficient and fault tolerant routing protocol for mobile sensor network. In: IEEE ICC Conference, pp. 1–5 (June 2011)
13. Khazaei, E., Barati, A., Movaghar, A.: Improvement of fault detection in wireless sensor networks. In: CCCM 2009. ISECS Colloquium, vol. 4 (2009)
14. Kim, D.J., Prabhakaran, B.: Motion fault detection and isolation in body sensor networks. In: IEEE PerCom Conference, pp. 147–155 (2011)
15. Lau, B.C., Ma, E.W., Chow, T.W.: Probabilistic fault detector for wireless sensor network. Expert Syst. Appl. **8**, 3703–3711 (2014). Jun
16. Lee, M.H., Choi, Y.H.: Fault detection of wireless sensor networks. Comput. Commun. **31**(14), 3469–3475 (2008)
17. Liu, K., Ma, Q., Zhao, X., Liu, Y.: Self-diagnosis for large scale wireless sensor networks. In: Proceedings of IEEE INFOCOM (2011)
18. Lo, C., Liu, M., Lynch, J.: Distributive model-based sensor fault diagnosis in wireless sensor networks. In: IEEE DCOSS Conference (2013)
19. Lo, C., Liu, M., Lynch, J., Gilbert, A.: Efficient sensor fault detection using combinatorial group testing. In: International Conference on IEEE DCOSS (2013)
20. Lo, C., Lynch, J.P., Liu, M.: Pair-wise reference-free fault detection in wireless sensor networks. In: IPSN Conference, pp. 117–118. ACM, NY (2012)
21. Ma, Q., Liu, K., Miao, X., Liu, Y.: Sherlock is around: detecting network failures with local evidence fusion. In: INFOCOM 2012 (Mar 2012)
22. Mahapatro, A., Khilar, P.M.: Fault diagnosis in wireless sensor networks: a survey. IEEE Commun. Surv. tutorials **15**(4), 2000–2026 (2013)
23. Miao, X., Liu, K., He, Y., Liu, Y., Papadias, D.: Agnostic diagnosis: discovering silent failures in wireless sensor networks. In: IEEE INFOCOM (2011)
24. Nguyen, T.A., Bucur, D., Aiello, M., Tei, K.: Applying time series analysis and neighbourhood voting in a decentralised approach for fault detection and classification in WSNs. In: SoICT, pp. 234–241. ACM Press, New York, USA (2013)
25. Ni, K., Pottie, G.: Bayesian selection of non-faulty sensors. In: IEEE International Symposium on Information Theory (ISIT 2007), pp. 616–620 (June 2007)
26. Ni, K., Pottie, G.: Sensor network data fault detection with maximum a posteriori selection and bayesian modeling. ACM Trans. Sen. Netw. **8**(3) (2012)
27. Ni, K., Ramanathan, N., Chehade, M.N.H., Balzano, L., Nair, S., Zahedi, S., Kohler, E., Pottie, G., Hansen, M., Srivastava, M.: Sensor network data fault types. ACM Trans. Sen. Netw. **25**:1, 25:29 (2009)
28. Nie, J., Ma, H., Mo, L.: Passive diagnosis for wsns using data traces. In: IEEE 8th Conference on Distributed Computing in Sensor Systems (DCOSS) (2012)
29. Pai, H.T.: Reliability-based adaptive distributed classification in wireless sensor networks. IEEE Trans. Veh. Technol. **59**(9) (2010)
30. Paradis, L., Han, Q.: A survey of fault management in wireless sensor networks. J. Netw. Syst. Manage. **15**(2), 171–190 (2007). Jun
31. Ramassamy, C., Fouchal, H., Hunel, P., Vidot, N.: A pragmatic testing approach for wireless sensor networks. In: ACM Q2SWinet Conference, pp. 55–61 (2010)
32. Rodrigues, A., Camilo, T., Silva, J.S., Boavida, F.: Diagnostic Tools for Wireless Sensor Networks: A Comparative Survey, vol. 21 (Jun 2012)
33. Salem, O., Guerassimov, A., Mehaoua, A., Marcus, A., Furht, B.: Sensor fault and patient anomaly detection and classification in medical wireless sensor networks. IEEE ICC Conference, pp. 4373–4378 (2013)
34. Shakshuki, E.M., Xing, X., Sheltami, T.R.: An intelligent agent for fault reconnaissance in sensor networks. In: iiWAS Conference, pp. 139–146, ACM (2009)
35. Snoussi, H., Richard, C.: Wsn06-5: Distributed bayesian fault diagnosis in collaborative wireless sensor networks. In: IEEE GLOBECOM, pp. 1–6 (2006)

36. Taleb, A., Mathew, J., Pradhan, D.: Fault diagnosis in multi layered de bruijn based architectures for sensor networks. In: IEEE PERCOM Conference (2010)
37. Venkataraman, G., Emmanuel, S., Thambipillai, S.: A cluster-based approach to fault detection and recovery in wireless sensor networks. In: ISWCS (2007)
38. Warriach, E., Nguyen, T.A., Aiello, M., Tei, K.: A hybrid fault detection approach for context-aware wireless sensor networks. In: IEEE Conference on MASS (2012)
39. Yu, M., Mokhtar, H., Merabti, M.: A Survey on Fault Management in Wireless Sensor Networks (2007)
40. Zhuang, P., Wang, D., Shang, Y.: Distributed faulty sensor detection. In: IEEE GLOBECOM Conference, pp. 1–6 (2009)

Thinking Lifecycle as an Implementation of Machine Understanding in Software Maintenance Automation Domain

Alexander Toschev and Max Talanov

Abstract The main goal of our work is to test the feasibility study of automation of incident processing in Infrastructure as Service domain to optimize the operational costs of management services that are delivered remotely. This paper also describes a framework that authors have developed to deliver an integrated incident, problem solution and resolution approach as an event-driven Automated Incident Solving System, for Remote Infrastructure Management (RIM) Model. Current approaches are mainly automated scripts, but this is a specific approach for one specific problem. Those systems can't think. Our approach is a system that exploits a thinking model thus can think and can learn. In other words system is capable of recombining its knowledge to solve new problems. Based on Minsky [11] thinking model we have created a machine understanding prototype which is capable of learning and understanding primitive incident description texts.

Keywords Artificial intelligence · Machine understanding · Remote infrastructure management · NLP · Reasoning · Automation · Knowledge base · Intelligent agents

1 Introduction

Our inspiration for this work is MIT Metafor [8–10], a tool that generates Python classified according to sentences in natural language (shallow English). Our assumption is: human understanding is completely based on thinking. The goal of the project is to study and propose system architectural design that is totally based on a human concept thinking process that is highly optimized for a RIM, Cloud-based offering. We used Marvin Minsky mind model taken from "The emotion machine" book [11] to create a thinking/understanding system, that is capable of learning and as a consequence, solving a range of problems or incidents. We explored different approaches for machine understanding described; for example "Artificial Intelligence The Modern Approach" [14] and found that Minsky's approach that we consider as the

A. Toschev (✉) · M. Talanov
Kazan Federal University, Kazan, Russia

G. Jezic et al. (eds.), *Agent and Multi-Agent Systems: Technologies and Applications*, Smart Innovation, Systems and Technologies 38,
DOI 10.1007/978-3-319-19728-9_25

most promising [13]. We also performed an analysis of a dump with thousands of incidents taken from Incident Management Systems and found a lot of repeatable and trivial incidents like: “Please install Firefox”, “I can’t open PDF files” [3]. Also team composition on project in IT maintenance domain has a lot of low-level specialists [2].

However, in our work we do not limit ourselves to one solution and reuse several frameworks for reasoning, for example, OpenCog [5], NARS [12].

At the current level our framework has version 1.0. We start evolution for automation from Decision tree in version 0.1 and found that this solution is not flexible. (see Menta 0.1 http://tu-project.com/demo/). After that we use Genetic algorithms to found solution, but they are very slow (see Menta 0.2 http://tu-project.com/demo/). In version 1.0 we used Minsky approach and found it suitable for flexibility and speed.

2 Materials and Methods

For the automated handling of events (incident, problem tickets via the first line of Service Desk support) or in Incident Management Systems, the core principle that has been adopted is to classify the highest priority event into an appropriate “class” and take a simple corrective action in response when applicable (e.g., restart a failed process or an interface), followed by increasingly more obtrusive actions (reboot, reimage, mark-as-failed) if simple actions do not fix the fault.

Additional aspects of automation include the creation and resolution (when possible) of problem tickets describing the incident, and influencing the placement engine to prevent provisioning for example of VMs on faulty or overloaded hypervisors.

We developed a framework for incident event-based automation solution search using Sensitive Emotion Machine to understand infrastructure IT domain elements, enhanced with state persistence to maintain a machine self-holistic view of the structure of subject matter, maintenance of event-action history for improved system self-decision making, and fault tolerance for dependability, features typically unavailable in off-the-shelf management systems. The IPM framework is described below.

Artificial Intelligence Approach as a vital core of the presented framework to infrastructure elements servers (hardware and hypervisor software providing virtualization services), network components, and shared storage components as an IT infrastructure Knowledge Base . The current system prototype handles approximately 10 different events. Each instance of incoming incident is used to treat and understand by system a clients infrastructure element from “birth” (deployment) to “death” (removal), based on events received from the Service Desk ticket aggregation and correlation system. An event that represents incidents (faults) in a given IT element is acted upon based on the type of event, the current state of the IT element in the clients infrastructure environment, the history of past events and actions taken on that clients system, and the working policies (instruction) associated with the resulting state transition encoded in the IPM system.

At the early stage we performed research and analysis of incident dump and typical tasks of IT Support team from the large IT Outsourcing Company. We have analyzed an option to automate several actions of Infrastructure as service domain. Typical domains are:

- Windows Intel (Wintel) domain (Windows operating System)
- Microsoft Active Directory Management
- UNIX management
- Storage management
- We created and implemented architecture for Marvin Minsky model and finally implemented application of it.

We created and implemented architecture for Marvin Minsky model and finally implemented application of it. As a NLP parser we used modified NLP RelEx [6] from OpenCog [5] project. To increase results quality we used public API for grammar correction: After The Deadline [1], Google. We created learning mechanism based on Minsky [11] approach to be able to train system.

3 Related Work

There is no documented evidence of such complex systems that can be collected officially. Even though some vendors of IT support software work in this direction, there is still no data supporting this.

We used results of many researchers in artificial intelligence. For natural language processing we use OpenCog Relex [6]. As a reasoning mechanism OpenCog PLN is used [4].

In 2010 we start our work to solve Software Development automation problem [15], but we found that amount of complex problems in Software Development are much higher than in IT infrastructure maintenance domain. That was the reason to change target domain. However, we use genetics algorithms to solve problem, now the whole approach has been revisited. Previously we have tried different approaches. See Sect. 1.

Due to application context of the work it can be compared to IBM Analytics [7], where the same problem trying to be solved. Also, wolfram alpha contains such knowledge system [16].

4 Solution

We base our solution on the Minsky approach [11] and use *Critic− > Selector− > Way To Think* triplet, six level thinking model, but implemented only five:

1. Learned
2. Deliberative
3. Reflective

4. Self-Reflective
5. Self-Conscious

Our system supports two modes: learning and operating. To better understand our solution lets make 3 input examples:

1. User Request: "Please install Firefox"
2. Train request: "Firefox is a browser"
3. Complex Request: "Office 2007 installed but Office 2010 required"

According to the Minsky approach, level 1 is at the bottom and level 5 indicates a top thinking level.

Critic represents some kind of trigger. In everyday life individuals face problems and tasks, such as "Open the door", "Eat with a spoon". Critic is a reaction to this problem. In ITSM domain, for example, when an auto generated incident comes to the queue the Auto Generated Incident critic will be activated. After activating, Critic becomes a Selector. From another point of view activated Critic is a Selector.

Selector is capable of retrieving Resources (Critic or Way to think) from memory.

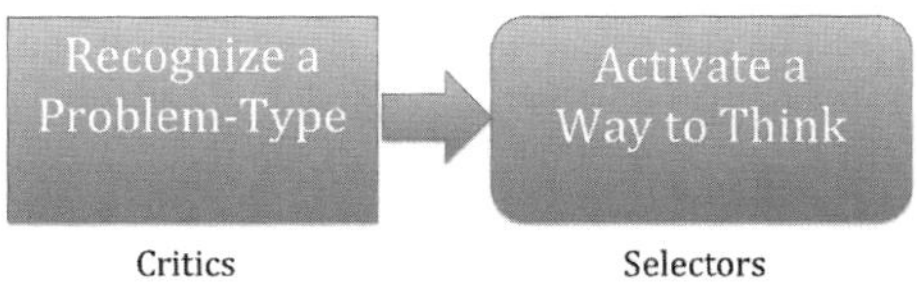

Fig. 1 Critic selector way to think collaboration

Selector can activate another Critic or a Way To Think (see Fig. 1).

The Way To Think according to Minsky is a way how a human thinks. For example, "If I know the problem, I use my experience in analogy situation to solve it". For example we have several types of way-to-think:

– Simulation
– Correlation
– Reformulation
– Thinking by analogy

Practical example 1. If an incident is automatically generated, the system should process it using instruction book A.

Practical example 2. If a system recognizes the problem, use analogy to solve it. Way To Think in current implementation is a worker that modifies short term memory.

4.1 Learned Level

On the first level the system uses previously obtained knowledge. The knowledge is obtained via training including communication with human expert (See for more info Sect. 4.7). On this level there are several types of components used:

1. Preprocess manager
2. Direct Instruction Analyzer
3. Problem With Desired State Analyzer
4. Problem Without Desired State Analyzer

Preprocess manager. This manager activates several Ways To Think to perform initial incident processing. The goal of this critic is to prepare incident description. There are several Way To Thinks:

- Auto Correction of spelling
- Synonymic search
- Annotation finding existing concepts in Knowledge Base

Direct Instruction Analyzer. This Critic is activated when direct instruction is detected in inbound request. For example, "Please install MS Office 2012" is a direct instruction for the system.

Problem With Desired State Analyzer. Critic is activated when a problem with desired state detected by the system. For example, I have Internet Explorer 8 installed, but finance software required is Internet Explorer 7. In other words, Problem With Desired State is a class of problems descriptions that contains current state and desired state descriptions.

4.2 Deliberative Level

On the deliberative level systems search for solution of the problem. The solution is stored in the KB as a link to recognized concepts. For example: "Install Firefox" recognized. "Install Firefox" has link to "InstallFirefoxSolution". However, there are could be several solution linked, especially in case of using induction or deduction during reasoning. System will try to apply different solutions, after successful application of a solution (The user estimates if solution successful or not) the score for this solution will be increased and in new request that match the same concepts of this solution will be used as priority one.

4.3 Reflective Level

On Reflective level system sets goals and monitors for SLA. On the reflective level the system sets processing goals, performs time control, solution completeness manager. Processing goals. Processing goals required to increase performance of incident processing. They have linked critics, way to think Main goal is a "Help User". There are other goals, that derive from it, for example:

- Resolve incident
- Understand incident type
- Model Direct Instruction

Time control. The time control watches across the thinking levels for time of incident processing. (SLA in terms of ITSM) Solution completeness manager. This manager searches for how-to and solution for current incident.

4.4 Self-Reflective Level

On this level the system controls results of lower levels such as initial context or start time control. All communication of a user is also regulated on this level by Do Not Understand Manager and communication with end-user. Human-user communication is activated when system faces unresolvable problem.

4.5 Self-Conscious

On the Self-conscious level the system traces its Emotional State. For example: reacting for long incident processing, system changes emotional state to allocate resources for processing.

4.6 Conceptual Process Stages

There are several conceptual stages in system in cooperation with Minskie's levels. The all concept stages in system processed on 2 levels.

1. NLP Processing (Learned level)
2. KB Annotating (Learned level)
3. Simulation (Learned level)
4. Reformulation (Learned level)
5. Solution Search (Deliberative)
6. Background processes
   (a) SLA control, local resource control (Reflective level)
   (b) User communication (Self-Reflective level)
   (c) Emotional state (Global Resource control) (Self-Conscious)

Background processes is not directly involved in processing stages, but they are very important. Emotional state is very important as a resource indicator. For example, in the "rush" emotional state there are insufficient resources. 6.a for local resource control for single request in comparison with 6.c where are global resource indicators. (System Dashboard).

4.7 Knowledge Base

Systems knowledges based on graph of concept. System has 2 base concepts: Action and Object. All other concepts can be trained through the training mode or communication with human expert. For example, in training mode we system can study base concepts.
H(Human Expert): Software is an object.
S(System): Ok.
H: Internet explorer is a browser.
S: I can't understand browser.
H: Browser is a software.
S: Ok.
H: Internet explorer is a browser.
S: Ok.

Such graph representation introduce possibility of applying different logic style:

- Deduction
- Induction
- Abduction
- e.t.c.

4.8 Memory

As a memory conception we used Short-term memory and Long-term memory. Short-term memory is a context for current request. Long-term memory is a global system memory. During the request systems acquires all necessary resources from Long-term memory and put them into Short-term memory context. During the request system can ask help from the human expert. Moreover, using the logic system can produce new knowledge. For example, system knows solution for "MSI software installation", system also knows that "MS Office is a MSI package". As a result system can create link and produce new knowledge "MS Office installation". New knowledge will be stored in Short Term memory (request context). After success request execution Short Term memory will be merged to Long Term memory and will be erased.

4.9 Sample

When first incident comes to the queue, Reflective level sets new goals and system starts processing from 1st level. It processes by NLP and Knowledge Base annotator on the Learned level (stages 1 and 2 See Sect. 4.6), which searches learned concept in local Knowledge Base and corrects NLP results. Knowledge Base annotator searches using the generalization and specialization mechanism. In other words, if the system knows browser concept and knows that "Firefox" is a "browser" it will find all concepts linked to browser.

After this system creates simulation for a complex problem (stage 3 and 4). In our example this is Case 3. Simulation creates problem description and finds that office 2010 required.

On the Deliberative level system searches for solution (stage 5).

In the second case Firefox is a browser system processes Stage 1 and 2. Then the system searches for known concepts and creates links between unknown and known concepts between Firefox and browser in our case.

In 3rd case "Office 2007 installed but Office 2010 required". System proceeds through Stage 1 to 5. Detailed example can be shown as dialog:
H(Human user): Please install firefox.
S(System): *Stage 1. Process request through NLP* (Italic means hidden for user actions).
S: *Stage 2. System finds concepts in database. For "please" - "form of politeness", for "install"-"install action", for "firefox"-"FirefoxConcept".*
S: *Stage 3, Stage 4 skipped because problem is direct instruction.*
S: *Stage 5. Find solution "InstallFirefoxSolution".*
S: Done. Does solution suitable for you?
H: Yes.

5 Conclusions

We prove an approach of automation in case of simple incidents. As a result, we have created architecture for Minsky model and created a proof of concept for the automated incident solving. We analyzed incidents in IT Outsourcing domain and found 3 types of typical incidents [2]:

1. Direct instruction (Please do something)
2. Problem with desired state (I have A but I need B)
3. Problem without desired state (I don't have B)

We processed 3 types of control incidents Fig. 2. However, the ability to learn can easily extend these types. Additional demo can be found at http://tu-project.com/demo/. The dump for the research contains 1895 incidents collected from IT maintenance company dump for 3 month. We studied dump and found 43 category.

All of categories are objects from base class of problem (see Sect. 4.1). The categories have been assigned by technical specialists from IT maintenance team. Top most:

- Invalid login
- Connectivity problem
- Can't find software
- Reinstall software
- Wrong application has been installed
- Insufficient disk space
- Shared disk connection required
- Invalid computer time
- Setup Wifi
- Add Alias for host
- No access to printer
- Insufficient rights
- Need additional info
- Clean profile after migration

We select 45 incidents over the all categories and process them using the prototype. The results presented on Fig. 2 (Table 1).

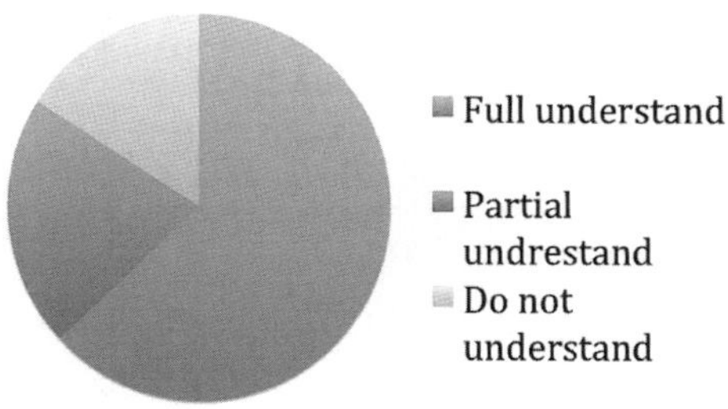

Fig. 2 Incident understanding diagram

Table 1 Incident processing results numbers

Processing result	Quantity	Summary
Full understand incidents	63 %	All concepts have been understood
Partial understand incidents	21 %	Some concepts do not understand correctly, but solution can be found
Do not understand	16 %	Most of the concepts haven't been understood, solution can't be found

5.1 Deployment Experience

The Automation Incident Handling component is currently deployed in VM infrastructure and can be presented for operational build in a six-eight months period. The scalability and persistence features of commercial version will meet the current and according road map design goals during steady state operations and as it based on business requirements to system under Remote Infrastructure Management business model.

However, implementation based on Java platform and can be easily run on different platform like:

- Windows
- Linux
- Mac OS X

Prototype implementation supports multi-threading as part of architecture and can be easily deployed on multiprocessor systems.

References

1. After the deadline, Web (04 2012). http://www.afterthedeadline.com/
2. Alexander, T.: Thinking model and machine understanding in automated user request processing. In: CEUR Workshop Proceedings, vol. 1297, pp. 224–226 (2014)
3. Alexander Toschev, M.T.: Analysis results for it outsourcing, Web (04 2013). http://tu-project.com/for-business/
4. Goertzel, B.: Probabilistic Logic Networks. Springer, A Comprehensive Framework for Uncertain Inference (2008)
5. Goetzel, B.: Opencog, Web (04 2012). http://opencog.org/
6. Goetzel, B.: Opencog relex, Web (04 2012). http://wiki.opencog.org/w/RelEx
7. IBM: Ibm watson, Web (00 2014). http://www.ibm.com/analytics/watson-analytics/
8. Lieberman, H., Hugo Liu, in H. Lieberman, F.P., Wulf, V.: Feasibility studies for programming in natural language. End-User Development (2005)
9. Liu, H., Lieberman, H.: Metafor: Visualizing stories as code. In: IUI 05–2005 International Conference on Intelligent User Interfaces (2005)
10. Liu, H., Lieberman, H.: Programmatic semantics for natural language interfaces. In: ACM Conference on Computers and Human Interaction (CHI-2005) (2005)
11. Minsky, M.: The Emotion Machine. Simon & Schuster (2006)
12. Pei, W.: Rigid Flexibility: The Logic of Intelligence. Springer, The Netherlands (2006)
13. Sloman, A., Chrisley, R.: Virtual machines and consciousness. J. Conscious. Stud. (2004)
14. Russell, S., Norvig, P.: Artificial intelligence. A Modern Approach. Pearson (2010)
15. Talanov, M., Krekhov, A., Makhmutov, A.: Automating programming via concept mining, probabilistic reasoning over semantic knowledge base of se domain. In: 6th Central and Eastern European Software Engineering Conference (CEE-SECR) (2010)
16. Wolfram: Wolfram alpha, web (00 2014). http://www.wolframalpha.com/

Towards Anthropo-Inspired Computational Systems: The P^3 Model

Michael W. Bridges, Salvatore Distefano, Manuel Mazzara, Marat Minlebaev, Max Talanov and Jordi Vallverdú

Abstract This paper proposes a model which aim is providing a more coherent framework for agents design. We identify three closely related anthropo-centered domains working on separate functional levels. Abstracting from human physiology, psychology, and philosophy we create the P^3 model to be used as a multi-tier approach to deal with complex class of problems. The three layers identified in this model have been named PhysioComputing, MindComputing, and MetaComputing. Several instantiations of this model are finally presented related to different IT areas such as artificial intelligence, distributed computing, software and service engineering.

Keywords Physiology · Psychology · Philosophy · Neuroscience · Layered model

M.W. Bridges
ICarnegie Global Learning, Pittsburgh, USA

S. Distefano (✉)
Politecnico di Milano, Milano, Italy

M. Mazzara
Innopolis University, Kazan, Russia

M. Minlebaev
Institut de Neurobiologie de la Méditerranée, Marseille, France

M. Minlebaev
Aix-Marseille University, Marseille, France

S. Distefano · M. Minlebaev · M. Talanov
Kazan Federal University, Kazan, Russia
e-mail: max.talanov@gmail.com

J. Vallverdú
Universitat Autònoma de Barcelona, Catalonia, Spain

G. Jezic et al. (eds.), *Agent and Multi-Agent Systems: Technologies and Applications*, Smart Innovation, Systems and Technologies 38,
DOI 10.1007/978-3-319-19728-9_26

1 Introduction

Emulation of functional aspects of living systems have been so far based on the perspective of linear interaction between sets of pre-programmed subsystems. As a consequence, the exhibited behavior is that of modular systems respectively responsible for specific functional domains.

Development of cognitive science has recently shown as living beings actually integrate several functionalities which operate in parallel and emergence of higher functions like consciousness may not be possible without the complex cooperation of such sub-systems.

In this paper, we introduce a three domain model providing a computational framework for an ecosystem of cognitive entities. This approach is then extended to computing systems and then instantiated to specific IT domains such as robotics, services or networks. The model has to be seen as a multi-views representation of phenomena of like artificial consciousness, artificial creativity and artificial intuition. Each view is best illustrated by an analogy with well-known anthropocentric domains: physiology, psychology, philosophy (P^3).

The framework can be seen as a set of perspectives in order for researchers to grasp a better understanding of a specific phenomena, starting from the lower level (i.e. hardware and software implementation) up to the higher (i.e. artificial intelligence (AI) or cognitive robotics).

2 The P^3 Anthropocentric Model

Human brain follows a non centralized subsumption architecture in which consciousness is a very important but not determinant managing layer. We must operate into dynamical environments out of our full control. It explains why, according to their internal states, memories or necessities, people faced to similar environmental conditions react differently. It can be justified by the fact that the topology of cognition is distributed all along the body, the cultural sphere and the brain. Up today AI attempts to reproduce or emulate human intelligence have failed in the achievement of similar adaptative and multilayered systems. Our aim is to produce more efficient, problem-solving and innovative AI systems providing a new architecture design.

We consider three domains as the real modules of any truly intelligent system which produces complex behaviors, reasonings and belong a consciousness, as shown in Fig. 1, bottom up: physiology, psychology and philosophy of P^3 model. This is the starting point towards a new approach where we consider these domains as layers of a layered model. Each layer implements specific functionalities and interacts with the adjacent ones through specific interfaces.

Fig. 1 The P^3 model

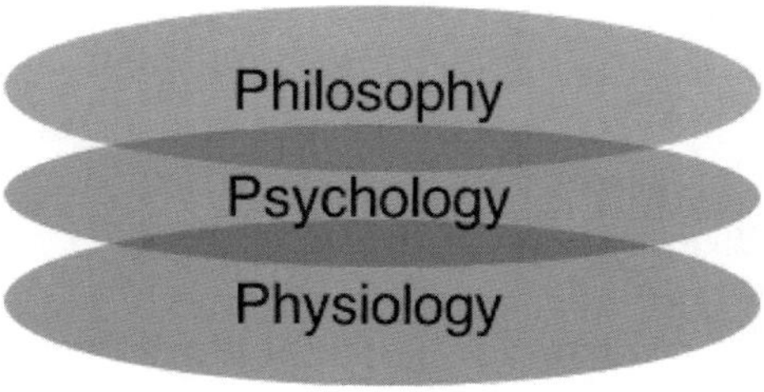

2.1 The Layers

Physiology Physiology is the study of basic normal function in a living system, focusing on organisms and on their main physical functions [1]. The subdomain of physiology that studies the central nervous system is neuroscience. The principal functional unit of our brain is neuronal cell, i.e. the "brick" of central nervous system. The principal difference of the neuron from major part of the cells is "excitability". That as well as an option to transfer the information between the neuronal cells allow to transmit the signal along multiple cells. The intercellular connections are selforganised that and neuronal chains form networks that serve specific particular functions. The simplest example is knee reflex, when light kick of the knee excites the receptor part of the sensory neuron that carries the sensory signal to the spinal cord, where it evokes the activity in intermediate cell that in its turn activates the motor neuron making the contraction of the hip muscles.

Thus, the neuronal network is an organized structure of the neurones, that serves specific function, i.e. "brick wall". However, presence of multiple factors i.e. different cell classes (inhibitory and excitatory) and diversity of the neuronal interconnections (axosomatic, axoaxonal, perisomatic, axodendritic etc.) and other factors (such as extracellular medium content, that affects electrical properties of the cells, neuronal plasticity of the synaptic strengths based on the previous experience of the cell) provides the enormous freedom for the functionality of our neuronal system for decision making even at the microscopic level. Considering the entire brain, the direct measurements of the immediate state of the small neuronal networks allow us to describe the general principles of the organisation of the central nervous system and its functioning but are hardly sufficient for analysis and prediction of the results of the concrete complex task, like decision making. We have to take into account the interactions between multiple functionally linked structures of the brain, previous experience (including acquired skills) and the temporary the state of mind to do that. This in its turn places us into completely another domain of science investigating our thinking brain — psychology.

Psychology Psychology studies the behavior and mental activity of human beings according to the brain processes with which are correlated [4]. Psychologists cover a broad range of interests regarding mental processes or behaviors: from the human being as an individual being (internal data), as well as considering the human as a node of a web of social, collective and ecological interactions (external). Although cultural variables are very important in order to understand the role of psychic events into human actions [7], we can affirm that psychology deals more directly with the several brain data processing mechanisms that operate at a presymbolic level. Consequently aspects like attention, visual field selection, interaction management, time perception, mental disorders, among others, belong to the field of this research area. Other very important mechanisms that fall under the general category of mental processes are attentional mechanisms, decision-making mechanisms, language processing mechanisms or social processing mechanisms [8]. As stated above, Psychology is not directly involved into the neural mechanisms that are present during mental actions,but on the functional mental managing of processed data. Let us use a metaphor: if we take brain as hardware, then psychology is the operating system, and philosophy is the software. Psychologists work with the ways brain transforms raw data into informational data, from several domains: cognition (thought), affect (emotion), and behaviour (actions). In combination, these fundamental units create human experience.

For instance, the visual sensory system may detect a large, dark, fury shape approaching quickly. Through a broad range of psychological processes and mechanisms, the shape might be interpreted as a viscous dog about to attack. The output would be a mental representation of a dangerous animal (cognitive), as well as trigger an emotion like fear (affect) and, finally, elicit a response such as running behavior). Alternatively, based on prior experience with this animal, the output might be a mental representation of a friendly animal (cognitive), which provokes an excitement (affect) and ask the human for a specific action like petting (behavior). The sensory input is the same, but the processing results in very different outputs. Highest level perspective of the psychology domain are formed of the concepts describing whole mind states and transitions like: ego and super-ego, personality types, tempers and so on that brings us closer to psychologically philosophical question "Who am I" and philosophy of cognition.

Philosophy Philosophy is the study of general and fundamental problems, such as those connected with reality, existence, knowledge, values, reason, mind, and language [9]. Philosophy is distinguished from other ways of addressing such problems by its critical, generally systematic approach and its reliance on rational argument, without supranaturalist influences [10]. Despite of specific cultural differences between Eastern and Western philosophical paradigms (whose boundaries are not so clear across regions and periods), philosophy looks for the rational analysis of reality and the obtention of metainformation. Philosophers not only know that they know, but also ask themselves how is it possible, which are the consequences of this process and other related questions. During this process, philosophers create heuristics of knowledge adapted to the surrounding socio-epistemic circumstances. This implies the work at the meta-level analysis. Our use of the concepts like: consciousness, emotions, mind, intuition, talking about a specific computational system domain, deals with the necessity of a meta-level of analysis. This implies to own the capacity of creating cognitive tool; it allows to extend the processing of information from the brain (and body) to external devices. Thus, humans are extended entities. And they evolve, culturally, creating meaning and ways to obtain meaning.

According to the previous explained notions, philosophy is a meta-level of data acquisition and processing. It's sphere belongs to the work at symbolic level, using semantic and syntactic mechanisms in order to create new information.

2.2 Integrating Domains

Previous sections sketched our holistic way to understand the multilayer nature of a cognitive process. The living and cognitive entity is based onto a physiological structure (the body) that by virtue of its specific architecture determines a strict way to interact with the environment as well as forces to follow the requirements of its own bodily system. Inside this body, several mechanisms deal with internal and external sets of data that must be analyzed in real time to produce adequate responses, if the organism that wants to survive and fulfil its intentional necessities (energy conservation, energy acquisition, data communication — as reproduction or cultural transmission —, maintenance of the system — playing, exercising, etc.).

This internal process of data management is handled by the brain, at the psychology level, and fully modelled and controlled by emotional mechanisms that allow the evaluation of acquired data as well as it's storage and application during decision processes. This allows to this entity to create a semantic view of the world and a second-order intentional approach to reality. At this point, all the processes are the result of a self-emergent data integration mechanism which makes possible the emergence of the binding activity called "consciousness". But it is still a passive process because it lacks a symbolic way to understand it, which leads us to the next level: philosophical.

Finally, this system creates symbolic ways to perform natural calculations. These bodily calculations are initially inspired by bodily requests such as calculating amounts or sets of objects (subitization, numerosity)or predicting possible outcomes of actions and events (the cause of events, the consequences of special events such as death, survival constraints, malfunctioning of the system, ...). From this binding process at several layers energes a metalevel: the conscious Self. It is a mechanism to integrate more efficiently and dynamically strategies of actions based on bodily and psychological constraints but also connected socially with other entities. At this level we enter into the realm of meta-information, created by the entity using externally shared tools like symbols, concepts or reasoning strategies/heuristics.

This architecture explains how the layers identified in the P^3 model interact to produce an activity,which is always implemented as a combination of actions, operations and interactions at the three layers and among them. The sensorimotor leads to psychological processing, which affects the conscious mind, although from a functional perspective the three domains are independent and directly fully controlled by none of the domains. They affect themselves but each one reigns on a different informational level.

2.3 Abstraction

The P^3 model can be abstracted from a human perspective into a more general system view. This way, the physiology layer, at the bottom, is mainly focused on the *main (physical) mechanisms* implementing and providing the *basic (mandatory) functionalities* required for the system operation. These basic functionalities are used at higher level to implement more complex functions, policies and strategies. Specifically, the psychology layer mainly implements individual reasoning and behaviour by just taking into account an *introspective* (internal, from the inside) view. In the system view this corresponds to advanced functionalities and policies operating *locally* to a system node or component, taking into account this local perspective. On the other hand, philosophy provides a wider, *introspective* or external view, by which the individual reasons from an external perspective as just an element of a wider global context, a kind of estrangement characterizing a meta-layer/view. This way, from the system perspective, the philosophy layer provides advanced functionalities and policies operating at a *global* view, considering the overall systems and its interaction with the external environment.

Highlighting both horizontal and vertical relationships different interpretations are possible: any block (physiology, psychology, philosophy) is independent, and therefore independently mapped into the application domain, or as a whole. Layers are connected and interdependent in layered way, complex functions at higher layer are based on simpler ones provided at lower layer(s).

The following example describes this layered model and its abstraction in the neuroscience area.

Physiology Example of high level abstraction of this domain could be artificial living organism or artificial living spiking pseudoneuronal network. This implies artificial life as main concept of each building block of the system. On the other hand main property of artificial living system could be the adaptation, self-evolving, and self-organizing abilities that defines overall behaviour of the system. The lowest level abstraction is artificial living cell or an object. Artificial living cell (ALC) — basic element of the system with self adaptation mechanisms to ever changing environment in some boundaries. One of the types of ALC is artificial pseudoneuron — mainly targeted on management and information processing of the artificial living organism (ALO). ALC should be self adaptable for the environmental changes in some boundaries. Mid level is represented by artificial organs or devices that are created via ALCs in form of ALC networks. An artificial living organism (ALO) is the combination of artificial organs which provide the functions set necessary for the artificial life.

Psychology Most of complex computational heuristics run by AI experts deal at a certain level with the implementation of processes that could be labelled as 'psychological' if they were run by human brains. We are thinking of heuristics of decision under poor informational environments, learning processes (supervised or unsupervised), artificial creativity, artificial vision, semantic data mining, etc. But in no case, these implementaions have been done once clarified the main architecture of the system, as we've been doing now. Cognitive architectures like SOAR, CHREST, CLARION, ICARUS, DUAL, and Psi are based on a full symbolic level [2, 3].

There is not a bottom up increasing complexity that justifies the interaction between the hard-physio and the upper levels of data processing (like psychological or also meta-heuristic ones). Nevertheless, the cited cognitive architectures have succeeded on simulating specific functionalities of the human mind. A good approach should be grounded as well as embodied, allowing a true merging of layers of interaction of a system with their environment as well as with other intelligent systems.

Philosophy *Philosophical corollaries about thinking*

Philosophical research has not an unit of study. Historically, at least for Western thought, the syntax coherence of written thoughts was considered the basis of philosophical analysis; that is, logical coherence, from Aristotle's syllogistic to contemporary non-monotonic or fuzzy logics. But also in parallel, there is another stream of studies devoted to the internal feeling about the world (phenomenology) and the corresponding debates on the existence or not of a perceiving unit (*I, Me, Myself*). Curiously, with the recent possibility of analyzing neural correlates *in vivo*, the internal states of mind are now partially discretizable. In both cases, the epistemic and experiential relationship with reality belong to a symbolic level of action. And in all philosophical systems is studied where are the limits of the own system and how we can know that our knowledge is correct. In a nutshell: it implies a meta-analysis level. Is in that sense that the philosophical level can be understood from two different perspectives: as a specific way to think on the value of symbolic processes as well as ac activity to think about the own existential experience ("to be conscious of"). Our aim is not to create artificial consciousness or artificial systems that think about their existence at this point but to create possible mechanisms of innovative ways to deal symbolically with information. This is the meta-level value of the notion of philosophy with which we are working to.

This meta-skill is mainly defined in the philosophy layer and we think on how meta-levels of significance are achieved and the mechanisms by which humans improvise new ways to understand and define their relevant aspects of reality.

3 A P^3 Computing System Model

Starting from the P^3 anthropocentric model of Fig. 1 in this section we map the three layers there identified into the corresponding layer for an abstract computing system. From our perspective, a computing system is a physical and/or abstract system, such as computer machines, devices, robots, networks, distributed systems, software, services, algorithms, workflows, data, in computer science and engineering. This way, a new, paradigmatic approach for computing is specified in Fig. 2 where three layers are identified: MetaComputing, MindComputing, PhysioComputing. These layers interact each other and implement functionalities of the corresponding P^3 model layers, i.e. physiology, psychology and philosophy, respectively, keeping relationships, interactions and basic mechanisms of the former model, just specialized in computer systems.

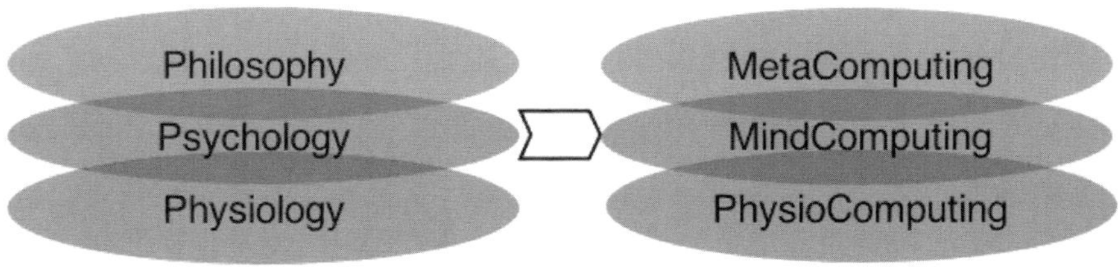

Fig. 2 The P^3 computing model

3.1 PhysioComputing

PhysioComputing implements the *physiology of computing*. It is therefore usually related to the hardware and the low level software, e.g. the firmware, providing the basic functionalities for a computing system to properly work. From an abstract point of view PhysioComputing provides the basic mechanisms implementing the main functionalities for a computing system to work. The scope of PhysioComputing is on structural, functional, organizational and communicational aspects of computing systems.

3.2 MindComputing

MindComputing is the *psychology of computing*. It should consider the high level details of phenomena taking into account broader and more conceptual approaches of the artificial mind operation based on the concepts, mechanisms, and functionalities introduced and provided at lower level by the PhysioComputing. From an abstract computer systems perspectives, PhysioComputing implements advanced and enhanced, mechanisms, policies, and strategies for locally optimizing the system, just considering the system introspectively. Scope of the MindComputing is therefore to implement enhanced services or systems on top of basic mechanisms, processes and functionalities provided by the lower level.

3.3 MetaComputing

MetaComputing is the *philosophy of computing*. Based on the MindComputing and PhysioComputing concepts, mechanisms and policies, it provides a further abstraction from a more general viewpoint, i.e. considering the interaction of the overall computing system with the environment, from outside. This is a meta-level, where the computing system is considered as part of the environment and its interactions with the other systems or elements are taken into account. The scope of MetaComputing includes definitions and high level, meta- views on the problems, thus providing meta-solutions to be enforced as policies and mechanisms at lower MindComputing and PhysioComputing layers.

4 Applications

To get the narration to more practical perspective we provide several examples of possible applications of introduced above framework P^3. We describe social network and cognitive robotics domains represented in the three layered perspectives of philosophy, psychology, physiology. This trifocal view on the complex phenomena could be beneficial in several domains that are demonstrated below.

4.1 Social Network

One of the most significant applications of the concepts described so far is in the social networks domain. Internet and the Web constitute a global artificial organisms/ecosystem on which actual life, in the broad sense we intend it, can be built. At the *physiology (PhysioComputing)* level this organism (node of the network) represents the basic "building block" of the living ecosystem that can be built on top of it. Applications like Facebook allow users to create profiles and self-identification in the living organisms. Facebook profiles precisely represent, at the *psychology (MindComputing)* level, the subjective self characterization of users in relation to other individuals populating the network. The metaphor solidly applies to the *philosophy (MetaComputing)* level too, where meta-analysis on the organism activities is performed. Here is where, for example, analysis of big data or opinion mining stands. The system has the peculiar ability to reason about itself at this level, and this can be done by the users themselves [5].

4.2 Cognitive Robotics

One more application of MetaComputing, MindComputing and PhysioComputing to the cognitive robotics domain invokes three new emerging domains: MetaBotics, MindBotics, HardBotics. Where in HardBotics we could identify following concpts: artificial living systems that cold be capable of reproduction thus regeneration via universal living "bricks/cells". Specifically neuronal systems could be capable of reconfiguration of their connections/"synapses" and generation of new neurons via "artificial neurogenesis". MindBotics contains definitions of several phenomena: affects, high-level emotions,temper, psychotypes, consciousness etc. In its turn a robotics system could fit in to the philosophical "model of 6" [6] by Marvin Minsky using emerging effects of living "artificial bricks/cells" like pattern matching, predictions, associative learning, deliberations, reflections, self-consciousness. Thus trifocal approach could be mapped and used for the benefit of robotic systems providing extended reflections of phenomena and their processes for more complete and exhaustive picture of robotics systems and their environment.

5 Conclusions and Future Works

The basis of any biological system is to be alive and keep living. Viruses (for which is still under debate whether they belongs to the realm of living entities), also have feeding and reproduction as their fundamental aims. Once these systems acquire tools to process bigger arrays of data, they can create more complex patterns of interaction. This base is well understood by neurobiologists, but is hard to implemented into current AI applications.

The approach presented in this work aims at providing a more coherent framework for agents design identifying three closely related domains, though working at separate functional levels.

- The foundation of artificial living system stands in the idea of *PhysioComputing*;
- Self-emergent characters, based on self-organization of connections and signals as processed by living cells (neurons) is the main aim of *MindComputing*. Consciousness, for example interpreted as a self-emergent property of neuronal networks;
- From a functional perspective, consciousness is the result of a multi-integration of data, but the feeling of being performing a conscious experience relies completely on the symbolic level providing tools that create elaborated semantic frameworks. Symbolic processing lies a basement for *MetaComputing*.

References

1. Hall, J.: Guyton and Hall Textbook of Medical Physiology (12th ed.). Saunders/Elsevier (2011)
2. Lin, J., Spraragen, M., Zyda, M.: Computational models of emotion and cognition. Adv. Cogn. Syst. **2**, 59–76 (2012)
3. Marsella, S., Gratch, J., Petta, P.: Computational models of emotion. In: Scherer, K., Bnziger, T., Roesch, E. (eds.) A Blueprint for a Affective Computing: A Sourcebook and Manual. Oxford University Press, Oxford (2010)
4. Matsumoto: The Cambridge Dictionary of Psychology. Cambridge University Press (2009)
5. Mazzara, M., Biselli, L., Greco, P.P., Dragoni, N., Marraffa, A., Qamar, N., de Nicola, S.: Social networks and collective intelligence: a return to the agora. Social Network Engineering for Secure Web Data and Services. IGI Global (2013)
6. Minsky, M.: The Emotion Machine: Commonsense Thinking, Artificial Intelligence, and the Future of the Human Mind. Simon & Schuster (2007)
7. Nisbet, R.: The Geography of Thought. The Free Press, New York (2003)
8. Putnam, H.: Reductionism and the nature of psychology. In: Haugeland, J. (ed.) Mind Design: Philosophy, Psychology, Artificial Intelligence, pp. 205–219. MIT Press, Cambridge (1981)
9. Rationality: The Cambridge Dictionary of Philosophy. Cambridge University Press (1999)
10. Tomberlin, J. (ed.): Philosophical Perspectives 13: Epistemology. Blackwell, Oxford (1999)

Part VI
IS: The Design and Implementation of Intelligent Agents and Multi-Agent Systems

Keyword Search in P2P Relational Databases

Tadeusz Pankowski

Abstract We discuss the way of using keyword search in the case of P2P connected relational databases. Each peer is then an agent deciding about translation and propagation of a keyword query, as well as about the way of merging partial answers. The way of merging (partial or total) is essential for both efficiency of execution and information contents in the final answer. We prove a theorem stating a necessary condition for deciding about the strategy of merging made by a peer.

Keywords Keywords queries · Relational databases · P2P data integration

1 Introduction

Keyword search is the most popular in information retrieval systems, where queries often are boolean or regular expressions over keywords, and information sources are text documents [4]. Lately, we observe a widespread application of keyword search paradigm to structured and semistructured data sources [6–8], in particular to relational databases [2, 11, 13].

In this paper, we focus on answering keyword queries issued against a P2P system populated with relational databases. The user expects that an answer will contain all relevant data stored in databases reachable from the chosen peer directly or indirectly. A keyword query is propagated in the P2P system and the peers (agents) translate it according to their database schemes into a relational database query. Partial answers are appropriately merged and sent back to a target peer. A crucial issue is then the strategy of query propagation and the way of merging partial answers. A peer can apply a *partial merge* (involving only answers returned to the peer) or a *total merge* (besides answers also the entire database stored on the peer is taken into account). The decision influence both the cost of the execution and the information contents

T. Pankowski (✉)
Institute of Control and Information Engineering, Poznań University of Technology, Poznań, Poland
e-mail: tadeusz.pankowski@put.poznan.pl

G. Jezic et al. (eds.), *Agent and Multi-Agent Systems: Technologies and Applications*, Smart Innovation, Systems and Technologies 38,
DOI 10.1007/978-3-319-19728-9_27

in the final answer. Application of the total merge may infer some *missing values*, which otherwise were not discovered in the final answer. We propose and discuss some necessary condition that specifies whether it will be beneficial to apply the total merge.

In Sect. 2, we review some information about relational databases and keyword queries. Motivation and basic concepts, as well as the running example, are included in Sect. 3. P2P keyword query answering is discussed in Sect. 4. A necessary condition concerning making decision about total merge is formulated in Sect. 5. Section 6 concludes the paper.

2 Preliminaries

In this paper, to avoid overwhelming notation, we will consider only one-attribute integrity constraints in relational databases. Let **Att** be a set of *attributes*. We assume that a *relational database schema* is a triple $\mathcal{R} = (\mathbf{R}, att, \mathbf{IC})$, where:

- $\mathbf{R} = \{R_1, \dots, R_p\}$ – a set of relation names,
- $att : \mathbf{R} \to 2^{\mathbf{Att}}$ – a function assigning to each relational name $R \in \mathbf{R}$ a set $att(R) \subseteq \mathbf{Att}$ of attributes referred to as the *type* of R,
- $\mathbf{IC} = \mathbf{FD} \cup \mathbf{PK} \cup \mathbf{FK}$ – a set of integrity constraints, where: **FD** is a set of *functional dependencies*, $\mathbf{FD} = \{R.A \to R.A' \mid A, A' \in att(R)\}$; **PK** is a set of *primary keys* (for each relation name there is at most one primary key), $\mathbf{PK} = \{R.A \mid A \in att(R)\}$; **FK** is a set of foreign keys, $\mathbf{FK} = \{R.A \to R'.A' \mid A \in att(R), R'.A' \in \mathbf{PK}\}$.

Let **Const** and **Id** be, respectively, sets of *constants* (strings, for simplicity) and *identifiers* (or *surrogate keys*). A *tuple* of type R is a set $r = \{(A_1 : a_1), \dots, (A_n : a_n)\}$, where $\{A_1, \dots, A_n\} = att(R)$, $a_i \in \mathbf{Const} \cup \mathbf{Id}$. A *fact* is an expression $R(r)$, where r is a tuple of type R. A *database* with schema $\mathcal{R}$ is a finite set of facts, $DB = \{R(r) \mid R \in \mathbf{R},\ r \text{ is a tuple of type } R\}$. A database is *consistent* if satisfies all integrity constraints in **IC** [1]. Further on, we will consider only consistent databases.

A *keyword query* is a set of keywords, $Q = \{q_1, \dots, q_k\}$, where each $q \in Q$ is a constant. A keyword searches tuples interconnected by means of foreign keys and selects tuples containing the keywords (all or some of them). A formula $contain(r.A, q)$ is true for a tuple r if the value of attribute A contains a keyword q. In general, a keyword can occur in any string-valued column as well as any subset of keyword from Q can be taken into account. We will restrict ourselves to the situation described below.

We assume that for a keyword $q \in Q$, there is a pair (R_q, A_q) such that q can be searched in the column $R_q.A_q$. A keyword query Q is translated into a *tuple relational calculus* [1] query $trc(Q)$:

$$trc(Q) = \{(r_1, \dots, r_m) \mid \Phi(r_1, \dots, r_m) \wedge \Psi(r_1, \dots, r_l)\}, \quad (1)$$

where: (1) $l \leq m$; (2) for each tuple variable r_i, a range formula $R_i(r_i)$ is in Φ; (3) Ψ is a conjunction of formulas of the form $contain(r_j.A_q, q)$, where $R_q = R_j$; (4) for each i, $1 \leq i < m$, $R_i.A \rightarrow R_{i+1}.A' \in \mathbf{FK}$, for some $A \in att(R_i)$, $A' \in att(R_{i+1})$, and the join condition $R_i.A = R_{i+1}.A'$ is in Φ.

3 Motivation and Basic Concepts

Further on in this paper, we will present databases either in a tabular form or as sets of facts. Let us consider three relational databases DB_1, DB_2 and DB_3 (Fig. 1), storing information about films and actors playing in those films. These databases are semantically homogeneous, but their structures differ slightly (in fact, only in presence or absence of some columns). Moreover, film identifiers (*FilmId*) have only local meaning and may be different in different databases (e.g., X1 in DB_2 and Y1 in DB_3). So, values of *FilmId* are from **Id**, values of other attributes are from **Const**.

DB_1 :

Film1

FilmId	*Title*	*Director*

PlaysIn1

Name	*Country*	*FilmId*	*Year*

DB_2 :

Film2

FilmId	*Title*
X1	The Pianist

PlaysIn2

Name	*Country*	*FilmId*
A. Brody	USA	X1
E. Fox	England	X1

DB_3 :

Film3

FilmId	*Title*	*Director*
Y1	The Pianist	R. Polanski

PlaysIn3

Name	*FilmId*	*Year*
E. Fox	Y1	2002

Fig. 1 Databases DB_1 (with the empty instance), DB_2 and DB_3

Let us consider a keyword query of size 2, $Q = \{Brody, Polanski\}$. We expect that an answer to the query should contain all relevant data from DB_1, DB_2 and DB_3. Let *DB* be a database created from DB_1 by renaming *Film*1 to *Film*, and *PlaysIn*1 to *PlaysIn*. Then the translation of Q to $trc(Q)$ with respect to a *DB* is:

$$trc(Q) = \{(p,f) \mid PlaysIn(p) \wedge Film(f) \wedge p.FilmId = f.FilmId \\ \wedge contain(p.Name, \text{“}Brody\text{”}) \wedge contain(f.Director, \text{“}Polanski\text{”})\}. \tag{2}$$

Now, let us consider a naive way of answering Q by applying $trc(Q)$ to *DB*. First, data from DB_1, DB_2, and DB_3 are transformed to *DB*, as follows:

$$Film = Film1 \cup Film2 \cup Film3, \\ PlaysIn = PlaysIn1 \cup PlaysIn2 \cup PlaysIn3. \tag{3}$$

It is easy to see that the answer produced by *Q*(*DB*) (or by *trc*(*Q*)(*DB*)) is empty. It contradicts our intuition since we expect that the answer should be: *(R.Brody, USA,* 2002, “*The Pianist*”, *R.Polanski)*, that can be deduced from the database. To this order, the following problems must be solved: (a) dealing with identifiers: X1 and Y1 denote in fact the same entity but they are different; (b) during the transformation some values in target tables are *missing*, i.e., can be represented by NULLs; (c) constraints from **IC** may be used; (d) direct transformation of entire databases i inefficient and should be avoided.

4 P2P Keyword Query Answerin

In this section, we discuss a way for solving problems enumerated in the end of Sect. 3. First, we will show how the theory of data exchange [10] may be used to deal with identifiers, null values and constraints. Next, a way utilizing P2P environment will be shown.

4.1 Query Answering Using Data Exchange with Constraints

We apply the theory of data exchange developed by Fagin et al. [9, 10]. In relational database data exchange, a source database with a schema $\mathcal{R}_s$ is transformed to a target database with a schema $\mathcal{R}_t$, by means of a set Σ of *source-to-target dependencies* (STDs) and a set Δ of *equality-generating-dependencies* (EGDs).

We assume that in a target database: (1) a source constant value from **Const** different than NULL, is represented by itself; (2) a source identifier (such as a value of *FilmId*) is mapped to a variable (*a labeled nulls*) in a set $\mathsf{Var}_{\mathrm{Id}}$; (3) a missing value and NULL value is mapped to a variable in a set $\mathsf{Var}_{\mathrm{V}}$. **Const**, $\mathsf{Var}_{\mathrm{Id}}$ and $\mathsf{Var}_{\mathrm{V}}$ are pairwise disjoint. Mapping between identifiers and variables in $\mathsf{Var}_{\mathrm{Id}}$ is made by means of a Skolem function *SkId*(), that assigns to an identifier a unique and fresh variable from $\mathsf{Var}_{\mathrm{Id}}$. *SkId*() is defined for any source database separately, i.e., equal identifiers in the same databases are mapped to the same variable, but equal identifiers in different databases are mapped to different variables.

The following set Σ of STDs defines transformation of union of source databases DB_1, DB_2 and DB_3 to a target database DB:

$$\Sigma = \{ \begin{array}{ll} Film1(x,y,z) \wedge v = SkId_1(x) & \Rightarrow Film(v,y,z), \\ PlaysIn1(x,y,z) \wedge w = SkId_1(z) & \Rightarrow PlaysIn(x,y,w), \\ Film2(x,y) \wedge v = SkId_2(x) & \Rightarrow \exists z Film(v,y,z), \\ PlaysIn2(x,y,z) \wedge w = SkId_2(z) & \Rightarrow PlaysIn(x,y,w), \\ Film3(x,y,z) \wedge v = SkId_3(x) & \Rightarrow Film(v,y,z), \\ PlaysIn2(x,z) \wedge w = SkId_3(z) & \Rightarrow \exists y PlaysIn(x,y,w)\}. \end{array} \tag{4}$$

A transformation process in data exchange is referred to as *chase* [3, 10]. In our case, in the result of chasing $\{DB_1, DB_2, DB_3\}$ with respect to Σ, and into a database with the schema of DB, denoted $chase_{\Sigma}^{DB}(\{DB_1, DB_2, DB_3\})$, we obtain a database:

$DB' =$

Film

FilmId	*Title*	*Director*
X_1	The Pianist	V_1
X_2	The Pianist	R. Polanski

PlaysIn

Name	*Country*	*FilmId*	*Year*
A. Brody	USA	X_1	V_3
E. Fox	England	X_2	V_4
E. Fox	V_2	X_2	2002

In the schema of DB', we have the following set of functional dependencies:

$$FD_{Film} = \{FilmId \rightarrow Title, Director; Title \rightarrow FilmId, Director\}$$
$$FD_{PlaysIn} = \{Name \rightarrow Country; FilmId \rightarrow Year\}$$

FD implies the following set of *equality-generating-dependencies* (EGDs):

$$\begin{aligned} \Delta = \{ \ & Film(x, y_1, z_1) \wedge Film(x, y_2, z_2) \Rightarrow y_1 = y_2 \wedge z_1 = z_2 \\ & Film(x_1, y, z_1) \wedge Film(x_2, y, z_2) \Rightarrow x_1 = x_2 \wedge z_1 = z_2 \\ & PlaysIn(x, y_1, z_1) \wedge PlaysIn(x, y_2, z_2) \Rightarrow y_1 = y_2 \wedge z_1 = z_2\}. \end{aligned} \tag{5}$$

Now, we chase DB' with respect to Δ, and obtain $DB'' = chase^{DB}_{\Delta}(DB')$, where EGDs generates equalities: $X_2 = X_1$, V_1="R. Polanski", V_2="England":

$DB'' =$

Film

FilmId	*Title*	*Director*
X_1	The Pianist	R. Polanski

PlaysIn

Name	*Country*	*FilmId*	*Year*
A. Brody	USA	X_1	2002
E. Fox	England	X_1	2002

Then the answer to Q is (variables in tuples belonging to the result are discarded):

$$\begin{aligned} Q(\{DB_1, DB_2, DB_3\}) = Q(chase^{DB}_{\Sigma\cup\Delta}(\{DB_1, DB_2, DB_3\})) = Q(DB'') = \\ \{(A.Brody, USA, 2002, The\ Pianist, R.Polanski)\}. \end{aligned} \tag{6}$$

4.2 Query Answering in P2P System

Now we assume that databases depicted in Fig. 1 are stored in three peers forming a P2P data integration system (Fig. 2). We will discuss how the keyword query Q can be propagated and answered in the system, and how the way of the propagation influence the contents of the answer to the query.

In rewriting procedures in P2P query answering, we will need the following translation. Let $Q = \{q_1, \ldots, q_k\}$ be a keyword query, then

$$dlSet(Q) = \{dl(q_1), \ldots, dl(q_k)\}, \tag{7}$$

where $dl(q)$ is a datalog query over a database schema $\mathcal{R}$ and returning a database with a schema being a subset of $\mathcal{R}$.

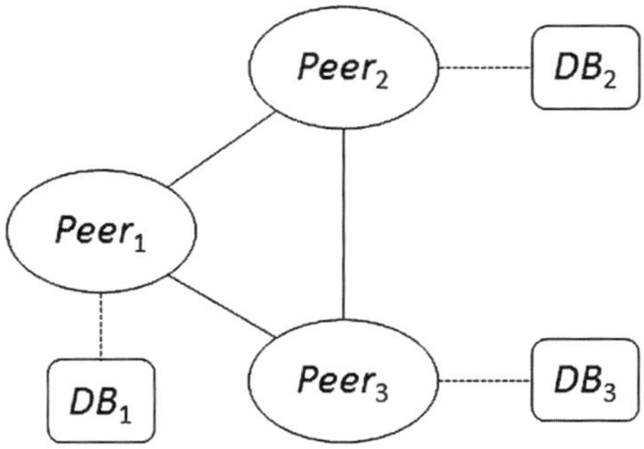

Fig. 2 A sample P2P integration architecture with three peers (agents) and three local databases stored in peers

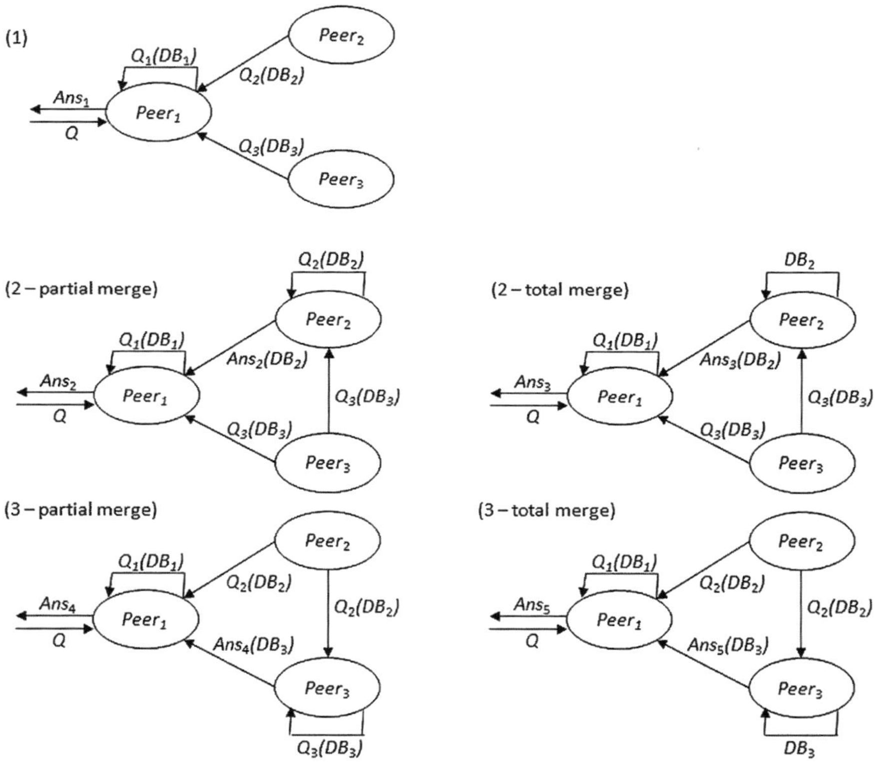

Fig. 3 Query answering in P2P system with different propagation strategies and with partial and total merge

Query Q can be answered using different strategies. We will consider three strategies depicted in Fig. 3. Moreover, in strategies (2) and (3), two different ways of merge are applied: in strategy (2) $Peer_2$ applies either partial or total merge, similarly – $Peer_3$ uses partial or total merge in strategy (3).

First, a keyword query $Q = \{Brody, Polanski\}$ is sent to $Peer_1$ that rewrites it to Q_1 and translates to $dlSet(Q_1)$ using the knowledge included in the schema of its database. Next, $Peer_1$ propagates Q to its partners and waits for answers. Finally, because $Peer_1$ is the target peer, it is responsible for producing the final answer. This is done by executing the query $trc(Q_1)$ against the result of the partial or total merge over collected answers at $Peer_1$.

Every partner, $Peer_2$ and $Peer_3$ in this case, behaves in the same way, i.e., rewrites and translates the query, propagates it to its partners and waits for answers. Additionally, a peer decides about the way of merging answers obtained from partners: in the partial merge only answers are taken into account, in the case of total merge besides of answers also the peer's entire database is taken into account. The decision on this stage is significant since influence both the cost and information contents of the final answer. In rest of this section, we will illustrate this issue.

We have the following translations:

$$\begin{array}{l} Q_1 = \{Brody, Polanski\},\ \ Q_2 = \{Brody\},\ \ Q_3 = \{Brody, Polanski\}; \\ dlSet(Q_1) = dlSet(Q_3) = \{dl(Brody), dl(Polanski)\}, \\ dlSet(Q_2) = \{dl(Brody)\}, \end{array}$$

where, for example (prefixes *.tgt* and *.src* distinguish target tables from the source ones):

$$\begin{array}{l} dl_{DB_2}(Brody) = \\ \quad tgt.PlaysId2(n,c,f) \leftarrow src.PlaysIn2(n,c,f), contain(n, \text{“}Brody\text{”}) \\ \quad tgt.Film(f,t) \leftarrow tgt.PlaysIn(n,c,f), src.Film2(f,t); \\ dl_{DB_3}(Polanski) = \\ \quad tgt.Film3(f,t,d) \leftarrow src.Film3(f,t,d)\}, contain(d, \text{“}Polanski\text{”}). \end{array}$$

Analogously, for $dl_{DB_1}(Brody)$, $dl_{DB_1}(Polanski)$, and $dl_{DB_3}(Brody)$.

We have:

$$\begin{array}{l} Q_1(DB_1) = dlSet(Q_1)(DB_1) = \emptyset; \\ Q_2(DB_2) = dlSet(Q_2)(DB_2) = \{PlaysIn2(Name : A.Brody, Country : USA, \\ \qquad FilmId : X1), Film2(FilmId : X1, Title : The\ Pianist)\}, \\ Q_3(DB_3) = dlSet(Q_3)(DB_3) = \{Film3(FilmId : Y1, Title : The\ Pianist, \\ \qquad Director : R.Polanski)\}. \end{array}$$

In the following, we denote: $Ans_j(DB_i)$ – a result of a merge by means of the chase procedure restricted to the schema of DB_i; Ans_j – the final result obtained by executing $trc(Q)$ over $Ans_j(DB_1)$ (since in our example $Peer_1$ is the final peer).

Attribute names are omitted. Additionally, in the final result all variables (labeled nulls) are discarded.

Strategy (1)

Q is sent to $Peer_1$, and $Peer_1$ propagates it to $Peer_2$ and $Peer_3$ (Fig. 3(1)). All received answers: $Q_1(DB_1)$, $Q_2(DB_2)$, and $Q_3(DB_3)$, are merged:

$$\begin{aligned}&Ans_1(DB_1) = chase^{DB_1}_{\Sigma\cup\Delta}(Q_1(DB_1), Q_2(DB_2), Q_3(DB_3))\\&= \{PlaysIn1(A.Brady, USA, X_1, V_1), Film1(X_1, The\ Pianist, R.Polanski)\}.\end{aligned}$$

The final result is (note that a value of *Year* is not given):

$$Ans_1 = (A.Brady, USA, The\ Pianist, R.Polanski).$$

Strategy (2)

Now, $Peer_2$ propagates Q to $Peer_3$ and waits for an answer $Q_3(DB_3)$, Fig. 3(2). Then decides about partial or total merge:

- Partial merge:

$$\begin{aligned}&Ans_2(DB_2) = chase^{DB_2}_{\Sigma\cup\Delta}(Q_2(DB_2), Q_3(DB_3))\\&= \{PlaysIn2(A.Brady, USA, X_1), Film2(X_1, The\ Pianist)\}.\end{aligned}$$

  Next, $Ans_2(DB_2)$ is sent to $Peer_1$, and merged there with $Q_1(DB_1)$ and $Q_3(DB_3)$. The final answer Ans_2 is equal to Ans_1.
- Total merge:
  The entire database DB_2 is merged with $Q_3(DB_3)$. However, the result is the same as by the partial merge, i.e.,

$$Ans_3(DB_2) = chase^{DB_2}_{\Sigma\cup\Delta}(DB_2, Q_3(DB_3)) = Ans_2(DB_2).$$

  The final answer Ans_3 is again equal to Ans_1.

Strategy (3)

Now, $Peer_3$ propagates Q to $Peer_2$ and waits for $Q_2(DB_2)$, Fig. 3(3). Then decides about partial or total merge:

- Partial merge:

$$\begin{aligned}&Ans_4(DB_3) = chase^{DB_3}_{\Sigma\cup\Delta}(Q_3(DB_3), Q_2(DB_2))\\&= \{PlaysIn3(A.Brady, X_1, V_1), Film3(X_1, The\ Pianist, R.Polanski)\}.\end{aligned}$$

  Again, the final answer Ans_4 is equal to Ans_1.

– Total merge:

$$Ans_5(DB_3) = Q_3(chase^{DB_3}_{\Sigma\cup\Delta}(Q_2(DB_2), DB_3)) =$$
$$= \{PlaysIn3(A.Brady, X_1, 2002), Film3(X_1, The\ Pianist, R.Polanski)\}.$$

Now, the final answer has additionally value 2002 of *Year* inferred in the integration process:

$$Ans_5 = (A.Brady, USA, 2002, The\ Pianist, R.Polanski).$$

5 Strategy Choice Criterion

In the previous section, we discussed some ways (strategies) for propagation, and answering keyword queries in P2P systems. It turns out that both costs of the execution and information contents of the final result may be different. Here we focus on the way of merging (partial or total) that is to be decided in a peer. It follows from the considered example in Sect. 4.2 that:

1. In the case of strategy (2), in the both ways of merging the results were the same, i.e., $Ans_2 = Ans_3$. Of course, applying of the total merge is much more costly than applying the partial merge. So, the total merge should be applied only if there is a chance to discover some missing values (i.e., to replace variables by constants).
2. In strategy (3), applying of the total merge was profitable. We were able to discover (infer) a value of *Year*, i.e., that Brody played in "The Pianist" in year 2002. Thus, $Ans_5 \supset Ans_4$.

Now, we will discuss a necessary condition for deciding the total merge in a peer.

Definition 1 *Let Ans be an answer to a query Q, $R(r) \in Ans$, and $(A : v) \in r$, where $v \in \mathsf{Var}_V$ and A is an attribute of R. Then we call v a missing value of A in Ans.*

For example, V_1 is a missing value of *Year* in $Ans_4(DB_3)$, however, X_1 in this answer is not a missing value since X_1 is in Var_{Id} – not in Var_V.

Theorem 1 *Let v be a missing value of A in Ans, where $(A : v) \in r$ and $R(r) \in Ans$. The necessary condition for discovering the value of v in a DB with a schema* $(\mathbf{R}, \mathbf{IC})$*:*

– *there is $A' \in att(R)$ such that $A' \to A \in \mathbf{IC}$, and*
– *there is $q \in Q$ such that: $R = R_q$, $A_q \in att(R)$, $contains(A_q, q)$ occurs in $trc(Q)$, and $A' \to A_q \notin \mathbf{IC}$.*

Then we say that the missing value of v can be discovered by means of the constraint (functional dependency) $A' \to A$.

For example, *Brody* $\in Q$, and *contains*(*Name*, }}*Brody*″) occurs in *trc*(Q). Moreover, *FilmId* $\rightarrow$ *Year* $\in$ **IC** and *FilmId* $\rightarrow$ *Name* $\notin$ **IC**. Thus, we see that the promise of Theorem 1 is satisfied for discovering the missing value of V_1 in $Ans_4(DB_3)$ by means of the constraint *FilmId* $\rightarrow$ *Year* (referring to DB_3).

Proof (*of Theorem 1*) Assume that the promise of the theorem is satisfied. Let $(A : v) \in r$, where: $R(r) \in Ans$, $R \in \mathbf{R}$. Additionally, let us assume that $(A' : c_1) \in r$ and there is a tuple r' of type $att(R)$, such that:

- $R(r') \in DB$,
- $(A' : c_1) \in r'$,
- $(A : c_2) \in r'$,
- $(A_q : c_3) \in r'$,
- $contains(A_q : q)$ is false in r'.

By the assumption $A' \rightarrow A \in \mathbf{IC}$, so it follows from the corresponding EGD:

$$r.A' = r'.A' \Rightarrow r.A = r'.A,$$

that $v = c_2$. Moreover, since $contains(A_q : q)$ is false in r', the above substitution cannot be inferred by means of the partial merge. □

6 Conclusions

We discussed the problem of answering keyword queries issued against a P2P-connected relational databases. In such a system, there are many autonomous peers (agents), which collaborates in the process of producing answers to queries. The system is flexible and peers can enter and leave the system dynamically. A query posed to a peer is propagated to its partners along semantic paths defined by mappings, those partners propagate the query to their partners, etc. Answers to the query flow back in the opposite directions. The target peer merge the answers producing the expected answer. We focused on translating and propagating keyword queries, as well as on the way of merging partial answers by a peer. The proper decision significantly influence both the quality of the answer as well as efficiency of processing. The discussed strategy was inspired by the SixP2P system originally designed for integrating XML data [5, 12].

Acknowledgments This research has been supported by Polish Ministry of Science and Higher Education under grant 04/45/DSPB/0136.

References

1. Abiteboul, S., Hull, R., Vianu, V.: Foundations of Databases. Addison-Wesley, Reading, Massachusetts (1995)
2. Agrawal, S., Chaudhuri, S., Das, G.: Dbxplorer: a system for keyword-based search over relational databases. ICDE **2002**, 5–16 (2002)
3. Arenas, M., Barceló, P., Libkin, L., Murlak, F.: Relational and XML Data Exchange. Morgan & Claypool Publishers, Synthesis Lectures on Data Management (2010)
4. Baeza-Yates, R., Ribeiro-Neto, B.: Modern Information Retrieval. Addison Wesley Publishing Company (1999)
5. Brzykcy, G., Bartoszek, J., Pankowski, T.: Schema mappings and agents' actions in P2P data integration system. J. Univers. Comput. Sci. **14**(7), 1048–1060 (2008)
6. Chaudhuri, S., Das, G.: Keyword querying and ranking in databases. PVLDB **2**(2), 1658–1659 (2009)
7. Chen, Y., Wang, W., Liu, Z., Lin, X.: Keyword search on structured and semi-structured data. ACM SIGMOD **2009**, 1005–1010 (2009)
8. Dalvi, B.B., Kshirsagar, M., Sudarshan, S.: Keyword search on external memory data graphs. PVLDB **1**(1), 1189–1204 (2008)
9. Fagin, R.: Equality-generating dependencies. In: Liu, L., Özsu, M.T. (eds.) Encyclopedia of Database Systems, pp. 1009–1010. Springer, US (2009)
10. Fagin, R., Kolaitis, P.G., Miller, R.J., Popa, L.: Data exchange: semantics and query answering. Theor. Comput. Sci **336**(1), 89–124 (2005)
11. Hristidis, V., Papakonstantinou, Y.: DISCOVER: Keyword Search in Relational Databases. VLDB
12. Pankowski, T.: Query propagation in a P2P data integration system in the presence of schema constraints. In: Data Management in Grid and P2P Systems DEXA/Globe'08, LNCS 5187, pp. 46–57 (2008)
13. Yu, J.X., Qin, L., Chang, L.: Keyword search in relational databases: a survey. IEEE Data Eng. Bull. **33**(1), 67–78 (2010)

Intelligent Social Agent for the Development of Social Relations Based on Primary Emotions

Arnulfo Alanis Garza, Lenin G. Lemus Zuñiga, María del Rosario Baltazar, Bogart Yail Marquez, Carlos Lino Ramírez and Karina Romero

Abstract This article shows the experimentation with emotions in a scenario with a specific task, where the main goal is to see the behavior of emotions to the task given to them that based on the level of empathy that exists between these emotions, all this work is done within a Social Multi-Agent System, in which it is intended that two or more robots can present a profile of personality and emotion for the search of empathy between them to make a team.

1 Introduction

More than 90 years have passed since the publication of the first psychological study on facial asymmetry and its attribution, and the issue remains as controversial as it was then. Although the initial interest in facial asymmetry dates back to the

A.A. Garza (✉) · B.Y. Marquez · K. Romero
Computing and Systems Department, Technology Institute of Tijuana, Tijuana, BC, México
e-mail: alanis@tectijuana.edu.mx

B.Y. Marquez
e-mail: bogart@tectijuana.edu.mx

K. Romero
e-mail: Karina_romero@lycos.com

L.G. Lemus Zuñiga
Universitat Politècnica de València, Valencia, Spain
e-mail: lemus@upv.es

L.G. Lemus Zuñiga
RIS-Itaca, Universitat Politècnia de València, Valencia, Spain

M. del Rosario Baltazar · C.L. Ramírez
Research Division, Technology Institute of Leon, Tijuana, BC, México
e-mail: charobal@yahoo.com

C.L. Ramírez
e-mail: carloslino@itleon.edu.mx

G. Jezic et al. (eds.), *Agent and Multi-Agent Systems: Technologies and Applications*, Smart Innovation, Systems and Technologies 38,
DOI 10.1007/978-3-319-19728-9_28

second half of the nineteenth century [14], it was J. Hallervorden, who, in 1902, first examined the possibility of divergent emotional expressions on the right and left hemifaces [17, 18]. Hallervorden's conclusion was thought-provoking: the right side of the face is apperceptive and active, whereas the left side is affective and "having dark unformed content" [17, 25]. In 1933, Werner Wolff [31, 32] introduced the photographic technique used by Hallervorden to the English speaking world of psychology, and his name has subsequently become synonymous with this methodology ("Wolff's split-face technique"). As with his predecessor, Wolff found that the right hemiface shows vitality and power, representing the public and the individualistic aspects of personality, whereas the more passive left side shows features representing the unconscious and collective characteristics.

The focus of the universality issue has been on the panculturality of recognition and expression of facial emotion. Primary questions of interest that have been examined in the literature include: (a) Are facial emotions recognized similarly across all cultures?, (b) To what extent does culture influence the expression of emotion?, and, (c) Are there any culture-specific emotions?

2 Concepts

In this section we mention some existing research around the world related to the project we are developing, so we are sure it is a relevant and innovative in the scientific community in some laboratories are focused on the study of human beings as memory short and long term, identification of emotions in the brain, etc., due to great interest in how the brain works.

2.1 Representation of Personality

One legacy of the studies of Hallervorden and Wolff has been the interest in differential representation of personality in each hemiface. However, sporadic research has yielded only inconsistent results. Wolff [31, 32, 33] found the left hemiface passive and dead like, and Lindzey et al. [22], in more systematic research, obtained similar results: the left hemiface is more passive and the right one is more vital (see also [27]). In contrast, Karch and Grant [21, 30] found subjects to rate men's left side (neutral expression) as stronger, harder, and more masculine. Finally, Stringer and May [29, 51] (cited in [16]) failed to replicate previous findings. Nonetheless, they did find the left hemiface to be more receptive and intuitive, whereas the right hemiface was not attributed with the opposing traits.

2.2 Emotional Expression

Researchers dealing with facial asymmetry have shifted their efforts in the last two decades into the arena of asymmetry in emotional expression. Their principal objective has been the exploration of brain mechanism in general and hemispheric specialization in particular. Many studies have focused on the proposition that each hemiface, constantly or temporarily, exhibits emotion of different valence, or the same ones but in different magnitudes. Their basic hypothesis maintains that the face reflects brain activity. In other words, stronger expressions in one hemiface indicate a greater involvement of the opposite cerebral hemisphere due to the crossover of pathways from the motor cortex to the facial nucleus. The extensive research on brain specialization for emotion has yielded two "grand" theories. The first one, called the valence hypothesis, postulates that the left brain hemisphere is specialized for positive emotions and the right hemisphere for negative emotions [1, 28]. One variant of this theory maintains that there is less hemispheric asymmetry for positive emotions [10], while a second variant maintains that there is a differential specialization for expression based on valence, but not for perception of emotions [9]. The second theory, the right- hemisphere hypothesis, postulates that the right hemisphere dominates overall in perception and expression of emotion, regardless of their valence [11, 20].

Several studies have suggested that a primary distinction should be made between neutral expressions found in resting faces and emotional expressions. It has been argued that whereas the former expressions are easily and naturally generated without cortical inference, the latter expressions stem from various often difficult to determine sources [8, 29]. Dealing with the face at rest, Karch and Grant [21] and Campbell [5] found right-handers' left and left-handers' right composites to be more expressive (miserable). In a subsequent larger sample, Campbell [6] (cited in [16, 26]) obtained these results only for females. Still, in another study, Campbell [7] found left-handers' right-side composites to be more expressive (happier or more miserable). Several studies found the left hemiface to show greater "intensity". Borod et al. [2] reported the neutral expression of elderly male subjects to be "significantly left-sided", and similar results were obtained by Moreno et al. McGee and Skinner [23] also reported more frequent attributions of emotion adjectives to the left hemiface, and likewise, Bennett et al. found the left hemiface to be rated as more happy (cheerful, excitable). Finally, Rhodes and Lynskey [24].

2.3 EEG-Based Emotion Recognition the Influence of Visual and Auditory Stimuli [2]

In the area of human-computer interaction information about the emotional state of a user becomes more and more important. For instance, this information could be used to make communication with computers more human-like or to make computer learning environments more effective [12, 13].

This paper proposes an emotion recognition system from electroencephalographic (EEG) signals. Emotional states were induced by pictures from the international affective picture system. EEG data was recorded either from 16 electrodes with a standard EEG-cap according to the 10–20-system or by using a headband with covers only four frontal electrodes but is much more comfortable to wear. For emotion recognition two types of classifiers support vector machines (SVMs) and hidden Markov models (HMMs) were investigated. In experiments where the three emotional states pleasant, neutral, and unpleasant were discriminated, with the HMM based system mean recognition rates of 46.15 % were achieved, whereas with the SVM based system we obtained a mean recognition rate of 62.07 % [3, 15].

This study shows that electroencephalographic signals are feasible for emotion recognition and that SVMs seem to be better suited for emotion recognition than a sequence-based approach with HMMs.

3 Development

We begin by mentioning the work that we are building in the area of psychology.

3.1 Cattell Personality Theory

The definition of personality best suited to our work would be to say that Raymond Cattell's personality is the determinant of behavior in a given situation. The basic components of personality are the features. Cattell believed that there is an underlying personality structure of the language that describes the features, represented as follows

$$R = f(S, P) \tag{1}$$

R = nature and magnitude of the response or a person's behavior is a function of S, a situation that is stimulating, and P or nature of his personality.

The trait is the basic structural element of the theory of Cattell, defined as a tendency, relatively permanent and comprehensive, to react in a certain way. It assumes a certain regularity of behavior over time and situations [2].

Emotions are reactions to the information (knowledge) that we receive in our relations with the environment. The intensity of the reaction based on subjective assessments that we make about how information received will affect our wellbeing.

What is involved in such evaluations is subjective prior knowledge, beliefs, personal goals, perceptions, environmental challenge, and so on.

An emotion depends on what is important to us. If the emotion is intense can cause intellectual or emotional disorders (phobia, stress, depression) [3, 19].

We can define emotion as nonverbal expressions that allow the human being to show the mood in which it is located.

In psychology the way a person can communicate their emotions and they can be understood by another person is related to the concept of empathy. In the literature we can find various definitions of empathy which took only two for this investigation are more representative:

"The action and the ability to understand are aware, being sensitive or alternative way to experience the feelings, thoughts and experiences of another, without these feelings, thoughts and experiences have been reported in an objective or explicit" [5].

"Mental function that allows us to not be focused on yourself and see things from another point of view. Through them we may be partakers of others' experiences and develop common experiences" [25].

4 Social Multi-agent System

In this module design and analysis was conducted for this we use two tools for development, UML and Gaia respectively include the completion of Social Multi-Agent System (SMAS) is based on FIPA standards for better utilization and greater possibility of extension. In this work we got the facts and rules that make up the knowledge base of the agents below show an example in Table 1.

Table 1 Facts awareness agent

Agent name	Facts
Awareness agent (AAwareness)	AAwareness_p.AskEmotionMAS-SOE.ACOR AAwareness_p.ReceiveEmotionMAS-SOE.ACOR AAwareness_p.ReceiveFactorsAF_m AAwareness_p.NvosValues[EvaluaFIS(emocion, degree,16factors)]

Where we can see each of the processes that were obtained from the analysis of multi-agent system, this procedure was performed for each of the agents that make up the SMAS as a first step in getting the BC or the Knowledge Base.

5 Results

The implementation of our Social Multi-Agent System was done using the Jade container as it is the most suitable for the development of intelligent agents.

Jade (Java Agent Development framework) is the development environment for creating, communicating and interacting intelligent agents Can be considered a "middleware" that implements agent:

- An agent platform (runtime)
- A development framework (class library).

The experiments were performed using the Maze shown in Fig. 1, varying the number of entities involved in the simulations four experiments using 6 robots, is the explanation that ach experiment is simulated in the same conditions, Table 2.

Fig. 1 Maze

Table 2 MASS experiments

Experiment #1 6 robots	Robot5.Anger robot1. Fear robot4.Disgust	Experiment #3 6 robots	Robot1.Joy robot3.Surprise robot3.Fear robot4.Sadness
Experiment #2 6 robots	Robot4.Joy Robot6.Fear robot3. Sadness	Experiment #4 6 robots	Robot2.Anger Robot3.Fear robot4.Disgust Robot1.Sadness

6 Conclusions

The surprise emotion appears in both classifications because it is considered that is able to interact in the same way with positive emotions as negative. The behavior observed in the conduct of the experiments was that emotions tend to cluster with the emotions of the same type i.e. positive emotions with positive emotions and vice versa, every emotion was a time-and behavior in different path depending on emotion and that tended to the same choice when they were the same type.

References

1. Ahren, G.L., Schwartz, G.E.: Differential lateralization for positive versus negative emotion. Neuropsychologia **17**, 693–698 (1979)
2. Borod, J.C., Kent, J., Koff, E., Martin, C., Alpert, M.: Facial asymmetry while posing positive and negative emotions: Support for the right hemispheric hypothesis. Neuropsycholoqia **26**, 759–764 (1988)
3. Bisquerra, R.: Educación emocional y bienestar. Praxis, Barcelona (2000)
4. Cattell, R.B., Krug, S.E.: The number of factors in the 16FP: a review of the evidence with special emphasis on methodological problems. Educ. Psychol. Measur. **46**(3), 509–522 (1986)
5. Campbell, R.: Asymmetries in interpreting and expressing a posed facial expression. Cortex **14**, 327–342 (1978)
6. Campbell, R.: Cerebral asymmetries in the interpretation and expression of a posed expression. Unpublished doctoral dissertation, University of London (1979)
7. Campbell, R.: Left handers' smiles: Asymmetries in the projection of a posed expression. Cortex **15**, 571–579 (1979)
8. Carpenter, M.B.: Core Text of Neuroanatomy, 2nd edn. Williams & Wilkins, Baltimore (1978)
9. Davidson, R.: Affect, cognition, and hemispheric specialization. In: Izard, C.E., Kagan, J., Zajonc, R. (eds.) Emotion, Cognition, and Behavior, pp. 320–365. Cambridge University Press, Cambridge (1984)
10. Ehrlichman, H.: Hemispheric asymmetry and positive-negative affect. In: Ottoson, D. (ed.) Duality and Unity in the Brain, pp. 194–206. Macmillan, Hampshire (1987)
11. Gainotti, G.: Emotional behavior and hemispheric side of lesion. Cortex **8**, 41–55 (1972)
12. Gabarre, J.: El rostro y la personalidad: rostro, cerebro y conducta. Ediciones flumen. Barcelona, 4ª edición ampliada (2007)
13. Goleman, D.: La inteligencia Emocional en la Empresa. In: Vergara, J. (ed.) Barcelona, p. 26 (1999). González de Rivera, J.L.: Crisis Emocionales. Espasa-Calpe, Madrid 5 (2005). González de Rivera, J.L.: Ecpatía y empatía. Psiquis, **25**(6), 243–245 (2004)
14. Gunturkun, O.: The Venus of Milo and the dawn of facial asymmetry research. Brain Cognit. **16**, 147–150 (1999)
15. Gruebler, A., Suzuki, K.: A wearable interface for reading facial expressions based on bioelectrical signals. In: International Conference on Kansei Engineering and Emotion Research 2010. Paris, 2–4 March 2010
16. Hager, J.C.: Asymmetries in facial expression. In: Ekman, P. (ed.) Emotion in the Human Face, pp. 318–352
17. Hallevorden, J.: Eine neue Methode experimenteller Physiognormk [A new method of experimental physiognomy]. Psychuu, Neuroloq. Wochensch. 4:309–311 (1902)
18. Hallevorden, J.: Rechts- und Links-handigkeit und Gesichtausdruck [Right- and left-handedness and facial expression]. Zentralbl. qes. Neuroloq. Psychiat. **53**, 560 (1929)
19. Best, C.T., Womer, J.S., Freya Queen, H.: Hemispheric asymmetries in adults' perception of infant emotional expressions. J. expo Psychol.: Hum. Percept. Perform. **20**, 751–765 (1994)
20. Heilman, K.M., Bowers, D.: Neuropsychological studies of emotional changes induced by right and left hemispheric studies. In: Stein, N.L., Leventhal, B., Trabasso, T. (eds.) Psychological and Biological Approaches to Emotions, pp. 97–113. Lawrence Erlbaum, Hillsdale, NJ (1990)
21. Karch, G.R., Grant, C.W.: Asymmetry in perception of the sides of the human face. Percept. Mot. Skills **47**, 727–734 (1978)
22. Lindzey, G., Prince, B., Wright, H.K.A.: A study of facial asymmetry. J. Person. **21**, 68–84 (1952)
23. McGee, A., Skinner, M.: Facial asymmetry and the attribution of personality traits. Br. J. Soc. Psychol. **26**, 181–184 (1987)

24. Rhodes, G., Lynskey, M.: Face perception: attributions, asymmetries and stereotypes. Br. J. Soc. Psychol. **29**, 375–377 (1990)
25. Borod, J.C.: Cerebral mechanism underlying facial, prosodic, and lexical emotional expression: a review of neuropsychological studies and methodological issues. Neuropsychology **7**, 445–463 (1993)
26. Sackeim, H., Weiman, A.L., Forman, B.D.: Asymmetry of the face at rest: size, area and emotional expression. Cortex **20**, 165–178 (1984)
27. Seinen, M., van der Werff, J.J.: The perception of asymmetry in the face. Net. J. Psychol. **24**, 551–555 (1969)
28. Silberman, E.K., Weingarten, H.: Hemispheric lateralization of functions related to emotion. Brain Cognit. **5**, 322–353 (1986)
29. Stringer, P., May, P.: Attributional asymmetries in the perception of moving, static, chimeric, and hemisected faces. J. Nonoerb. Behav. **5**, 238–252 (1981)
30. Van Gelder, R.S., Van Gelder, L.: Facial expression and speech: neuroanatomical considerations. Int. J. Psychol. **25**, 141–155 (1990)
31. Wolff, W.: The experimental study of forms of expression. Charact. Person. **2**(82), 168–176 (1933)
32. Wolff, W.: The Study of Personality. Harper & Row, New York (1943)
33. Wolff, W.: The experimental study of forms of expression. Charact. Person. **2**, 168–176 (1933)

Recognition of Primary Emotions Using the Paradigm of Intelligent Agents for the Recognition of Subtle Facial Expressions

Enrique Aguirre, Arnulfo Alanis Garza, María del Rosario Baltazar, Lenin G. Lemus Zuñiga, Sergio Magdaleno Palencia and Carlos Lino Ramírez

Abstract The subtle expressions represent is a discrete way to externalize emotions and feelings about situations or conditions that occur suddenly. In this document will be established the proposal of how approach the study of them using computational algorithms implemented in intelligent agents. Implemented computational logic that allows making decisions based on anatomical and psychological models related to the expressions. So in this way to build a more comprehensive model proposed to identify subtle expressions.

Keywords Subtle expressions · Intelligent agents · Cognitivist · Emotions · Feelings · Facial muscles

---

E. Aguirre (✉)
Technological Institute of Tijuana, Master of Science in Engineering, Tijuana, BC, Mexico
e-mail: eaguirre@tectijuana.edu.mx

A.A. Garza · S.M. Palencia
Computing and Systems Department, Technological Institute of Tijuana, Tijuana, BC, Mexico
e-mail: alanis@tectijuana.edu.mx

S.M. Palencia
e-mail: jmagdaleno@tectijuana.edu.mx

M. del Rosario Baltazar · C.L. Ramírez
Research Division, Technology Institute of Leon, Monterrey, Gto, Mexico
e-mail: charobal@yahoo.com

C.L. Ramírez
e-mail: carloslino@itleon.edu.mx

L.G. Lemus Zuñiga
Universitat Politècnia de València, Valencia, Spain
e-mail: lemus@upv.es

L.G. Lemus Zuñiga
RIS-Itaca, Universitat Politècnia de València, Valencia, Spain

G. Jezic et al. (eds.), *Agent and Multi-Agent Systems: Technologies and Applications*, Smart Innovation, Systems and Technologies 38,
DOI 10.1007/978-3-319-19728-9_29

1 Introduction

In the scientific study of facial expressions should be mentioned as antecedent the book "The Expression of the Emotions in Man and Animals" [1]. This book has contributed significantly to gather evidence that some emotions are universal facial expression.

More recent research as "Emotions Revealed Recognizing face and feelings to improve increase communication and emotional life" [2]. Dr. Paul Ekman, from the University of San Francisco California have focused on studying the issues concerning expressions, micro expressions and subtle expressions.

Emotions are a complex concept, but they can be seen as affective states related to an event, an object. In many cases, facial expressions, by contractions of the facial region corresponding to the muscles, results in the manifestation of temporarily deformed features such as eyelids, eyebrows, nose, lips and skin.

Recently, it has sparked an interest in the area of computer vision, focused on everything that has to do with matters related to image analysis. Including analysis of facial emotion recognition.

1.1 Understanding the Differences in the Types of Expressions

Below it is briefly explained the differences between the types of expressions.

(a) **Macro expressions**

The macro expressions are expressions that manifest in our daily interact with people, nature, and objects. We are so familiar to use that somehow is able to detect highly common and usually identify them. Dr. Ekman [2] considers 6 basic emotions: Anger or rage, disgust, fear, joy, sadness and surprise, the duration of its peak can oscillate generally between 0.5 and 5 s [2].

(b) **Micro expressions**

The micro expressions are expression less noticeable. Usually occur when an involuntary facial expression is generated for a short period of time, which usually happens when a person is consciously trying to hide all signs of how he or she feels, or when a person not consciously know he or she is feeling. It can be said to reflect on the face involuntarily a feeling. Micro expressions generally occur faster than macro expressions.

(c) **Subtle expressions**

Unlike the aforementioned types of expressions, a subtle expression usually occurs from an emotional response when a person begins to experience a feeling associated with an emotion. Unlike micro expressions subtle expressions are not associated with its reflection in the face for a length of time, but rather to the intensity of the emotion that is occurring. That is why his detection involves a more complex level of difficulty to detect them, because usually occur in the half the time of a subtle expression [2].

2 Concepts

2.1 *Emotions*

2.1.1 What Is an Emotion?

An emotion is defined by Darwin [1] as follows: "It is the response of the whole organism, involving: physiological arousal, expressive behaviors, and conscious experiences."

2.1.2 What Factors Influence an Emotion?

Emotions are part of what can be mentioned as complex psychological concepts.

It can be said that although many emotional characteristics shared similarities, these may vary from subject to subject due to training, education, culture and many others factors.

2.1.3 How Do People Can Perceive Emotions?

The emotions can be visualized when we are able to perceive one or more voluntary or involuntary expressions.

2.2 *Expressions*

2.2.1 What Is an Expression in Emotional Terms?

We have already mentioned how emotions can be perceived.

Now it's time to talk about expressions. Expressions are the physical reflection of a feeling that is projected.:

In this paper we use the definition made by Myers [3]: "They are not learned responses that you agree to a complex set of movements, particularly of the facial muscles".

3 Proposal

The following proposal aims to present a framework of analysis to identify subtle emotions through the use of intelligent agents. Below the modules that are contemplated in the proposed model are presented such purpose.

- **Real time video.**
- **Segmentation working area of interest.**
- **Database of video frames.**
- **Secondary Emotions BD.**
- **Mesh Algorithm.**
- **Facial regions segmentation work.**
- **Classifier.**
- **Agent IFB (indicator Biological Factors).**
- **Q-Learning Module.**
- **Module for sensing biological factors.**
- **Results Interface.**

3.1 How Does the Proposed Model Would Work?

The initial part of the model consists of the capturing via real time video images of people under study. Once the video is captured we can proceed to segment the section of interest with which to work with in video frames.

Once the above is done we can proceed to store the frames in a database with information about the people under study. Then apply the algorithm to each frame mesh that will be obtained according to their sequence, this will allow assigning points on the facial regions for analysis of movements.

Once the mesh algorithm is assigned to the study subject's face, we will proceed to divide the face into facial regions, so it will be able to group the points awarded for areas sharing each facial muscles.

In the following process the classifier will contain the relationships between regions in the areas work and the points assigned to them. So it may relate points of facial regions with an expression. It is important to note that this part of the process needs support from experts in the area of psychology whose feedback will have a significant weighting in the way that the classifier determines the relationships between points-regions and regions-expressions that serve are used to give the result of more subtle expression according to facial movements.

Subsequently the result obtained from the classifier will be channeled to the agent IFB to process the information acquired while the other channel will receive the results of monitoring biological factors of temperature and pressure of the subject that will serve to complement the information related to the changes made in a period of time. Thus the information obtained from both results is evaluated on a new analysis to get a final result.

The Q-learning module as already mentioned is used for reinforcement learning with the information obtained from the IFB agent.

In a similar way the sensing module biological factors, will serve to monitor temperature and EEG signals from people under study.

Finally, we find the results interface where we can visualize the result of the above processes and all work performed. We will find 2 frames of facial movement and a highlighted area on the face where the subtle expression, quantity or weighting on the subtle expression that are acknowledging reflected.

The information displayed on the interface results will be fed back to the database with which we work in order to keep track of the results already obtained.

Having described the above, the proposed model is presented, as shown in Fig. 1.

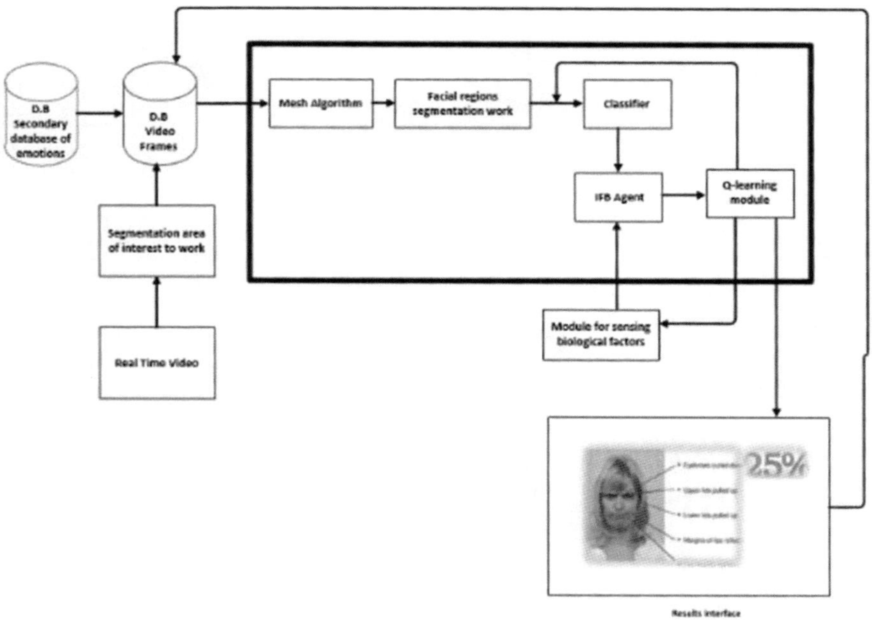

Fig. 1 Proposed model

Once the classifier is finished whit the analysis process he will send the information to the agent IFB for processing.

While on the other information of the agent will be receiving the data from sensing module biological factors which will be evaluated by the information previously received from the classifier to thereby assign an evaluation.

Once obtained THE corresponding evaluation result, this agent will be sent to Q-learning module to verify that the required target is met and evaluated. This information will be sent to the agent to be deployed in the user interface.

Below the IFB agent shown in Fig. 2.

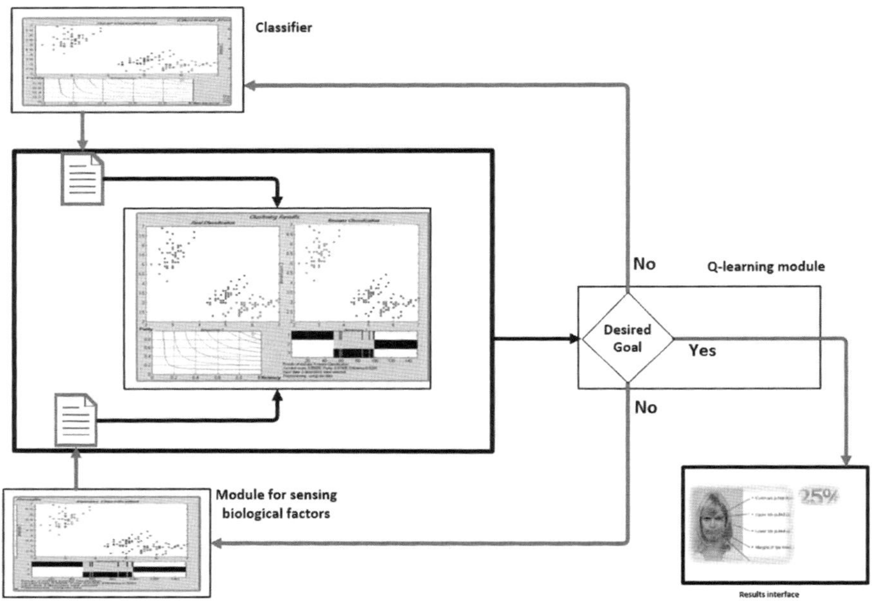

Fig. 2 Agent IFB

4 Conclusions

Until now the results are not as expected due to multiple factors involved in the process of obtaining successful integration of various disciplines such as psychology, sociology and medicine needed to achieve a more efficient identification and detection model expressions. However it is expected that the following test phases can obtain more complete results and a more accurate recognition.

References

1. Darwin, C. (ed.): The Expression of the Emotions in Man and Animal. Oxford University Press Inc, New York (1998)
2. Ekman, P. (ed.): Emotions Revealed Recognizing Face and Feelings to Improve Increase Communication and Emotional Life. Times Books, New York (2003)
3. Myers, D. (ed.): Psychology. In: Emotions, Chap. 13, pp. 500–502. Medical Panamericana, Buenos Aires, ARG (2005)
4. Paniagua, E., Sánchez, J., Martin, F.: Computational Logic. In: The Intelligent agents and logic, Chap. 6, pp. 191–192 (2003)
5. Vrancx, P. (ed.): Decentralised Reinforcement Learning in Markov Games Brusselas. Vubpress, Belgium (2010)
6. Ross, A., Jain, A.K.: Information Fusion in Biometrics. Pattern Recogn. Lett. (Special Issue on Multimodal Biom)

Assessment and Intervention with Wii Fit in the Elderly

Lenin Guillermo Lemus Zúñiga, Natalia Fernández Pintos, Esperanza Navarro Pardo, Arnulfo Alanis Garza and José Miguel Montañana Aliaga

Abstract Nowadays, the number of elderly people is constantly increasing, and the prospective is that this collective will continue to grow in the future. As aging is a process that implies many changes and consequences for the person, it is important to focus the research on the improvement of the quality of life of this increasing age group. Psychologists dedicate a great amount of time to design programs that improve the health of the elderly, and that includes gathering information and statistically analyze it to interpret and provide recommendations for an improved design. Expert systems are a good solution to reduce the amount of work that these professionals have to dedicate, automatizing tasks.

Keywords Physical exercise · Psychological intervention · Psychological assessment · Cognitive impairment · Expert system

L.G. Lemus Zúñiga (✉)
RIS-Itaca, Universitat Politècnia de València, Valencia, Spain
e-mail: lemus@upv.es

L.G. Lemus Zúñiga · N.F. Pintos · E.N. Pardo
InterTech Interdisciplinary Modelling Group, Valencia, Spain

N.F. Pintos · E.N. Pardo
University of Valencia, Valencia, Spain
e-mail: naferpin@alumni.uv.es

E.N. Pardo
e-mail: esperanza.navarro@uv.es

A.A. Garza
Instituto Tecnológico de Tijuana, Tijuana, México
e-mail: alanis@tectijuana.edu.mx

J.M. Montañana Aliaga
DACYA, Universidad Complutense de Madrid, Madrid, Spain
e-mail: jmontanana@fdi.ucm.es

G. Jezic et al. (eds.), *Agent and Multi-Agent Systems: Technologies and Applications*, Smart Innovation, Systems and Technologies 38,
DOI 10.1007/978-3-319-19728-9_30

1 Introduction

Population aging is a phenomenon that nowadays is increasing its importance. Concretely, the World Health Organization declared that about 8 % of the world population in 2010 was aged 65 years or older (an estimated 524 million people), and remarked that the prospective is that this number will continue to increase, reaching a 16 % of the world population in 2050 (approximately 1.5 billion) [1].

The aging process implies many changes on the physical characteristics of the human being, as many areas of the physical performance are affected by it. One of the current greatest concerns that is derived by this changes, is the increasing number of falls that are experienced worldwide as the elderly population grows, as they have many consequences for the health of this group [2]. Concerning this problem, participation in physical exercise activities can be a good way of decreasing the risk of suffering falls [2], as it improves physical features that contribute to better mobility, balance and reaction time [3].

However, it is frequent that the elderly living in nursing homes engage in little physical exercise, as most of the institutions have established traditional physical activity programs that the elderly might find not very appealing, and without an inner desire to engage in this kind of activities, it is not probable that the elderly will become physically active in the future [4]. As a result, besides from the participation in these established programs, the elderly do not normally involve in other physical exercise activities on their behalf, so it is important that they have resourceful environments with diverse activity related equipment, as these supportive environments encourage their participation in this kind of activities, promoting an active lifestyle [4, 5].

In the last years, different gaming consoles have created videogames that are based on the performance of physical exercise, but focusing on the recreational aspects of these activities. These games constitute an appealing alternative to the traditional physical exercise routines that are generally implemented in nursing homes and day care centers for the elderly. In this way, many commercial videogames have proved to produce improvements in different areas in the elderly population, such as self-esteem, social interaction, confidence, quality of life and physical features, as the mobility, movement time and also reaction time [6, 7].

One of these games is Wii Fit, which is exclusive of the Nintendo Wii™ gaming system. This commercial videogame is based on the performance of physical exercise, and has the objective of improving the physical status of the user by engaging in different kinds of exercises and mini-games. It requires a peripheral to be used, the "Wii Balance Board", a board that serves as a scale and allows the videogame software to know how the user is moving, by using the measures provided by its four pressure points. This videogame has proved to produce good results in the elderly population, as it improves the muscle strength, walking speed and balance, and as a consequence, decreases the fear of falling experienced [8–12]. Moreover, the beneficial effects of physical exercise itself in the elderly include reduction in the risk of falls [13, 14], and improvements in physical features, as well as psychological attributes [15], so it is expected to find these benefits too with the engagement in a new and more appealing physical exercise routine.

On the other hand, Expert Systems (ES) are part of the computer science research area called Artificial Intelligence. The ES are also known as knowledge based systems. An ES is a computer application that contains the knowledge of one or more human experts in a specific domain. The first expert system aim is to capture that knowledge from a human expert and make it available to the users [16].

The knowledge base can be generated by using various methods which can include methodical interviews and the repertory grid technique from a trained knowledge engineer. Often the expert knowledge area is "fuzzy" in nature and contains a great deal of procedural knowledge, so the knowledge engineer must be an expert in the process of knowledge elicitation. Fundamentally, a knowledge-based system consists of some representation of expertise, or a problem to be solved, and some mechanisms to apply the expertise to a problem in the form of rules [17].

The second ES aim is to emulate an expert in a specialized knowledge domain [16].

Expert Systems are used in many areas, in the case of psychology several Expert Systems have been developed for automatic emotion recognition [18, 19], the simulations of social interaction within groups [20], phobias therapies, computer aided treatment in psychiatry [21], electronic inquires and automatic evaluation and diagnostic generation of disabilities. As an example of these diagnostic systems, it has been proposed to use intelligent agents for the assessment of changes in aging [22]. And also, they have been used to offer solution strategies for learning disabilities [23], performing its classification of disabilities based on their past experiences.

The objective of this paper is to propose the architecture of an expert system on the assessment of an intervention program. That program was evaluated within a group of institutionalized elderly using Wii Fit. In this experiment, many quality of life variables were assessed, both physical and psychological, such as self-esteem, fear of falling, perceived efficacy for avoiding a fall, general health symptoms, physical symptoms, anxiety, depression, social dysfunction, gait and balance. Concretely, it is planned to propose an expert system to help the psychologist perform the following tasks that are designed for the intervention:

1. Efficiently gather the information provided by the different assessment instruments (physical and psychological tests) and the intervention device (Wii Fit).
2. Statistically analyses of the previously gathered information.

In the future, when a high number of elderly will be evaluated and the knowledge base of the expert system would have enough data to start making inferences, the system will help psychologists to: Establish conclusions based on the performed analyses.

2 Methods

We will start introducing the original intervention performed on the elderly.

2.1 Subjects

The participants from the Wii Fit study were elderly subjects living in a nursing care facility, who met the inclusion and exclusion criteria that was established to filter the subjects that were able to perform the activities required for the intervention. Concretely, these were the requirements that were redacted for the sample selection:

- Aged 65 years or more
- Not have used Wii Fit with WBB before
- Ability to stand on their behalf, or with an assist device
- Willing to participate in the study
- Score 0–4 in the Spanish version of the Short Portable Mental Status Questionnaire (SPMSQ) [24, 25].

2.2 Design

A randomized case-control design was implemented, dividing the sample in different groups. The SPMSQ scores were used to separate the subjects in two different groups according to their cognitive status. Those who scored 0–2 points in the questionnaire were assigned to the "Non Cognitive Impairment" group, and the ones who had a 3–4 points score, entered the "Mild Cognitive Impairment" group. Both of these groups were divided in half, creating four subgroups, randomly designating them to the experimental and control conditions.

2.3 Materials

Physical and Psychological Assessment Materials. Two different evaluations were required for the comparison of the groups before and after the intervention. Thus, all the participants were assessed with the following tests.

Physical Assessment. This evaluation included the following measurements:

- The Spanish version of Tinetti's Balance and Gait Instrument [26], as a measure of the physical features of the participants.
- The Wii Fit Age value that the videogame gives in every session, which is calculated from the age, weight, center of balance and performance in balance tests.
- The Clinical Opinion about the physical state of the elderly, given by the physiotherapist of the center, who was not aware of which group every participant was assigned to. This value was measured in a scale from 0 to 10, being 0 a very poor physical status for their age, and 10 an optimum physical status for their age.

Psychological Assessment. For the psychological variables that are evaluated, Spanish versions of the following tests were administered:

- The Falls Efficacy Scale [27], an instrument designed to estimate the self-efficacy of the person to avoid falling.
- A single item question: "Are you afraid of falling?" as a simple measure of the fear of falling. The item was chosen according to different studies that focus on the advantage of measuring this variable through a direct and simple question [28].
- Rosenberg's Self-esteem Scale [29, 30], which offers a measure of the self-esteem of the participant.
- The Philadelphia Geriatric Center Morale Scale [31, 32] was used in the study to assess the quality of life satisfaction of the elderly.
- Goldberg's General Health Questionnaire [33, 34] was the instrument used to evaluate the anxiety, depression, social dysfunction and physical symptoms, as well as the general health of the users.

Intervention Materials. For the sessions, a closed room with enough space for the participants to move comfortably was prepared with a 32 in. computer display, the Wii Fit gaming console and the Wii Balance Board, placed 2 meters away from the screen.

2.4 Procedure

First Stage. The inclusion and exclusion criteria were used to select the subjects that were eligible to participate in the study, who were included in two lists (Non Cognitive Impairment or NCI and Mild Cognitive Impairment or MCI), according to their SPMSQ score. Then, the subjects from these lists were randomly assigned to the experimental and control subgroups, thus creating the four groups for the experiment: NCI experimental, NCI control, MCI experimental and MCI control.

Second Stage. The physical and psychological pre-intervention individual assessments were performed with all the groups to obtain the baseline for the study. Also, the clinical opinion of the physiotherapist about their physical status was obtained. .

Third Stage. The participants from the experimental conditions performed Wii Fit sessions every week, for a period of two months of intervention, continuing too with their usual physical exercise routine programmed in the nursing home. In the other hand, the subjects from the control conditions did not have any contact with the Wii Fit, and continued their usual physical exercise routine too.

The Wii Fit sessions consisted of an initial physical status test that delivered the Wii Fit Age value, and was followed by four physical exercise activities, according to the session planning shown in Table 1. They were performed in the morning, and lasted about 30–45 minutes. The users were always accompanied to prevent possible falls. All the scores obtained were recorded.

Fourth Stage. After the intervention, the assessments were performed again with all the participants, and the clinical opinion from the physiotherapist was obtained once more.

2.5 Statistical Analyses

The physical and psychological data that was obtained was statistically analyzed using Mixed Analyses of Variance (ANOVA), for the variables that were assessed in both experimental and control groups. However, the data that was only obtained from the experimental groups, that is, the Wii Fit Age and the different exercise scores, were analyzed using a Repeated Measures Student's T test that compared the first and last Wii Fit session.

Table 1 Activity planning for the Wii Fit sessions

Session number	Exercises
First session	Deep breathing/half moon/torso twists/soccer heading
Second session	Deep breathing/torso twists/hula hoop/soccer heading
Third session	Deep breathing/half moon/lunge/ski slalom
Fourth session	Deep breathing/sun salutation/torso twists/ski slalom
Fifth session	Deep breathing/sun salutation/lunge/ski slalom
Sixth session	Deep breathing/sun salutation/ski slalom/table tilt
Seventh session	Deep breathing/soccer heading/ski slalom/table tilt
Eight session	Deep breathing/half moon/torso twists/soccer heading

Also, the scores from every session were used to create line charts that showed the evolution of the performance with every session. The first line chart was created with the mean of the Wii Fit Age values obtained by each participant; and the second was drawn with the sum of the score obtained in the four exercises that every session consisted of. In the event that the session included the "Ski Slalom" balance mini-game, which delivers the final score as a measure of minutes and seconds needed to perform the exercise, the value obtained had to be converted to a 0–100 score system, as the other exercises use.

3 Proposed Expert System

The Expert System was designed in a non-intrusive way. In other words, the target participants (Sect. 2.1) and the experiment design (Sect. 2.2) used by the psychologist does not need to be modified.

Next, it is described the additional materials, the procedure and the architecture of the Expert System.

3.1 Materials

Additionally to the materials of Sect. 2.3. It is necessary the following: A tablet running the Android® OS version 4.3 (API 18). And a personal computer permanetly connected to Internet, running the Windows 7 OS® with the following characteristics: An Intel® Core™ i7@3,4GHz Processor. And 16 GBytes of RAM DDR-3@1,666GHz.

The needed software is;

- XAMPP® with PHP 5.6.3 version. XAMPP is a completely free Apache distribution containing MySQL®, PHP®, Tomcat® and Perl.
- SPSS® version 20.
- Java® Development Kit (JDK).
- Android® SDK bundle for Android Studio.
- Android Studio®. Android Studio® is the official IDE for Android application development, based on IntelliJ IDEA.
- Sublime Text editor version 3.0.

3.2 Procedure

The procedure is the same as previously described on Sect. 2.4. Though the psychologist should store the information of the assessments in the database, as follows:

- Once the selection stage (First Stage) has finished, the socio-demographic data should be stored in the system database using an internet connection. Besides, it is recommended to store in the database information about who is the psychologist responsible of each one of the participants.
- It is recommended that the psychologist stores in the database the scores of each one of the participants after they finish every session

The Fig. 1 shows how the psychologist should proceed when using the expert system.

3.3 Architecture of the Expert System Under Development

The ES is conformed of four blocks. Each one them are defined below.

- Perception: This block receives the data of each one of the participants and analyze each one of the actions made by the participant in order to store them.
- Agent of action's selection: The agent is in charge of selecting and provide recommendations about the actions that one participant should do.
- Knowledge base: This block stores all the data and also stores the learning rules of the IS.
- Human-machine interface: This interface is used by the psychologist to input data and consult the data stored in the system.

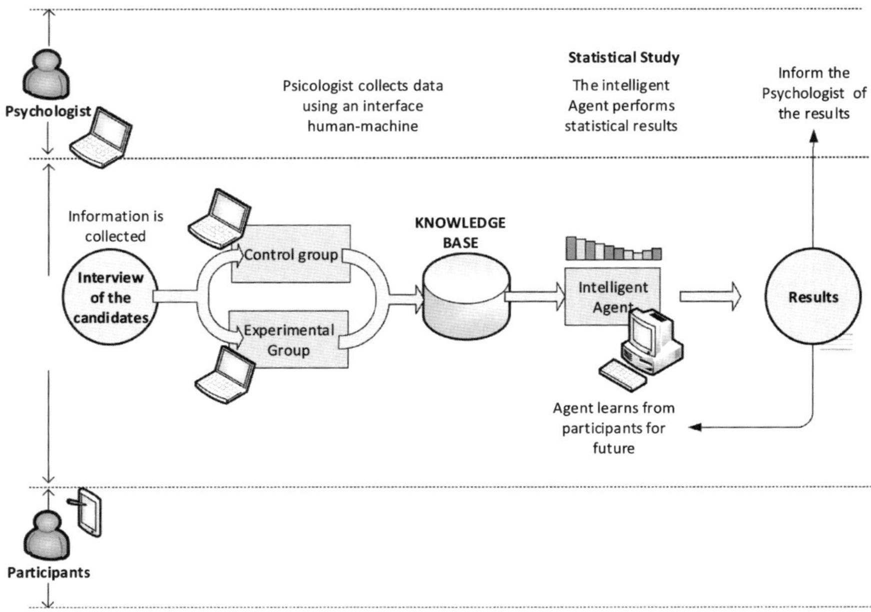

Fig. 1 Procedure to use the expert system

3.4 Statistical Analyses

The statistical Analysys is madewith SPSS. For testing porpouses, it has been verified with synthetic data.

3.5 Human Interface

Concerning to the Human-Machine interface, we currently have developed an application for Android mobile devices (Smartphones and tablets).

Below, screen captures of the mobile App are shown.

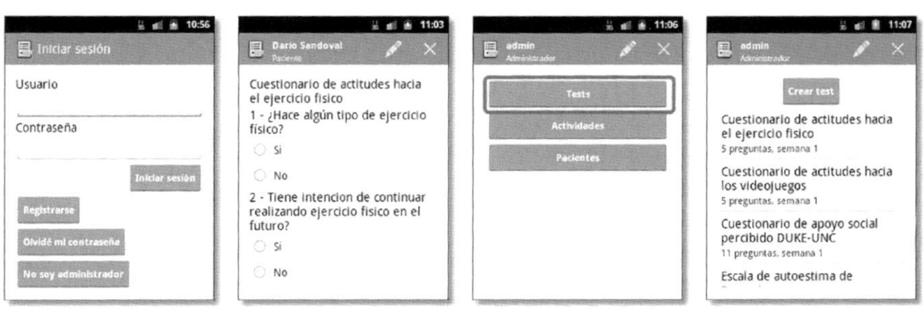

Fig. 2 Different views of the Android application for smartphones and tablets

4 Discussion and Conclusions

The original procedure has been improved. Now, the method includes the upload of the data using Internet, and its analysis through automatized procedures.

The original intervention was made with a sample of 14 people aged 65 years or more. Seven of them were randomly assigned to the control group and the other seven to the experimental group. Unfortunately, the ANOVA analyses show that this sample size is not big enough to obtain a representative statistical analysis.

This new procedure, based on the expert system, has not been used with real participants. The ES has been tested with synthetic data and with the values obtained in the original intervention program.

It has been proposed the architecture of an Expert System to assist psychologist on assessment of an intervention program.

The system has been designed to be non intrusive.

In the near future, we are establishing collaborations with nursing homes for the elderly located in Mexico, in order to recreate this design using the new procedure based on the expert system.

As future work we have to: (i) Verify the rules to obtain recommendations based on the statistical data analyses. (ii) Enhance the human interface.

References

1. World Health Organization: Global Health and Ageing. WHO & US National Institute of Aging, Geneva (2011)
2. World Health Organization: WHO Global Report on Falls Prevention in Older Age. WHO, Geneva (2007)
3. Day, L., Fildes, B., Gordon, I., Fitzharris, M., Flamer, H., Lord, S.: Randomised factorial trial of falls prevention among older people living in their own homes. BMJ **325**, 1–6 (2002)
4. Chen, Y.M.: Perceived barriers to physical activity among older adults residing in long-term care institutions. J. Clin. Nurs. **19**, 432–439 (2010)
5. World Health Organization: Active Aging: From evidence to Action. WHO, Geneva (2000)
6. Keogh, J.W.L., Power, N., Wooller, L., Lucas, P., Whatman, C.: Can the NINTENDO WIITM Sports game system be effectively utilized in the nursing home environment? J. Commun. Inf. **8**(1) (2012)
7. Schoene, D., Lord, S.R., Delbaere, K., Severino, C., Davies, T.A., Smith, S.T.: A randomized controlled pilot study of home-based step training in older people using videogame technology. Plos One **8**(3) (2013)
8. Nitz, J.C., Kuys, S., Isles, R., Fu, S.: Is the Wii Fit a new-generation tool for improving balance, health and well-being? A pilot study. Climacteric **13**(5), 487–491 (2010)
9. Padala, K.P., Padala, P.R., Burke, W.J.: Wii-Fit as an adjunct for mild cognitive impairment: clinical perspectives. J. Am. Geriatr. Soc. **59**(5), 932–933 (2011)
10. Young, W., Ferguson, S., Brault, S., Craig, C.: Assessing and training standing balance in older adults: a novel approach using the Nintendo WiiTM' Balance Board. Gait Posture **33**, 303–305 (2011)
11. Agmon, M., Perry, C.K., Phelan, E., Demiris, G., Nguyen, H.Q.: A pilot study of Wii Fit exergames to improve balance in older adults. Geriatr. Phys. Ther. **34**, 161–167 (2011)

12. Bainbridge, E., Bevans, S., Keeley, B., Oriel, K.: The effects of the Nintendo Wii FitTM on community-dwelling older adults with perceived balance deficits: a pilot study. Phys. Occup. Ther. Geriatr. **29**(2), 126–135 (2011)
13. Campbell, A.J., Robertson, M.C., Gardner, M.M., Norton, R.N., Tilyard, M.W., Buchner, D.M., et al.: Randomised controlled trial of a general practice programme of home based exercise to prevent falls in elderly women, BMJ **315** (1997)
14. Province, M.A., Hadley, E.C., Hornbrook, M.C., Lipsitz, L.A., Miller, J.F., Mulrow, C.D., et al.: The effects of exercise on falls in elderly patients: a preplanned meta-analysis of the FICSIT trials. JAMA **273**(17), 1341–1347 (1995)
15. Fiatarone Singh, M.A.: Exercise comes of age: rationale and recommendations for a geriatric exercise prescription. J. Gerontol. **57A**(5), M262–M282 (2002)
16. Dennis, R.: Artificial intelligence. McGraw-Hill, New Delhi (1996)
17. Engelbrecht, A.P.: Computational Intelligence: An Introduction. Wiley (2002)
18. Datcu, L.R.: Facial expression recognition in still pictures and videos using active appearance models: a comparison approach. In: Proceedings of the 2007 International Conference on Computer Systems and Technologies (2007)
19. Pantic, M., Rothkrantz, L.: Expert system for automatic analysis of facial expressions. Image Vis. Comput. (2000)
20. Nygren, E.: Simulation of user participation and interaction in online discussion groups. In: Proceedings on Modeling and Mining Ubiquitous Social Media, Springer, pp. 138–157 (2011)
21. Liao, S.-H.: Expert system methodologies and applicationsa decade review from 1995 to 2004. Elsevier **28**(1), 93–103 (2005)
22. Kaye, J.A., Maxwell, S.A., Mattek, N., Hayes, T.L., Dodge, H., Pavel, M., Jimison, H.B., Wild, K., Boise, L., Zitzelberger, T.A.: Intelligent systems for assessing aging changes: home-based, unobtrusive, and continuous assessment of aging. J. Gerontol. Ser. B Psychol. Sci. Soc. Sci. **66B**(S1), 180–190 (2011)
23. Elsayed, K.N.: Diagnosing learning disabilities in special education by an intelligent agent based system. Int. J. Adv. Comput. Sci. Appl. **4**(3) (2013)
24. Pfeiffer, E.: A short portable mental status questionnaire for the assessment of organic brain deficit in elderly patients. J. Am. Geriatr. Soc. **23**, 433–441 (1975)
25. García-Montalvo, J.L., Rodríguez, L., Ruipérez, I.: Valoración del cuestionario de Pfeiffer y la escala de incapacidad mental de la Cruz Roja en la detección del deterioro mental en los pacientes externos de un servicio de geriatría. Rev. Esp. Geriatr. Gerontol. **27**, 129–133 (1992)
26. Tinetti, M.E.: Performance-oriented assessment of mobility problems in elderly patients. J. Am. Geriatr. Soc. **34**, 119–126 (1986)
27. Tinetti, M.E., Richman, D., Powell, L.: Falls efficacy as a measure of fear of falling. J. Gerontol. **45**(6), 239–243 (1989)
28. Legters, K.: Fear of falling. Phys. Ther. **82**(3), 264–272 (2002)
29. Rosenberg, M.: Society and the Adolescent Self-Image. Princeton University Press, Princeton (1965)
30. Atienza, F.L., Moreno, Y., Balaguer, I.: Análisis de la dimensionalidad de la Escala de Autoestima de Rosenberg en una muestra de adolescentes valencianos. Rev. Psicol. Univ. Tarrac. **22**(1–2), 29–42 (2000)
31. Lawton, M.P.: The Philadelphia Geriatric Center Morale Scale: a revision. J. Gerontol. **30**, 85–89 (1975)
32. Montorio Cerrato, I.: Evaluación psicológica de la vejez: instrumentación desde un enfoque multidimensional (doctoral thesis). Universidad Autónoma de Madrid, Madrid, Spain (1990)
33. Lobo, A., Pérez-Echeverría, M.J., Artal, J.: Validity of the scaled version of the General Health Questionnaire (GHQ-28) in a Spanish population. Psychol. Med. **16**, 135–140 (1986)
34. Goldberg, D.: The Detection of Psychiatric Illness by Questionnaire. Oxford University Press, London (1972)

A Multiagent System Proposal for 30 Day Readmission Problem Management

M.A. Mateo Pla, L. Lemus Zúñiga, J.M. Montañana, J. Pons Terol and S. Tortajada

Abstract Thirty day readmission rate is an important quality estimator for hospitals. Confident tools that forecast this risk for each patient before hospital discharge are needed by medical staff in order to delay the discharge and to plan additional home care interventions. This paper presents a proposal for a multi agent system that will evaluate this risk by integrating information from each patient and historical data from local and remote medical histories. This systems will not only help with the hospital discharge decision, but it will integrate a basic telecare system in order to reduce readmissions and increment patient quality of life by detecting problems arisen after hospital discharge.

Keywords Multiagent system · Learning agents · Decision support systems in healthcare · Machine learning

1 Introduction

When a patient is discharged from a hospital, there is always a risk of an unplanned readmission after a short period of time. One of the most used time period when evaluating this risk is 30 days. Readmissions are a costly process [1, 2] that could

M.A. Mateo Pla (✉) · J.P. Terol
DISCA, Universitat Politècnia de València, Valencia, Spain
e-mail: mimateo@disca.upv.es

J.P. Terol
e-mail: jpons@disca.upv.es

L. Lemus Zúñiga
RIS-Itaca, Universitat Politècnia de València, Valencia, Spain
e-mail: lemus@upv.es

J.M. Montañana
DACYA, Universidad Complutense de Madrid, Madrid, Spain
e-mail: jmontanana@fdi.ucm.es

S. Tortajada
IBIME-Itaca, Universitat Politècnia de València, Valencia, Spain
e-mail: vesaltor@upvnet.upv.es

G. Jezic et al. (eds.), *Agent and Multi-Agent Systems: Technologies and Applications*, Smart Innovation, Systems and Technologies 38,
DOI 10.1007/978-3-319-19728-9_31

be used as a quality index for health care delivery. In fact it is has been used in USA from 2012 [3–5] to penalize those hospitals with excess readmissions.

The medical staff need tools and validated indexes to determine the 30 day readmission risk of a patient. Several validated indexes have been proposed in the last years, some for all patients [6, 7] and others for specific kind of patients [8, 9]. One problem with many of these indexes is their locality, i.e., the indexes have been validated to a set of data that is collected more or less locally: inside a city, region, state or country. When trying to use these indexes in other contexts, the results have not always been the expected [10].

The validated indexes are not only useful in the discharge decision. Basic home care intervention can reduce the risk of readmission. This is true for elderly Congestive Heart Failure patients [13] and we think that these results could be extrapolated to other kind of patients.

On the other hand, Multi Agent Systems (MAS) are well suited both to model large-scale distributed complex systems and to ease the development of novel application related to those systems [11]. Health care systems are good candidates to use MAS technologies as they are clear examples of large-scale, complex and decentralized systems. A good review can be found in [12] about the use of MAS in health care, defining five main fields of application of MAS in health care. Among these fields, the MAS proposed in this paper is mainly a Decision Support System (DSS). Furthermore, some Medical Data Management tasks are needed and basic Remote Care applications are planned to be developed.

The proposed DSS will produce an index based in the well-known LACE+ [7]. Although recent research [14] concludes that LACE (which is the base for LACE+) could not accurate predict unplanned readmission, the presented system will include more data than LACE. One of the reasons is that in the available data there can be found a set of variables different from the ones used in LACE+ validation. In any case, our DSS will use an alternative approach to compute the index that will retain LACE+ meaning.

Our approach to evaluate the 30 day readmission risk combine the use of incremental learning agents to calculate this index and the use of discharged patient additional information as source for the learning process.

In the following sections, first different system requirements and working scenarios are defined, then the proposed system architecture is presented, and finally the conclusions and future work are outlined.

2 Working Scenarios and Requirements

There are three different scenarios that our system hast to deal with: the learning and validating scenario, the decision support scenario and the monitoring scenario. Each of this scenarios has its own set of requirements and all must be fulfilled in order to proper system work.

2.1 The Learning and Validating Scenario

This scenario happens at system installation and needs access to historical medical records. In this stage decision-selection agents are created and trained with the available data using one of the available methods. At this moment the available methods are the learning algorithm described in [15, 21].

To train the decision-selection agent the data-filter-agents will be used to extract information from the available medical data, coming from historical offline records, available Electronic Health Record (EHR) and other relevant information e.g. weather information or local festivities dates.

In this scenario the main users are the hospital computer department staff. They will install the software and hardware related with the involved agents and launch the initial learning process.

The requirements of this stage are mainly related to data quality and data access. An adequate data quality is needed to correctly train and validate learning models [22], and run them after. Also, medical information access has been regulated in many countries, so access to this kind of information is granted only to medical staff.

2.2 The Decision Making Scenario

This is the main functionality of the proposed system, determining the risk of readmission for each patient in the hospital in daily basis. The agents involved in this process belong to one of these three classes: decision-selection-agents, data-filter-agents and the EHR-update-agent. Obviously, medical staff always has the last word about the discharging decision.

A fourth class of agents, incremental-learning-agents, is defined to allow an incremental learning process in selected decision-selection agent based on the work described in [15, 21].

The users in this scenario are the medical staff, who interact with data-filter-agents and use the decision-selection-agent results in their decisions.

In this stage the requirements are related to MAS scalability and connectivity. In the connection side, the EHR system has to be modified in order to include the index in the patient record and to grant the data-filter-agents access to its records. On the other hand, the observed patient number growth could generate problems if the system is not scalable enough to cope with it.

2.3 The Monitoring Scenario

The system accepts new information for discharged patients that will be used in two ways: first, to fire patient health state related alarms, and second, to improve incremental learning processes in decision-selection-agent and incremental-learning-agents.

Users of this scenario are the medical staff (alarm management) and the people responsible for patient care (source of data) or the patients themselves if they are able to use the tools needed to send the information.

The collection and validation of this new data are information-retrieval-agents specific tasks. The collected data is added to the data-filter-agents information sources.

Alarm system management needs to be flexible and configurable in order to adapt to a wide range of diseases and patient characteristics. Medical staff will be responsible of the definition of alarm fire rules and for this task a tool has be developed or adapted from related work [16].

3 Architecture and Implementation

The architecture has four main blocks: a multiagent platform, a web user interface for medical staff, a set of web and mobile applications for monitor discharge patient evolution and databases. Figure 1 shows the components relationships.

In order to increase privacy, external components can access external databases containing only anonymous data, that is, the monitor subsystem cannot access internal databases. Medical staff web interface can only be accessed from within the hospital network. All the components of MAS will be located inside the medical care center.

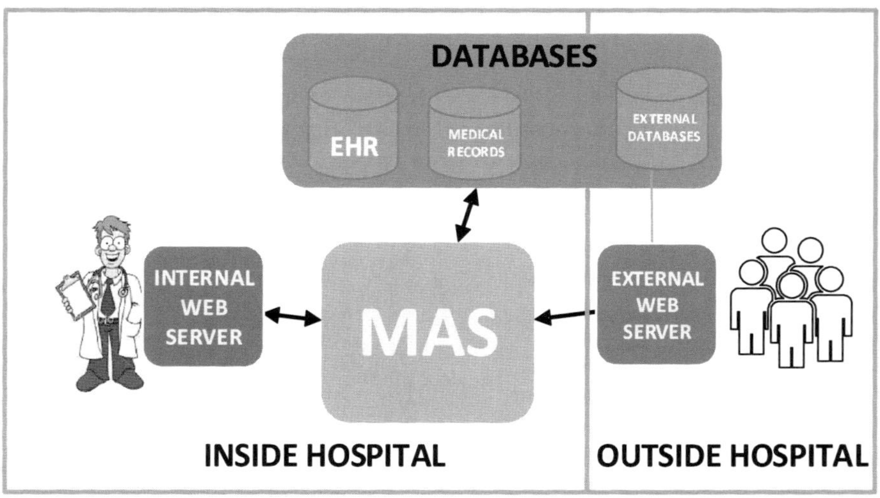

Fig. 1 Relationship between main system components

3.1 Multiagent System

The MAS is being implemented using Open Source tools, specifically JADE [17] has been selected because it is a complete multiagent Java platform based on FIPA standards [18].

The main characteristics of MAS agents are:

- Data-filter-agents (DFA) are responsible of selecting the relevant information related to a patient from the available data sources, and represent it using a standardized agent communication language with a fixed and known ontology. In this way, the whole system can be easily adapted to new hospitals using different data sources.
- Decision-selection-agents (DSA) are specialized agent that performs inference with the data retrieved by data-filter-agents. Although one DSA could be enough in a primary deployment, our idea is to develop DSA with various learning algorithms in order to increase prediction accuracy and overall fault tolerance.
- Incremental-learning-agents (ILAg) are used in two different ways. First, when creating new DSA with training needs, a ILAg is selected for doing this training and the corresponding validation. And second, DSAs with incremental learning capabilities create ILAgs to monitor and select data about previous DSA results and its accuracy. These DSA will use the messages sent by ILAgs to implement the incremental learning process.
- EHR-Update-Agent (EUpA) is the agent responsible of gathering results from DSA and selecting a value to be written in hospital EHR. Although more than one EUpAs could be deployed in the system, the current versions only allows one to be active at a time.
- Information-retrieval-agents (InfReA) act as an interface between external data gathering applications. These external applications could range from basic web based questionnaires to complex telecare systems with several biometrical sensors [16]. The use of these intermediate agents makes easier the adaptation of system to improvements or new environments.
- Alarm system management. This component will create and interact with INfReA in order to set new alarm types and conditions and read current patient and alarm status.

Figure 2 summarizes the relationship between these agents and the MAS input/output ports.

Being raised as a standalone system, it was not considered necessary to use an existing ontology in these early stages of the design, as the ones defined in the systems described in [12, 20]. The idea is to add more agents to provide points of external communication when needed after system evaluation.

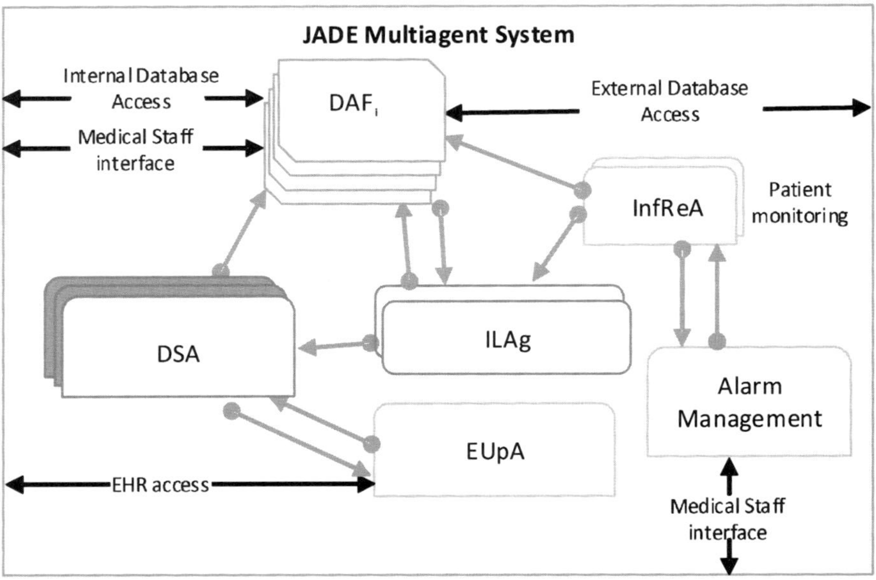

Fig. 2 Internal MAS components and their communications relations

3.2 Web User Interface for Medical Staff

Medical staff interacts with MAS using a web application. This applications could be hosted in an existing hospital network infrastructure or in a new server. In any case, as the web application accesses to internal information, an authentication method will be mandatory.

This web system allows any medical staff member to inquire about the process of calculating the 30 day readmission risk index for a given patient, including the ones already discharged. The information does refer not only to the current index value, but also to their development and potential components that determine its value. In case of more than one DSA in the system, the medical team also receives the information estimated for each of these DSA agents and how the overall value is obtained.

Furthermore, this web interface implements the alarm management system. This system will allow to define new alarms and edit, delete and check the status of the ones already defined ones. Initially, the system only allows you to set alarms such as 'greater than threshold' and 'less than threshold' with respect to the defined variables on the information provided by the remote monitoring system.

3.3 Monitoring Discharged Patients System

In the current approximation this system will be a web based questionnaire application that tries to gather relevant information from discharged patients and their environment.

If tele monitoring systems already exist in the hospital and if it is feasible to do so, interfaces will be developed to use those systems. In any case, the questionnaire design has been based on protocols and guidelines used in the referral hospital home care department.

This system will be outside the hospital to increase its scalability, as well as to not consume hospital external connection bandwidth. This externalization implies that the information related to the privacy of patients should be minimized. In any case, the system needs to implement techniques for anonymization and encryption of transferred information.

3.4 Databases

There are three types of databases: the external databases, the EHR and MAS internal database. The external databases will be used mainly for non-medical information relevant to the index calculation process. In this case we must make an earlier study of this information (as in [19] about season effect). Furthermore, they can be used as a scalable storage for information provided by outpatients, after being anonymized or encrypted.

The MAS internal databases are used to store information about the agent status, the alarm status and the results generated by the DFS. The idea of having a database not integrated into the EHR facilitates quick system installation and adaptation.

The EHR is an essential part of the current health system, so it is not easy to get full access to it for a new application. In our system we have tried to minimize access to it. On one hand, a read permission is required to access patient data. On the other hand, it is desirable that the calculated value for 30 day readmission risk is included in the EHR to facilitate its use.

4 Conclusions and Next Steps

The 30 days readmission rate is a measure of quality of medical care provided by hospitals. Several risk indexes have been proposed to help medical staff to reduce this value by not discharging those patients with high readmission probability. Among these indexes, LACE is one of the best known and different systems have been developed for its calculation.

Multi-agent systems have proven their suitability for building complex applications related to health, especially in the fields of decision making, remote monitoring and complex information interchange.

This paper presents a proposal for a multi-agent system that provides information about the 30 days readmission risks of hospital patients, including those who have already been discharged.

This system will integrate a DSS to generate risk indexes, an information retrieval system and a tele monitoring system. All three systems are combined to get improvements in the 30 days readmission rate.

In addition, DSS integrates incremental learning methods that will allow the system to adjust their predictions to changing and evolving situations. The implementation of DSS as a multi-agent system allows to include fault tolerance techniques, for example by replicating agents. Moreover, this flexibility allows the inclusion of specialized agents to obtain the risk index for patients with exceptional features.

Currently the system is in a preliminary stage of implementation awaiting that the necessary permissions to access the medical history of patients are granted.

References

1. Jencks, S.F., Williams, M.V., Coleman, E.A.: Rehospitalizations among patients in the Medicare fee-for-service program. N. Engl. J. Med. **360**(14), 1418–1428 (2009)
2. Ben-Chetrit, E., Chen-Shuali, C., Zimran, E., Munter, G., Nesher, G.: A simplified scoring tool for prediction of readmission in elderly patients hospitalized in internal medicine departments. Isr. Med. Assoc. J. **14**(12), 752–756 (2012)
3. Medicare Payment Advisory Commission. Report to congress: Medicare and the health care delivery system. Washington, DC: MedPAC (2011). http://www.medpac.gov/documents/jun11_entirereport.pdf
4. United States Congress House. Office of the Legislative Counsel: Compilation of Patient Protection and Affordable Care Act: as amended through Nov 1, 2010 including Patient Protection and Affordable Care Act health-related portions of the Health Care and Education Reconciliation Act of 2010. Washington: U.S: Government Printing Office (2010)
5. Report to Congress: national strategy for quality improvement in health care (2011) http://www.healthcare.gov/news/reports/quality03212011a.html
6. van Walraven, C., Dhalla, I.A., Bell, C., Etchells, E., Stiell, I.G., Zarnke, K., Forster, A.J.: Derivation and validation of an index to predict early death or unplanned readmission after discharge from hospital to the community. Can. Med. Assoc. J. **182**(6), 551–557 (2010)
7. van Walraven, C., Wong, J., Forster, A.J.: LACE+ index: extension of a validated index to predict early death or urgent readmission after hospital discharge using administrative data. Open Med. **6**(3), e80 (2012)
8. Amarasingham, R., Moore, B.J., Tabak, Y.P., Drazner, M.H., Clark, C.A., Zhang, S., Halm, E.A.: An automated model to identify heart failure patients at risk for 30-day readmission or death using electronic medical record data. Med. care **48**(11), 981–988 (2010)
9. Berman, K., Tandra, S., Forssell, K., Vuppalanchi, R., Burton, J.R., Nguyen, J., Chalasani, N.: Incidence and predictors of 30-day readmission among patients hospitalized for advanced liver disease. Clin. Gastroenterol. Hepatol. **9**(3), 254–259 (2011)
10. Cotter, P.E., Bhalla, V.K., Wallis, S.J., Biram, R.W.: Predicting readmissions: poor performance of the LACE index in an older UK population. Age Ageing **41**(6), 784–789 (2012)

11. Müller, J.P., Fischer, K.: Application impact of multi-agent systems and technologies: a survey. In: Agent-Oriented Software Engineering, pp. 27–53. Springer, Berlin Heidelberg (2014)
12. Isern, D., Snchez, D., Moreno, A.: Agents applied in health care: a review. Int. J. Med. Informatics **79**(3), 145–166 (2010)
13. Proctor, E.K., Morrow-Howell, N., Li, H., Dore, P.: Adequacy of home care and hospital readmission for elderly congestive heart failure patients. Health Soc. Work **25**(2), 87–96 (2000)
14. Wang, H., Robinson, R.D., Johnson, C., Zenarosa, N.R., Jayswal, R.D., Keithley, J., Delaney, K.A.: Using the LACE index to predict hospital readmissions in congestive heart failure patients. BMC Cardiovasc. Disord. **14**(1), 97 (2014)
15. Tortajada, S., et al.: Incremental gaussian discriminant analysis based on graybill and deal weighted combination of estimators for brain tumour diagnosis. J. Biomed. Inform. **44**(4), 677–687 (2011)
16. Traver, V., Monton, E., Bayo, J.L., Garcia, J.M., Hernandez, J., Guillen, S.: Multiagent home telecare platform for patients with cardiac diseases. In: Computers in Cardiology, pp. 117–120. IEEE (2003)
17. Bellifemine, F.L., Caire, G., Greenwood, D.: Developing multi-agent systems with JADE, vol. 7. John Wiley & Sons (2007)
18. Foundation for Intelligent Physical agent. http://www.fipa.org/
19. Park, L., Andrade, D., Mastey, A., Sun, J., Hicks, L.: Institution specific risk factors for 30 day readmission at a community hospital: a retrospective observational study. BMC Health Serv. Res. **14**(1), 40 (2014)
20. Isern, D., Snchez, D., Moreno, A.: Ontology-driven execution of clinical guidelines. Comput. Methods Programs Biomed. **107**(2), 122–139 (2012)
21. Tortajada, S., et al.: Incremental logistic regression for customizing automatic diagnostic models. In: Fernn-dez-Llatas, C., Garca-Gmez, J.M. (eds.) Data Mining in Clinical Medicine, Methods in Molecular Biology (2014)
22. Sez, C., et al.: Stability metrics for multi-source biomedical data based on simplicial projections from probability distribution distances. In: Statistical Methods in Medicine (2014). doi: 10.1177/0962280214545122

Part VII
IS: Serious Game and Business Intelligence Through Agent-Based Modelling

An Evaluation Model for Order-Decision Methods of Contents in Information-Providing Sites

Masato Mori and Setsuya Kurahashi

Abstract Web sites called an information-providing site provide information to people looking for a job, a home and so forth in Japan. Companies that want to display their products or services on the sites pay advertisement rates for website operators. In this study, we consider how to decide the posting order of products for publication on list pages of contents in information-providing sites. The order should be desirable not only for advertisers, but likewise for users. First, we show a result of a regression analysis between posting order and page views. Second, we define User Popularity Degree using the analysis. Finally, we employ agent-based modelling to simulate the websites in which web contents are defined as agents bidding advertisement rates for getting the optimized or effective posting orders autonomously.

Keywords Information-providing site • Posting order • Rank • Auction • Bid • User popularity degree • Reinforcement learning • Actor critic

1 Introduction

1.1 What Are Information-Providing Websites?

There is a certain type of website in Japan, which is called an information-providing website. This type of websites collects information on a particular industry, and charges online content published on this website. Representative examples include recruitment websites and real-estate websites. This study explores the posting order of online content on a list page where all the content of an information-providing

M. Mori (✉) · S. Kurahashi
University of Tsukuba, 3-29-1 Otsuka Bunkyo, 112-0012 Tokyo, Japan
e-mail: masatobar@yahoo.co.jp

S. Kurahashi
e-mail: kurahashi.setsuya.gf@u.tsukuba.ac.jp

G. Jezic et al. (eds.), *Agent and Multi-Agent Systems: Technologies and Applications*, Smart Innovation, Systems and Technologies 38,
DOI 10.1007/978-3-319-19728-9_32

website is displayed. This study also verifies how to determine this content display order which is desirable for both website visitors and advertisers and which brings about profit for website operators.

These websites provide products in a form where advertisers apply to publish their content for a certain period of time. Therefore, the content applied is published on the website for a specified contracted period, such as 4 weeks or 24 weeks. All products are packaged with the same predetermined number of photos and appealing items for each advertiser. However, each product is divided into two classes, the expensive product and the inexpensive product, where the expensive product can provide a greater number of photos and appealing items. In the Internet industry, this kind of content-providing advertisement business is referred to as an information-providing business. Additionally, those websites that are related to this business are referred to as information-providing websites.

The site structure of such information-providing websites is basically identical with that of recruitment websites and real-estate websites. Therefore, the following sections will describe real-estate websites.

Individuals who seek places to live access real-estate websites. On such a real-estate website, each individual visitor tries to narrow down their choice of property or places to live based on addresses, cost for rent, and layouts. The results are then displayed on a list page. This list page displays the name, photos, rent, and information about the nearest station for each property. If the visitor wants to check the details, they can click on the link for that particular property. After checking the detail page, the visitor will move to another page to send their personal information in order to inquire about the property. Completing the sending of the personal information is referred to as conversion.

1.2 About Posting Order

On a website, the upper level of the page is highly valuable per equivalent area when compared to the lower level of the page. This is because the uppermost level of the page is usually displayed when visitors access the website, and this level easily attracts attention from visitors.

On the list page of such an information-providing website, the upper level of the page easily catches visitors' attention when compared to the lower level, and at the same time has high effectiveness for advertising. When a page cannot display all the content requested on one page, a number of additional pages are used if needed. In this case too, the first page has a high advertising effectiveness when compared to the second page. Therefore, advertisers want to have their information posted on the upper level of the pages or pages near the beginning of the listed pages. It is impossible for all advertisers to post their information on the uppermost level of the page. Therefore, website operators make all kinds of efforts for adjusting posting order.

With respect to the content posting order type of information-providing websites, the most popular type is recent order. In this order type, the content is listed in order of the nearest date of posting from the top. Each of content portion has a new arrival period, and this provides the content with equivalent opportunities to be posted on the upper level of the page. However, this does not function with content posted on the same date. Additionally, how close the target property is to the visitor's desire is not considered.

Another order type that can be frequently seen is product-price order. In this order type, expensive products are posted and displayed on the upper level, while inexpensive products on the lower level. Advertisers usually can accept this arrangement for financial reasons. However, many information-providing websites have two or three classifications for packaged products. For this reason, many advertisers pay the same price; however, this type of order does not work for all advertisers. In addition, this product-price order type is the same as the resent order type from the point of view that the desired conditions of visitors are not taken into consideration.

Recommendation order is another order type. Popular advertisers might be posted and displayed on the upper level of the page. Machine learning could be utilized; however, the details are not disclosed and it is difficult to evaluate this order type.

These are the characteristics of posting order types of information-providing websites. The majority of such websites have adopted the use of the recent order and product-price order. So generally speaking we can say that this situation is neither suitable nor desirable for advertisers and website visitors. Given that, the goal of this study is to conduct an Agent-Based Model simulation in order to examine methods for determining posting order. In this simulation, advertising content is set as agents, while these agents conduct automatic bidding so as to determine the posting order. We examined not only order determination methods based on bidding price range, but also a model that defines the popularity stakes for all content and uses its popularity.

2 Literature Review

Web search is included as one of the services where posting order is considered to be important. Search Engine Optimization (SEO) aims to display the search result in the upper level of the search results. This measure is popular because the results displayed in the upper level easily attract attention of website visitors, while having a high probability to be clicked on by the visitor. On the search result pages, advertisement frames also exist. This is referred to as search ads where advertisers bid for search keywords. Multiple search ads frames are displayed on a page, which display order is determined based on the bidding price and advertisement quality [1]. The Generalized Second Price (GSP) auction method is adopted for search ads. This auction method was created by generalizing the sealed second price auction which was developed by Vickrey [2]. In the second price auction, the winning bidder pays the bidding price placed by the runner-up bidder.

Many studies have been made on search ads. Edelman et al. proposed a generalized English auction by expanding the English auction to the search auction [3]. This auction assigns the lowest order rank in order of lower bidders. Feldman et al. made an approach from the optimization problem of advertisers [4]. There is also another study that proposed an auction mechanism where only the uppermost winning bidder can have its own advertisement displayed by paying premiums. Sakurai et al. proposed the Generalized Second Price auction system with exclusivity based on the ordinary GSP auction [5].

The order determination method based on an auction is also utilized for advertisement distribution networks called AdNetwork.

Mookerjee et al. analyzed services provided by Chikita based in the United States [6]. Chitika offers a service where publishers can set thresholds of the average click through rate (CTR) of advertisements they display on their websites, while leaving the selection of dilemmas between the profit and the media value with the publishers. Ghosh et al. conducted a study on bidding strategies in AdNetwork [7]. They commented on the balance between the bidding strategy and the goal value. There is also an auction trade type, referred to as Real-Time Bidding (RTB), which is adopted by many AdNetworks. Nabeta et al. built this RTB as an Agent-Based Model and verified the results depending on the differences in bidding strategies [8]. They conducted experiments on a RTB auction in order to examine the bidding price and the number of successful bids depending on the differences in bidding strategies.

As described above, there are a number of studies regarding business models and bidding strategies related to search ads and AdNetworks. However, little attention has been given to information-providing websites. The goal of this study is to simulate the determination method of posting order for an information-providing website by utilizing the auction method.

3 Analysis of Information-Providing Websites

3.1 Data Used

When conducting the simulation, we prepared the following two different types of data. The first one is property data. We obtained daily information of a certain real-estate website by means of the web-scraping program. The second type was sampling data from the page views (PV) of the same website.

Combining these two types of data, we obtained the daily posting ranks and the number of PV. Figure 1 shows the distribution of the posting ranks of properties on a certain day and the number of PV. Here, we can see some values located significantly away from the center, which were significantly influenced by optional products. Optional products mean banner advertisements that are placed on other

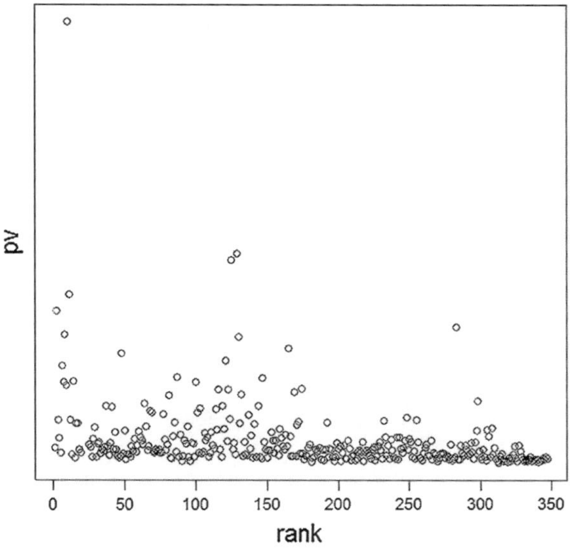

Fig. 1 Ranks and PV

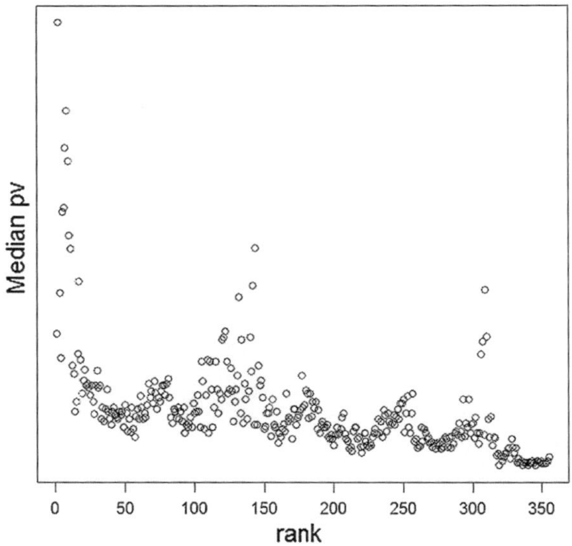

Fig. 2 Ranks and median PV

levels of the website. Clicking on one of these banner advertisements moves the page directly to the detail page. Additionally, unique factors of each property, such as the price and the site location, also had an influence on the number of website PV. In order to make it easier to confirm the influence of posting ranks on the number of website PV, we obtained the median value of the number of daily PV at each rank during one month. The result is shown in Fig. 2. This figure shows that when ranks drop, the number of page views also decreases. Chapter 1 mentioned that the contents posted on higher ranks have high advertising effectiveness by easily attracting much attention from visitors. This can also be demonstrated by the reduction of the number of page views.

3.2 Prediction of the Number of Page Views by Ranking Variation

In this study there are calculation equations necessary for the experiment described after next chapter. Under this simulation environment, posting ranks are determined by bidding done by agents, while the number of PV is given as reward. The number of PV varies depending on the posting rank; equations necessary for calculating the number of PV are defined as follows:

For each item of content there is certain property information. This information includes the address, the nearest train station, sell price, and layout. These are unique factors for each property, while only the posting rank varies. Given that, we tried to isolate the influence of posting ranks on the number of PV and the influence of other factors on the number of PV.

In Fig. 3, when we conducted regression analysis while setting posting ranks as explanatory variables and the PV values as objective variables, the determination coefficient of exponential regression became 0.5063. This value is more applicable than single regression where posting ranks or multiple regressions where posting positions and posting pages were set as explanatory variables. This is expressed by Eq. (1). Figure 3 shows the regression curve of the present case.

$$\text{Median}(\widehat{PV}) = \text{a} \cdot e^{b \cdot rank} \tag{1}$$

$$\text{* e : natural logarithm}$$

In addition, when we conducted another regression analysis while setting the absolute values of errors in Fig. 3 as objective variables and posting ranks as explanatory variables, the absolute error value increased exponentially (determination coefficient: 0.613).

$$\text{Abs}(\text{Median}(\widehat{Error})) = \text{c} \cdot e^{d \cdot rank} \tag{2}$$

$$\text{* e : natural logarithm}$$

Next, since the error expressed by Eq. (2) should differ depending on the property content, we tried to obtain the median error value from Eq. (1) according to each property. This error is between the actual number of daily PV for one month of each property and the prediction value of PV where the posting rank of each day was assigned to the Eq. (1).

$$Median_i\{(y_{i1} - \widehat{y_{i1}}), (y_{i2} - \widehat{y_{i2}}), \ldots, (y_{id} - \widehat{y_{id}})\} \tag{3}$$

$$* \; y_{id} \text{ : Actual PV of propaty i of the dth day}$$

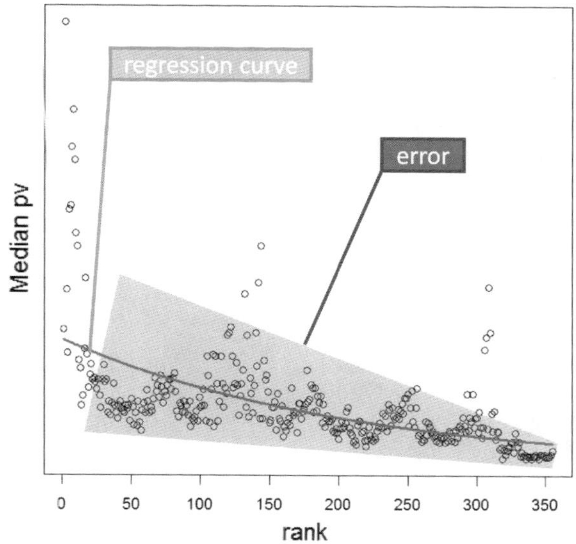

Fig. 3 Regression curve and error in median during one month data

The value obtained by Eq. (3) should increase exponentially similar to Eq. (2). With that, we conducted multiple regression analysis while setting Eqs. (1) and (2) * Eq. (3) as explanatory variables, and the actual number of PV during one month as objective variables. This produced the determination coefficient of 0.5396, resulting in deriving determination coefficient which is larger than where only the posting ranks were set as explanatory variables. Therefore, Eq. (4) was determined to be a PV prediction equation according to each property. Additionally, Eq. (3) was defined as the User Popularity Degree of each property.

$$\widehat{PV_i} = \mathrm{f}(\mathrm{a} \cdot e^{b \cdot rank}) + \mathrm{g}(\mathrm{c} \cdot e^{d \cdot rank}) Median_i\{(y_{i1} - \widehat{y_{i1}}), (y_{i2} - \widehat{y_{i2}}), \ldots, (y_{id} - \widehat{y_{id}})\} \quad (4)$$

4 Simulation Models

In this study, we built five experimental models. The experimental models we built are as follows: Model 1 determines order ranks based on the bid price, model 2 determines order ranks based on the bid price and the User Popularity, model 3 implements reinforcement learning agents in model 1, model 4 implements reinforcement learning agents in model 2, and model 5 implements imitation agents in model 1.

4.1 Simulation Flow

The models that were considered in this study move in a fashion shown in Fig. 4. First, visitors access to the website. This generates bidding requests. Each content that receives the request determines the bid price and bids. In other words, each content works as an agent. The goal of each agent is to maximize the number of its own PV.

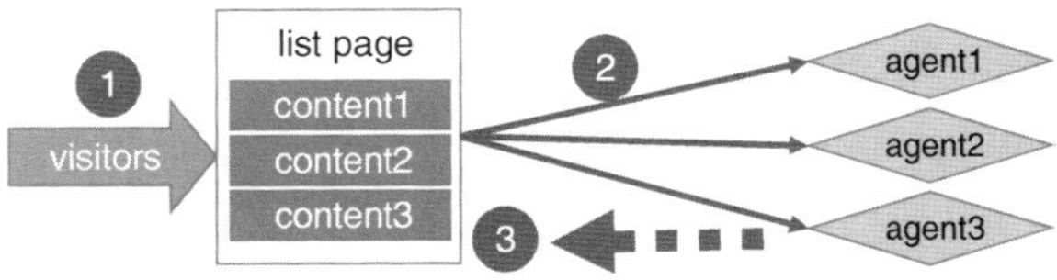

Fig. 4 Bid flow

The goal of this experiment is to maximize the number of detail PV for the entire website. The situation where there is a great number of PV is desirable for advertisers. This situation is also delightful for the website operator because the customer satisfaction level increases. From the standpoint of website visitors, this situation can be positive because a greater number of PV is actually achieved as a result of many visitors browsing a large number of real-estate properties with interest.

4.2 Model Explanation

Model 1 determines posting ranks based only on the bid price. First, a budget is equally allocated to each bid. The bid price is calculated by multiplying this equal price with an identically-distributed random number. Additionally, the posting ranks are given in order according to large bid prices. This posting order and the User Popularity are assigned to Eq. (4) in order to obtain the PV expectation. Finally, the bidding price is deducted from the budget, and the budget balance for the next bid is calculated.

Model 2 determines posting ranks based on the bid price and the User Popularity. By using the bid price similarly obtained as model 1 and the User Popularity, a Weight is calculated. This is calculated by Eq. (5). "α" indicates coefficient that determines the weight of Bidprice and Popularity, while "β" adjusts the scale of Bidprice and Popularity. The posting rank is given in order by a heavier weight that is calculated as above. The subsequent process is similar to that of model 1.

$$Weight_i = Bidprice_i \cdot \alpha + Popularity_i \cdot (1 - \alpha) \cdot \beta \quad (5)$$

The third model implements agents that determine bid prices based on reinforcement learning. The goal of this model is to determine whether the number of PV of the entire website can be maximized by having some agents conduct reinforcement learning or not. From a wide variety of reinforcement learning methods, we decided to use the Actor Critic method in this case.

Model 4 is based on model 2 where reinforcement learning agents similar to model 3 are implemented.

The last one, model 5, is the imitation type. This model also uses learning agents in order to imitate the behavior of agents that obtained the maximum reward.

5 Experimental Results and Discussion

There are 24 experimental patterns for the models described in Chap. 3. The basic four patterns are as follows: the status with each variable in the initial setting, the status where the dispersion range of identically-distributed random numbers that determine the bid price is narrowed down, the status where the influence given by the User Popularity to order ranks is increased, and the status where this influence is decreased. Next, there are three patterns, the pattern where reinforcement learning agents have high popularity, the pattern where they have low popularity, and the pattern where these agents are selected randomly. The final two patterns are the status where the learning rate is at the initial setting and the status where this rate is lowered. Combinations of these patterns provide 24 patterns in total.

Table 1 shows the ratio of the number of PV of the model 1's experimental result for the number of PV based on the actual rank order. Ten times of bids become 1 episode. The "first-half average" means the average value from episode 1 to episode 50, and the "second-half average" means the average value from episode 51 to episode 100. "Second/First" indicates the percentage where the second-half average is compared with the first-half average. This will be described in section about model 3 and model 4. The learning effect based on reinforcement learning was demonstrated within 50 episodes; therefore, this is a column in order to confirm learning effect.

The following section describes the experimental results of model 1. Model 1 showed almost the same result when compared to the actual posting order, even though at the program default status pattern and at the pattern where the dispersive range of the bid price is reduced. However, the number of PV slightly increased in total where the dispersion range was reduced.

In model 2, the number of PV increased in total (Table 2). As is the result of model 1, the number of PV increased when the dispersion of bid prices were reduced. Additionally, the number of PV increased when the influence of the User Popularity by Weight (Eq. (5)) was increased.

Table 1 The ratio of the PV of the model 1 for PV based on the actual rank order

Mode	Leaning agent	Pattern	Learning rate	100 episode average	First-half average	Second-half average	Second/first	1,000 episode average
1	–	default status	–	99.95 %	99.94 %	99.95 %	100.01 %	100.00 %
1	–	reduced dispersive range of the bid price	–	100.20 %	100.19 %	100.20 %	100.00 %	100.26 %

Table 2 The ratio of the PV of the model 2 for PV based on the actual rank order

Model	Leaning agent	Pattern	Learning rate	100 episode average	First-half average	Second-half average	Second/first	1,000 episode average
2	–	default status	–	102.45 %	102.43 %	102.46 %	100.02 %	102.51 %
2	–	reduced dispersive range of the bid price	–	111.36 %	111.37 %	111.36 %	99.99 %	111.42 %
2	–	increased the influence of the User Popularity	–	111.23 %	111.24 %	111.22 %	99.98 %	111.29 %
2	–	decreased the influence of the User Popularity	–	101.07 %	101.08 %	101.07 %	99.99 %	101.12 %

The experiment in model 3 is described by dividing the agents into three different types of reinforcement learning (Table 3). Among all 342 agents, 30 agents are reinforcement learning agents. In the case where reinforcement learning was conducted for popular agents, the number of PV increased in total. It is similar to model 1 or model 2 in that the number of PV increased when price dispersion was reduced. In addition, the increased rate of the number of PV decreased when the learning rate was reduced. When observing the column of "Second half/First half," we can see that the number of PV during the second half actually increased. This shows the reinforcement learning effect. Since the learning effect was demonstrated during 50 episodes, the number of PV increased in the "1000 episode average" rather than the "100 episode average." In the pattern where agents with low popularity had reinforcement learning, the number of PV increased slightly when compared to the actual model. This is because the agents with low popularity showed a low increase rate for the value expressed by Eq. (4) of Sect. 3.2 although such agents were placed in the upper posting ranks. The pattern where agents that conduct reinforcement learning were selected randomly showed the result just between the pattern where agents with high popularity conducted reinforcement learning and the pattern where agents with low popularity conducted reinforcement learning.

The experiment for model 4 was also described by dividing the agents into three different types of reinforcement learning (Table 4). Among all 342 agents, 30 agents are reinforcement learning agents. The first pattern is the case where agents with high popularity conducted reinforcement learning. The number of PV decreased in total. Since these agents with high popularity had already obtained high posting ranks before learning, when they conducted learning, negative results were shown. We can see this fact through the phenomenon that every "Second-half/First-half" value went below 100 %. On the other hand, in the pattern where agents with low popularity conducted reinforcement learning, the number of PV increased in total. The patterns where the price dispersion was reduced and where the influence of popularity to posting order showed the same tendencies as in model 2. The last pattern to be described is the one where agents that conduct reinforcement learning were selected in a random manner. In this pattern, the number of PV increased in total; however, this value of the second-half slightly decreased more than first-half. The pattern where price dispersion and the influence of popularity to posting order were varied, actually showed the same tendencies as model 2.

The final point is the imitation agent. The number of agents that took on the imitation behavior was set to 100 in order for agents with lower popularity. The bid prices to be imitated were the prices of 100 agents in descending order of the number of PV obtained in previous bidding. As a result, there were no significant differences in the number of PV when compared to the actual models (Table 5).

Table 3 The ratio of the PV of the model 3 for PV based on the actual rank order

Model	Leaning agent	Pattern	Learning rate	100 episode average	First-half average	Second-half average	Second/first	1,000 episode average
3	high popularity	default status	default status	101.95 %	101.11 %	102.79 %	101.66 %	102.99 %
3	high popularity	reduced dispersive range of the bid price	default status	102.38 %	101.79 %	102.98 %	101.17 %	103.05 %
3	high popularity	default status	reduced learning rate	101.75 %	100.88 %	102.63 %	101.73 %	102.66 %
3	high popularity	reduced dispersive range of the bid price	reduced learning rate	101.99 %	101.04 %	102.93 %	101.86 %	103.01 %
3	low popularity	default status	default status	100.29 %	100.30 %	100.28 %	99.97 %	100.14 %
3	low popularity	reduced dispersive range of the bid price	default status	100.17 %	100.27 %	100.07 %	99.79 %	99.21 %
3	low popularity	default status	reduced learning rate	100.29 %	100.30 %	100.28 %	99.97 %	100.14 %
3	low popularity	reduced dispersive range of the bid price	reduced learning rate	100.14 %	100.24 %	100.05 %	99.80 %	99.19 %
3	random	default status	default status	100.49 %	100.42 %	100.57 %	100.15 %	100.45 %
3	random	reduced dispersive range of the bid price	default status	100.40 %	100.38 %	100.42 %	100.03 %	100.46 %
3	random	default status	reduced learning rate	100.41 %	100.35 %	100.47 %	100.11 %	100.35 %
3	random	reduced dispersive range of the bid price	reduced learning rate	100.43 %	100.37 %	100.50 %	100.11 %	100.40 %

Table 4 The ratio of the PV of the model 4 for PV based on the actual rank order

Model	Leaning agent	Pattern	Learning rate	100 episode	First-half average	Second-half	Second/first	1,000 episode average
4	high popularity	default status	default status	96.88 %	98.74 %	95.02 %	96.23 %	94.90 %
4	high popularity	reduced dispersive range of the bid price	default status	100.96 %	102.58 %	99.34 %	96.83 %	99.09 %
4	high popularity	increased the influence of the User Popularity	default status	102.14 %	104.55 %	99.74 %	95.40 %	99.67 %
4	high popularity	decreased the influence of the User Popularity	default status	96.36 %	98.08 %	94.64 %	96.49 %	94.60 %
4	high popularity	default status	reduced learning rate	96.85 %	98.28 %	95.42 %	97.08 %	95.71 %
4	high popularity	reduced dispersive range of the bid price	reduced learning rate	101.04 %	102.19 %	99.89 %	97.74 %	99.61 %
4	high popularity	increased the influence of the User Popularity	reduced learning rate	101.54 %	103.66 %	99.43 %	95 92 %	99.55 %
4	high popularity	decreased the influence of the User Popularity	reduced learning rate	96.52 %	97.83 %	95.21 %	97.32 %	95.38 %
4	low popularity	default status	default status	102.57 %	102.57 %	102.58 %	100.00 %	102.61 %
4	low popularity	reduced dispersive range of the bid price	default status	110.72 %	110.76 %	110.68 %	99.93 %	110.43 %
4	low popularity	increased the influence of the User Popularity	default status	111.47 %	111.47 %	111.46 %	99.98 %	111.39 %
4	low popularity	decreased the influence of the User Popularity	default status	101.19 %	101.17 %	101.20 %	100.02 %	101.21 %

(continued)

Table 4 (continued)

Model	Leaning agent	Pattern	Learning rate	100 episode	First-half average	Second-half	Second/first	1,000 episode average
4	low popularity	default status	reduced learning rate	102.57 %	102.56 %	102.59 %	100.01 %	102 63 %
4	low popularity	reduced dispersive range of the bid price	reduced learning rate	110.68 %	110.74 %	110.62 %	99.88 %	110.11 %
4	low popularity	increased the influence of the User Popularity	reduced learning rate	111.47 %	111.50 %	111.45 %	99.95 %	111.41 %
4	low popularity	decreased the influence of the User Popularity	reduced learning rate	101.18 %	101.16 %	101.19 %	100.02 %	101 25 %
4	random	default status	default status	102.01 %	102.25 %	101.76 %	99.51 %	101 83 %
4	random	reduced dispersive range of the bid price	default status	110.07 %	110.30 %	109.83 %	99.58 %	109 39 %
4	random	increased the influence of the User Popularity	default status	110.45 %	110.88 %	110.03 %	99.23 %	109/9 %
4	random	decreased the influence of the User Popularity	default status	100.60 %	100.81 %	100.39 %	99.58 %	100.59 %
4	random	default status	reduced learning rate	102.21 %	102.39 %	102.04 %	99.65 %	101.95 %
4	random	reduced dispersive range of the bid price	reduced learning rate	109.31 %	109.57 %	109.06 %	99.52 %	10866 %
4	random	increased the influence of the User Popularity	reduced learning rate	110.46 %	110.83 %	110.09 %	99.33 %	109.88 %
4	random	decreased the influence of the User Popularity	reduced learning rate	100.49 %	100.74 %	100.25 %	99.50 %	100.48 %

Table 5 The ratio of the PV of the model 5 for PV based on the actual rank order

Model	Leaning agent	Pattern	Learning rate	100 episode average	First-half average	Second-half average	Second/first	1,000 episode average
5	–	default status	–	99.94 %	99.94 %	99.95 %	100.01 %	100.00 %
5	–	reduced dispersive range of the bid price	–	100.20 %	100.19 %	100.20 %	100.01 %	100.26 %

6 Conclusion and Future Issues

Through analysis of information-providing websites, we showed that the number of content detail page views would increase if posting order became higher. We also defined the User Popularity, and demonstrated that the page views (PV) increase rate would become significant when the content's popularity became higher. Through this study, we also found that the number of PV for the entire website would increase when content with high popularity would be placed in an upper level of the website. In addition, when reinforcement learning was conducted for content with low popularity while contents with high popularity were placed in an upper level of the website, the number of PV of the entire website would be significant.

However, this kind of situation would not be so favorable for advertisers that have content with low popularity. When considering the actual online advertising business, we need to establish certain bailout plans. Additionally, scramble for the number of PV among content with high popularity and heated competition in popularity are also a concern. When only content with high popularity are placed on a content list page, the content with higher popularity would obtain PV. As a result, the popularity of the content that could not obtain PV there will actually be a decrease in the relative popularity of all the content. Another issue we should consider in the future is to examine the standard for determining to what extent popular content can be placed on an upper level of a website.

References

1. Yahoo Japan Corporation. http://promotionalads.yahoo.co.jp/service/srch/features.html
2. Vickrey: Counterspeculation, auctions, and competitive sealed tenders. J. Financ. **16**, 8–37 (1961)
3. Edelman, B., Ostrovsky, M., Schwarz, M.: Internet advertising and the generalized second-price auction: selling billions of dollars worth of keywords. Am. Econ. Rev. **97**, 242–259 (2007)
4. Feldman, J., Muthukrishnan, S., Pal, M., Stein, C.: Budget optimization in search-based advertising auctions. In: Proceedings of the 8th ACM Conference on Electronic Commerce (EC'07), pp. 40–49 (2007)
5. Sakurai, Y., Yokoo, M., Iwasaki, A.: New keyword auction protocols for determining the appropriate number of slots. Comput. Softw. **25**(4), 60–67 (2008)
6. Mookerjee, R., Kumar, S., Mookerjee, V.S.: To Show or not show: using user profiling to manage internet advertisement campaigns at Chitika. Interfaces **42**(5) 449–464 (2012)
7. Ghosh, A., Rubinstein, B.I.P., Vassilvitskii, S., Zinkevich, M.: Adaptive bidding for display advertising. In: Proceedings of the 18th International Conference on WORLD WIDE Web, ACM, pp. 251 –260 (2009)
8. Nabeta, T., Yamamoto, G., Yoshikawa, A., Terano, T.: Study on the simulation model of a real-time auction in WEB advertising frame. 5th Symposium of Technical Committee on Social Systems, vol. 5, pp. 13–22 (2014)

Analysis of the Network Effects on Obesity Epidemic

Agent-Based Modeling with Norm-Related Adaptive Behavior

Kazumoto Takayanagi and Setsuya Kurahashi

Abstract The present paper aims at clarifying the effects of different network structures on the spread of obesity so as to examine the norm-related dynamics of obesity. While previous papers have demonstrated that social norms are relevant to the obesity epidemic, the issues on the key mechanisms of such operating norms still remains to be addressed. We attempt to construct an agent-based model (ABM) in which agents' adaptive behavior under social interactions plays a significant role, so that we may investigate how different network topologies influence the norm-related epidemic. Utilizing typical network-generating models, we will focus on several network indices (e.g. degree distribution, clustering coefficient, etc.) which might determine the social contagion as regards obesity. With the results of ABM-based simulations, we present our conclusions so far with respect to network structures which may affect obesity epidemic and suggest effective and feasible interventions either to prevent the epidemic or to treat obese persons.

Keywords Obesity epidemic • Agent-based modeling • Network topology • Social norm • Adaptive behavior

1 Introduction

1.1 Obesity Epidemic Throughout the World

Obesity is now a worldwide epidemic. Obesity rate has globally doubled since 1980 and has increased by more than 40 % over the past 10 years in a number of countries [1]. In 2012, the majority of the population is overweight or obese in 19 of

K. Takayanagi (✉) · S. Kurahashi
University of Tsukuba, Tokyo, Japan
e-mail: k.takayanagi1882@gmail.com

S. Kurahashi
e-mail: kurahashi.setsuya.gf@u.tsukuba.ac.jp

G. Jezic et al. (eds.), *Agent and Multi-Agent Systems: Technologies and Applications*, Smart Innovation, Systems and Technologies 38,
DOI 10.1007/978-3-319-19728-9_33

34 OECD nations [2]. Although the epidemic slowed down recently in several areas, obesity rates remain high in most of the developed countries where people continue to experience a large burden from chronic diseases associated with obesity.

Japan is not immune to the epidemic. While it shows the second lowest obesity rate in the OECD following Korea, the ratio of obese Japanese males has been steadily growing for the last several decades and eventually reached 30 % in 2007 [3]. Meanwhile, the increase of daily calorie intake per capita has not been confirmed, which contrasts sharply with the situation in the United States where the excessive energy intake is considered as the main cause of the world's highest rate of obesity.

Under these circumstances, this study aims to investigate how various network topologies affect the norm-related dynamics of obesity by means of simulations based on agent-based modeling (ABM). By integrating social interactions into agents' adaptive behavior, we attempt to examine how social norms influence the agents' weight statuses and the resulting obesity rates within a society. This paper is structured as follows. In the next section, we examine previous papers as regards (1) obesity spreading in a social network, (2) effects of network structure on the social contagion. In the third section, we present the overview of our agent-based model that focuses on the adaptive behavior of agents interacting with each other. In the fourth section, we will show the results of simulations carried out under various network structures, as well as present our conclusions so far together with the direction of future research.

2 Previous Research

2.1 Obesity Spreading in a Social Network

In their epoch-making paper, Christakis and Fowler have demonstrated that obesity may spread in social networks in a pattern which depends on the nature of social ties [4]. Evaluating an interconnected social network of 12,067 people assessed repeatedly over 32 years as part of the Framingham Heart Study, they used longitudinal logistic-regression models to examine whether weight gain in one person was associated with weight gain in his or her friends. They discerned clusters of obese persons in the network at all time-points and found that a person's chance of becoming obese increased significantly if he or she was connected in a network with a person who became obese in a given interval. These findings indicate, this paper says, that network phenomena appear to be relevant to the biologic and behavioral trait of obesity and obesity appears to spread through social ties. Furthermore, Christakis and Fowler have meaningfully stated that "the psychosocial mechanisms of the spread of obesity may rely less on behavioral imitation than on a change in an ego's general perception of the social norms regarding the acceptability of obesity." With the observation that geographic distance does not modify the effect of a friend's obesity, they have also inferred that norms may be particularly relevant because behavioral effects might depend on the frequency of contact, whereas norms might not.

If we assume that social norms are relevant to the mechanism of the obesity spreading, we will have to explore two important issues as regards the norm-related social dynamics: First, we need to examine how norms affect the agents' decision-making and adaptive behavior that eventually determine their weight statuses within a society. Second, we are also required to investigate the effects of different network topologies on the mechanism of social norms and the resulting weight status of each agent. We continue our inquiry by reviewing the discussion as to the "effective intervention" aimed at mitigating the obesity epidemic in the next section.

2.2 *Effects of Network Topologies*

Dispute Over the Effective Interventions

It is quite natural that, taking into account the network-based spread of obesity, one should consider exploiting the network effects to mitigate the obesity epidemic. Bahr et al. suggested network-driven interventions that target well-connected individuals at the edges of a cluster so as to halt the contagion of obesity [5]. They used different kinds of social networks (e.g., lattice, random, small-world, or scale-free) to test their hypothesis with respect to effective measures against the epidemic and found stable results across various network topologies. They concluded that weight loss among "friends of friends" was more important than that among "friends alone" in treating obese individuals, and that interventions among well-connected individuals would be more effective than those among individuals at random.

In contrast to the conclusion of Bahr et al., El-Sayed et al. demonstrated, with the results of simulations featuring the literature-based parameter, that interventions targeting highly networked individuals did not outperform at-random interventions [6]. However, based on agent-based counterfactual simulations featuring an artificially high parameter, they showed that targeted interventions aimed at preventing obesity outperformed the at-random ones. While the agents in their model were nested within a social network generated by a biased preferential attachment growth model creating a scale free (Barabási–Albert) network, they replicated each of their simulations by using a segregated Erdős–Rényi model and a clustered network so as to account for the model's potential sensitivity to network topology.

Notwithstanding their opposite conclusions, the robustness of these papers' results against network topologies might be analogously derived from the extreme simplicity of agents' algorithms in their simulation models (e.g. a majority rule in which an individual's probability of becoming obese was calculated from the obesity status among the majority of the individual's neighbors). When taking into consideration the underlying mechanism of the norm-related social contagion, a model that integrates the norm-linked social dynamics with agents' adaptive behavior would be necessary for the thorough examination into the effects of network topology over the spread of obesity.

Two Hypotheses Regarding Network Effects

Then, how does the network topology affect the social contagion process, when considering the underlying norm-related mechanism?

As for the spread of behavior, Centola tested two competing hypotheses regarding how network topology influences the diffusion [7]. The "strength of weak ties" hypothesis predicts that networks with many "long ties", e.g. "small world" structure, will spread social behavior farther and more quickly than a network in which ties are highly clustered. This hypothesis assumes that the spread of behavior can be considered as a simple contagion where a single contact might be sufficient to transmit the behavior. Long ties can reduce the redundancy of the diffusion process through connecting people whose friends do not know each other, which allows behavior to rapidly reach far areas in a network.

The other hypothesis argues that social contagion of behavior is a complex process in which people need contact with multiple sources of infection before becoming convinced to adopt the behavior. This hypothesis supposes that clustered networks with redundant ties, providing social reinforcement for adoption, should promote the behavioral diffusion across the population. Experimenting with a network-embedded population, he has investigated how the diffusion process is affected by different network topologies that these hypotheses assume. The results showed that topologies with greater clustering and a larger diameter were much more effective for spreading behavior. He concluded that, while locally clustered ties may be redundant for simple contagions such as information or disease, they can be highly efficient for the spread of behavior, not only because individual adoption is improved by reinforcing signals coming from clustered ties, but also because large-scale diffusion can spread quickly in clustered networks.

Noting the difference between behavior and social norm in the contagion process, we are not able to apply Centola's findings directly to the spread of obesity in which social norm plays a key role. We, therefore, try to assess the applicability of these two hypotheses by means of simulations based on agent-based model featuring different network topologies so that we may identify the network effects on obesity epidemic. We will present the overview of the model as follows.

3 Agent-Based Model

3.1 Time Discounting and Dual System Model

Time discounting appears to be closely connected with some kinds of tendencies observed among obese persons. Cutler et al. suggested that the lowered food price (preparation cost among others) caused by the technological progress promoted the time-inconsistent overeating, which could explain the vast increase of the obesity rate in the United States [8]. Ikeda et al. has reported that the body mass status of persons are expectedly related to their time discounting and that caloric intake and the resulting body mass status could be taken as determined by intertemporal decision-making with behavioral decision bias toward immediacy and/or toward aversion of future losses [9]. The time-inconsistency inducing obesity has been, in turn, correlated to the neural systems. McClure et al. have demonstrated that two separate systems are involved in a series of choices between monetary reward options varying by delay to delivery [10]. The limbic system associated with the midbrain dopamine system are preferentially activated by decisions involving immediately available rewards. In contrast, regions of the lateral prefrontal cortex and posterior parietal cortex are engaged uniformly by intertemporal choices irrespective of delay. The relative engagement of the two systems is directly associated with individuals' choices. We integrate this dual-system process into our agent-based model as the agents' algorithm consisting of the "affective system" and the "deliberative system" [11], the latter is based on a utility function [12].

3.2 Adaptive Behavior

The agents nested in a network, detecting the mean weight of their neighbors, vary the value of a parameter which represents the agent's time preference, so that their weight status might stay appropriate within the network. Each agent, for example, weakens the present-bias by inclining the dual system towards the deliberative system when they feel like being fatter than the others. Contrarily, aware of being lighter than others, the agent strengthens the present-bias by favoring the activation of the affective system.

3.3 Network Topologies

In this paper, we examine several types of network structure generated by Barabási–Albert (BA) model and Watts-Strogatz (WS) model so as to figure out the network effects over obesity epidemic [13]. As for BA model, it is characterized by the growth along with the preferential attachment described as (1),

$$\prod(k_i) = \frac{k_i}{\sum_j k_j}, \quad (1 \le i \le n) \tag{1}$$

where the probability that a new node is connected to the vertex v_i whose degree is k_i is denoted by $\prod(k_i)$. We test two different BA model networks which have the same value of average degree. They are different from each other in both the degree distribution and the value of global clustering coefficient, while indicating almost equal values of average path length (Table 1).

Table 1 BA model characteristics

Average degree	7.988	7.988
Average path length	2.88	3.57
Clustering coefficient	0.51	0.08

As regards WS model, a network is generated by rewiring edges of a regular graph by the probability p. As p increases from 0 to 1, the average path length decreases quickly and levels off while the clustering coefficient's value dwindles moderately as Fig. 1 shows. We compare two WS model networks either of which indicates the identical average degree. The first network is a "small world" which shows relatively high value of clustering coefficient while exhibiting the lowered level of average path length. The second one is "random graph" in which both the clustering coefficient as well as the average path length are levelled off (Table 2).

Table 2 WS model characteristics

Average degree	8	8
Average path length	5.93	5.28
Clustering coefficient	0.149	0.013

Furthermore, we explore the agents' weight status under network structures varying by 20 steps throughout the simulation, each of which is generated by WS model with a distinct p value. The result is compared with that of the simulation carried out with the invariable "small world" network.

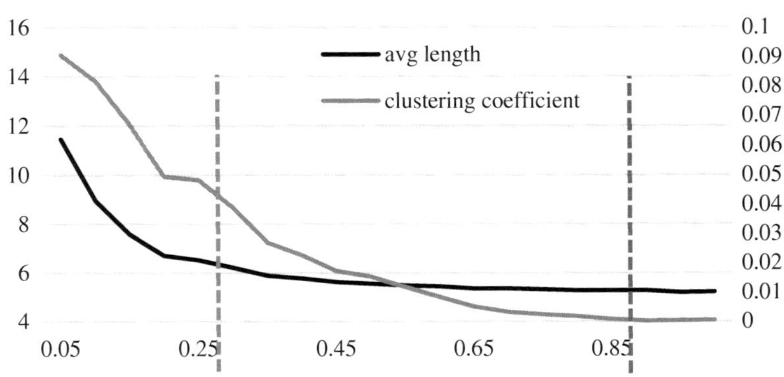

Fig. 1 Watts-Strogatz model

3.4 Simulations

We iterate the simulation 50 times under one of the network structures described above and obtain a data set consisting of agents' weight values at the end of the simulation. We also collect another data set comprised of agents' weight changes throughout the simulation. In order to compare the effects of different networks over the obesity epidemic, we examine the difference in the mean value between two data sets by carrying out the Wilcoxon rank sum test which is a nonparametric statistical hypothesis test. The null hypothesis is that two populations are the same.

4 Results and Discussion

4.1 Results

(See Tables 3, 4 and 5).

Table 3 Comparison between different networks. The value in the parentheses shows the global clustering coefficient

Network	BA (0.51)	BA (0.08)	WS (0.14)	WS (0.01)
Mean weight	74.25 kg	73.61 kg	73.55 kg	73.62 kg
Mean change	+12.97 kg	+12.45 kg	+12.22 kg	+12.42 kg

Table 4 Test results of Wilcoxon rank sum test as to WEIGHT

	BA (0.08)	WS (0.14)	WS (0.01)
BA (0.51)	$p < 0.001$	$p < 0.001$	$p < 0.001$
BA (0.08)		$p > 0.05$	$p > 0.05$
WS (0.14)			$p > 0.05$

Table 5 Test results of Wilcoxon rank sum test as to WEIGHT CHANGE

	BA (0.08)	WS (0.14)	WS (0.01)
BA (0.51)	$p < 0.001$	$p < 0.001$	$p < 0.001$
BA (0.08)		$p > 0.01$	$p > 0.05$
WS (0.14)			$p > 0.01$

BA Model

We compared two data sets each of which corresponds to one of the two networks generated by BA model. The results show that the difference in the mean value between the two data sets regarding agents' weight values is statistically significant

(P < 0.001). Equivalently, the difference in the mean value between the two data sets concerning agents' weight gains is also statistically significant (P < 0.001). We confirm also that the mean value of the agents' weight along with weight change is higher in the highly clustered network than that in another one.

Since the two networks differ with respect to the degree distribution as well as the global clustering coefficient, we inspect, furthermore, the correlation between the weight values and the network indices observed in each agent in order to investigate which network characteristic is critical. As Fig. 2 shows, the correlation between each agent's local clustering coefficient and its weight value varies notably according to the network in which the agents are nested. Under the highly clustered network whose global clustering coefficient is 0.51, the weight values of the agents whose local clustering coefficient is relatively low (from 0 to 0.2) extend widely over the level of obesity (red line).

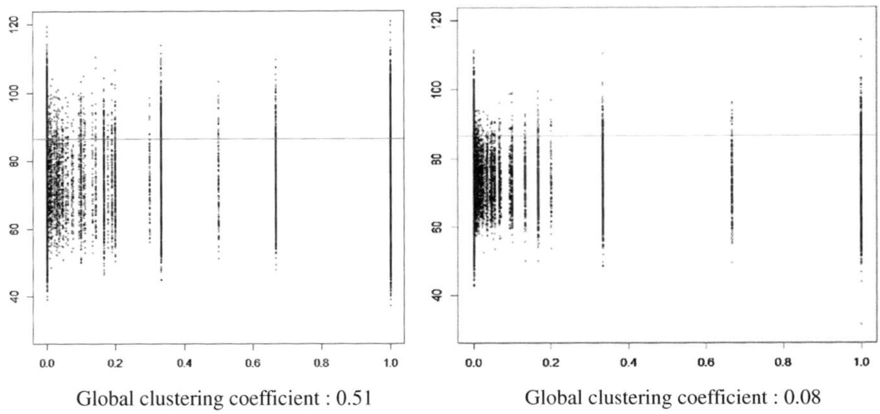

Fig. 2 Agent's local clustering coefficient (horizontal) and weight (vertical)

In contrast, in the other network whose clustering coefficient is 0.08, the weight values of the agents with the relatively low level of clustering coefficient converge onto the middle range (from 60 to 80) beneath the obesity level.

Equivalently, the association of each agent's degree with its weight also differs in accordance with the network, albeit in a subtle way as Fig. 3 indicates. The weight statuses of the agents having not so many edges (under 50) spread above the obesity level more remarkably in the highly clustered network than in the other network. The convergence of the agents' weight values onto the middle level below the obese line is also recognized where the range of each agent's degree is comparatively low in the network not highly clustered.

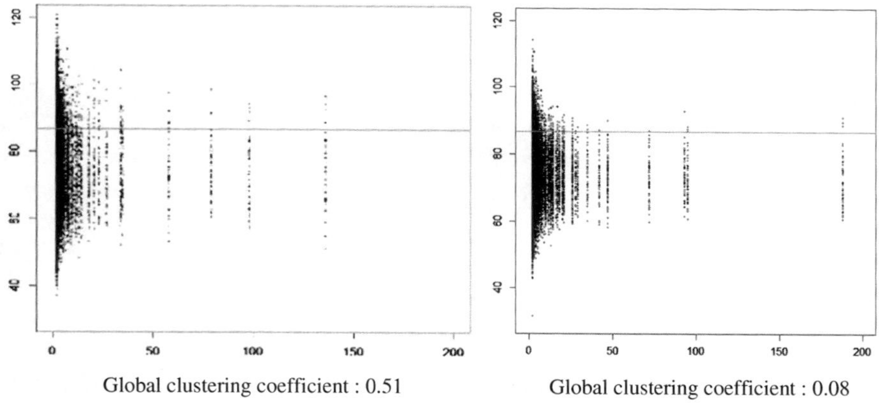

Fig. 3 Agent's degree (horizontal) and weight (vertical)

WS Model

Under the networks generated by WS model, the difference in the mean value between the two data sets is not statistically significant ($p > 0.05$), whether the data set consists of weight changes or eventual weight values. As Fig. 4 shows, the correlation between the clustering coefficient and the weight value seems to be unaffected by the difference in the network structure. Even when comparing the weight values between the network unchanged and the one varying by 20 steps throughout the simulation, the difference of the two data sets has not been statistically confirmed.

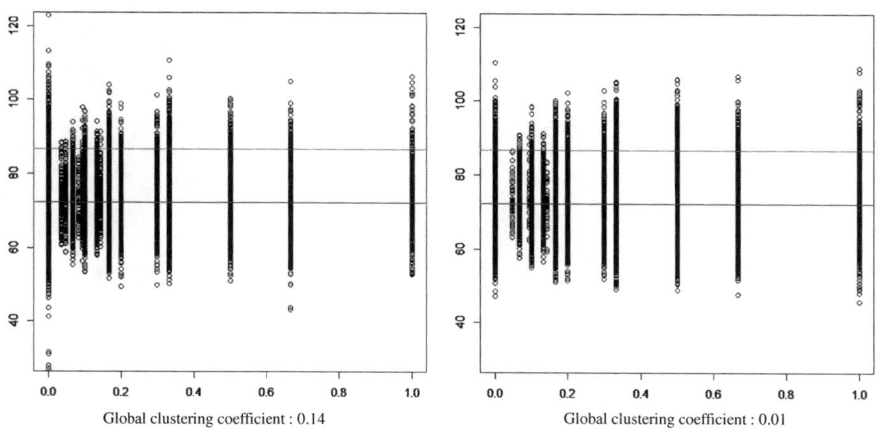

Fig. 4 Local clustering coefficient (horizontal) and weight (vertical)

5 Discussion

With the BA model networks, the difference in the global clustering coefficient or in the degree distribution resulted in the difference in the agents' mean weight. The simulation results also demonstrated that the agents with relatively low degree or clustering coefficient seem to be more biased toward becoming obese under one of the two networks than under the other one. Of the two networks, the one indicating the higher value of global clustering coefficient and the more right-skewed power law distribution provides such a bias for agents. We may conclude, therefore, that agents nested in such a highly clustered network have a higher probability of becoming obese.

The results of the simulation under WS models showed that the difference in the clustering coefficient is not sufficient to bring about the difference of the agents' weight, even when it is rather large. Consequently, we seem to be able to conclude that the degree distribution of the network is really critical because the agents observed within the low range of degree or clustering coefficient are influenced to gain weight when the degree distribution is right-skewed.

In order to confirm our conclusion, we will have to investigate further issues regarding the network effects over the spread of obesity. First, we need to identify the role of the highly-networked persons (hubs) in order to specify the effects of the highly clustered network over obesity described above. If the hub infects others with obesity and easily brings about the epidemic in such a highly clustered network, the network effects should be reduced to the mechanism of the hub system. Contrarily, the key effects inherent to the network would be found if it is proved that the hub does not play a significant role in a highly clustered network. Second, we will attempt to investigate other network models than BA or WS so that we may figure out the factors determining the network effects over obesity. Although we concluded that the clustering coefficient seem to be less important than the degree distribution with respect to the network's influence, our conclusion needs to be affirmed under various types of network model in order to be robust.

References

1. World Health Organization: http://www.who.int/topics/obesity/
2. OECD Obesity Update 2014: http://www.oecd.org/health/obesity-update.htm
3. National Health and Nutrition Survey Japan: http://www.mhlw.go.jp/bunya/kenkou/kenkou_eiyou_chousa.html
4. Christakis, N.A., Fowler, J.H.: The spread of obesity in a large social network over 32 years. New Engl. J. Med. **357**, 370/379 (2007)
5. Bahr, D.B, Browning, R.C., Wyatt, H.R., Hill, J.O.: Exploiting social networks to mitigate the obesity epidemic. Obesity, **17**, 723/728 (2009)
6. El-Sayed, A.M., Seemann, L., Scarborough, P., Galea, S.: Are network-based interventions a useful antiobesity strategy? An application of simulation models for causal inference in epidemiology. Am. J. Epidemiol. **178–2**, 287/295 (2013)

7. Centola, D.: The Spread of behavior in an online social network experiment. Science, **329**, 1194/1197 (2010)
8. Cutler, D.M., Glaeser, E.L., Shapiro, J.M.: Why have Americans become more obese? J. Econ. Perspect. **17–3,** 93/118 (2003)
9. Ikeda, S., Kang, M.I., Ohtake, F.: Hyperbolic discounting, the sign effect, and the body mass index. J. Health Econ. **29**, 268/284 (2010)
10. McClure, S.M., Laibson, D.I., Loewenstein, G., Cohen, J.D.: Separate neural systems value immediate and delayed monetary rewards. Science, **306**, 503/507 (2004)
11. Loewenstein, G.F., O'Donoghue, T.: Animal spirits: affective and deliberative processes in economic behavior. Working Paper, 04/14, Center for Analytic Economics, Cornell University (2004)
12. Strulik, H.: A mass phenomenon: the social evolution of obesity. J. Health Econ. **33**, 113/125 (2014)
13. Masuda, N., Konno, N.: Complex Network—From Fundamentals to Applications. Sangyo Tosho (2010)

A Health Policy Simulation Model of Smallpox and Ebola Haemorrhagic Fever

Setsuya Kurahashi and Takao Terano

Abstract This study proposes a simulation model of a new type of infectious disease based on smallpox, Ebola haemorrhagic fever and a health policy Game. SIR (Susceptible, Infected, Recovered) model has been widely used to analyse infectious diseases such as influenza, smallpox, bioterrorism, to name a few. On the other hand, Agent-based model or Individual-based model begins to spread in recent years. The model enables to represent behaviour of each person in the computer. It also reveals the spread of an infection by simulation of the contact process among people in the model. The study designs a model based on Epstein's model in which several health policies are decided such as vaccine stocks, antiviral medicine stocks, the number of medical staff to infection control measures and so on. Furthermore, infectious simulation of Ebola haemorrhagic fever, which has not yet any effective vaccine, is also implemented in the model. As results of experiments using the model, it has been found that preventive vaccine, antiviral medicine stocks and the number of medical staff are crucial factors to prevent the spread. In addition, a health policy game against a new type of infectious disease is designed as a serious game.

Keywords Infectious disease · Smallpox · Ebola haemorrhagic fever · Health policy game · Serious game

1 Introduction

Infectious diseases have been serious risk factors in human societies for centuries. Smallpox has been recorded in human history since more than B.C 1100. People have also been suffering from many other infectious diseases such as malaria, cholera,

S. Kurahashi (✉)
Graduate School of Business Sciences, University of Tsukuba,
3-29-1 Otsuka, Tokyo, Bunkyo, Japan
e-mail: kurahashi.setsuya.gf@u.tsukuba.ac.jp

T. Terano
Computational Intelligence and Systems Science, Tokyo Institute of Technology,
Nagatsuda-Cho, Yokohama, Midori-ku 4259-J2-52, Japan
e-mail: terano@dis.titech.ac.jp

G. Jezic et al. (eds.), *Agent and Multi-Agent Systems: Technologies and Applications*, Smart Innovation, Systems and Technologies 38,
DOI 10.1007/978-3-319-19728-9_34

tuberculosis, typhus, AIDS, influenza, etc. Although people have tried to prevent and hopefully eradicate them, a risk of unknown infectious diseases including SARS, a new type of influenza, as well as Ebola haemorrhagic fever have appeared on the scene.

A model of infectious disease has been studied for years. SIR (Susceptible, Infected, Recovered) model has been widely used to analyse such diseases based on a mathematical model. After an outbreak of SARS, the first SIR model of SARS was published and many researchers studied the epidemic of the disease using this model. When an outbreak of a new type of influenza is first reported, the U.S. government immediately starts an emergency action plan to estimate parameters of its SIR model. Nevertheless the SIR model has difficulty to analyse which measures are effective because the model has only one parameter to represent infectiveness. For example, it is difficult for the SIR model to evaluate the effect of temporary closing of classes because of the influenza epidemic. The agent-based approach or the individual-based approach has been adopted to conquer these problems in recent years [1–4]. The model enables to represent behaviour of each person. It also reveals the spread of an infection by simulation of the contact process among people in the model.

In this study, we developed a model to simulate smallpox and Ebola haemorrhagic fever based on the infectious disease studies using agent-based modelling. What we want to know is how to prevent an epidemic of infectious diseases not only using mechanisms of the epidemic but also decision making of health policy [5]. Most Importantly, we should make a decision in our modern society where people are on the move frequently world wide, so we can minimise the economic and human loss caused by the epidemic. Therefore we developed a new serious game in which authorities of areas or countries need to make a prompt decision regarding health policies such as vaccine stocks, antiviral medicine stocks, the number of medical staff in order to achieve effective control measures. In the game, decision-makers need to focus on many kinds of policy variables and control them in an appropriate manner to minimise human loss. Policies have inevitable costs to operate them. Moreover, it is extremely difficult to control an epidemic in a single country in the modern world; therefore cooperation between countries is tantamount. We developed a health policy game designed as a serious game against a new type of infectious disease. Through this game, we aim to analyse how complex decision-making of policies are, and confirm the effectiveness as an educational tool of policy decision-making.

2 Cases of Infectious Disease

2.1 Smallpox

The smallpox virus affects the throat where it invades into the blood and hides in the body for about 12 days. Patients developed a high fever after that, but rashes do not appear until about 15 days after the infection. While not developing rashes, smallpox virus is able to infect others. After 15 days, red rashes break out on the face, arms and legs, and subsequently they spread over the entire body. When all rashes generate pus, patients suffer great pains; finally 30 per cent of patients succumb to the disease. For thousands of years, smallpox was a deadly disease that resulted in thousands of deaths.

In 1796, an English doctor, Edward Jenner, developed vaccination of smallpox had been destroyed in the world. In 1977, after the last known smallpox patient in the East Somali, smallpox was eradicated. Still now, there is a substantial risk of bioterrorism using the smallpox virus as a biological weapon, because people do not have any tolerance to smallpox.

2.2 Ebola Haemorrhagic Fever

A source of Ebola infection is allegedly by eating a bat or a monkey, but it is unknown whether the eating these animals is a source of the infection. The current epidemic, which began in Guinea in Dec. 2013, 23 people have died. The authorities of Guinea, Liberia and Sierra Leone have each launched a state committee of emergency and have taken measures to cope with the situation. The prohibition of entry over the boundary of Guinea is included in these measures.

There is a risk that a cough and a sneeze includes the virus, so the infection risk is high within 1 m in length of the cough or sneeze. The incubation period is normally 7 days, and then the person gets infected after showing the symptoms. The symptoms in the early stage are similar to influenza. They are fever, a headache, muscular pain, vomiting, diarrhoea, and a stomachache. The fatality rate is very high; 50–90 %. There is no effective medical treatment medicine confirmed officially and several medicines are currently being tested. According to a guideline of WHO, the serum of a recovered patient is one of most effective treatments.

3 Related Work

3.1 Smallpox and Bioterrorism Simulation

Epstein [6] [7] made a smallpox model based on 49 epidemics in Europe from 1950 to 1971. In the model, 100 families from two towns were surveyed. The family includes two parents and two children thus the population is each 400 from each town. All parents go to work in their town during the day except 10 per cent of adults who go to another town. All children attend school. There is a communal hospital serving the two towns in which each 5 people from each town work. This model was designed as an agent-based model, and then simulation of infectious disease was conducted using the model. As results of experiments showed that 1) in a base model in which any infectious disease measures were not taken, the epidemic spread within 82 days and 30 per cent of people died, 2) a trace vaccination measure was effective but it was difficult to trace all contacts to patients in an underground railway or an airport, 3) a mass vaccination measure was effective, but the number of vaccinations would be huge so it was not realistic, 4) epidemic quenching was also effective, and reactive household trace vaccination along with pre-emptive vaccination of hospital workers showed a dramatic effect.

3.2 Individual-Based Model for Infectious Diseases

Ohkusa [8] evaluated smallpox measures using an individual-based model of infectious diseases. The model supposed a town including 10,000 habitants and a public health centre. In the model, one person was infected with smallpox virus at a shopping mall. They compared between a trace vaccination measure and a mass vaccination measure. As a result of simulation, it was found that the effect of trace vaccination dropped if the early stage of infection was high and the number of medical staff is small, while the effect of mass vaccination was stable. Therefore timely and concentrate mass vaccination is required when virus starts spreading. The estimation about the number, place and time of infection is needed quickly and the preparation of an emergency medical treatment and estimation system is required for such occasions.

3.2.1 Summary of Related Work

From these studies, the effectiveness of an agent-based model has been revealed, yet these are not sufficient models to consider a relationship between vaccination and antiviral medicine stocks, and the number of support medical staff and medicine from other countries. In addition, authorities need to make a decision regarding blockade, restrictions on outings including cars and railways while considering economic loss of the policy. This study takes into account these extensions.

4 A Health Policy Simulation Model of Infectious Disease

We designed a health policy simulation model of infectious disease based on Epstein's smallpox model. The model includes smallpox and Ebola haemorrhagic fever.

4.1 A Base Model of Smallpox

We assume all individuals to be susceptible which means no background of immunity. 100 families live in two towns Fig. 1. The family includes two parents and two children. Therefore the population is each 400 in each town. All parents go to work in their town during the day except 10 percent of adults commute to another town. All children attend school. There is a communal hospital serving two towns in which 5 people from each town work. Each round consists of an interaction through the entire agent population. The call order is randomised each random and agents are processed or activated, serially. On each round, when an agent is activated, she identifies her immediate neighbours for interaction. Each interaction results in a contact. In turn, that contact results in a transmission of the infection from the contacted agent to the active agent with probability.

The probability of contact at an interaction is 0.3 at a workplace and a school, while 1.0 at a home and a hospital. The probability of infection at a contact is 0.3 at a workplace and a school, while 1.0 at a home and a hospital. In the event the active agent contracts the disease, she turns blue to green and her own internal clock of disease progression begins. After twelve days, she will turn yellow and begins infecting others. Length of noncontagious period is 12 days, and early rash contagious period is 3 days. Unless the infected individual is vaccinated within four days of exposure, the vaccine is ineffective. At the end of day 15, smallpox rash is finally evident. Next day, individuals are assumed to hospitalize. After eight more days, during which they have a cumulative 30 per cent probability of mortality, surviving individuals recover and return to circulation permanently immune to further infection. Dead individuals are coloured black and placed in the morgue. Immune individuals are coloured white. Individuals are assumed to be twice as infectious during days 1 through 19 as during days 12 through 15.

4.2 A Base Model of Ebola Hemorrhagic Fever

In the event the active agent contracts the disease, she turns blue to green and her own internal clock of disease progression begins. After seven days, she will turn yellow and begins infecting others. However, her disease is not specified in this stage. After three days, she begins to have vomiting and diarrhoea and the disease is specified as Ebola. Unless the infected individual is dosed with antiviral medicine within three days of exposure, the medicine is ineffective. This is an imaginary medicine to play the policy game. At the end of day 12, individuals are assumed to hospitalize. After four more days, during which they have a cumulative 90 per cent probability of mortality, surviving individuals recover and return to circulation permanently immune to further infection. Dead individuals are coloured black and placed in the morgue. Immune individuals are coloured white. Other settings are the same as smallpox.

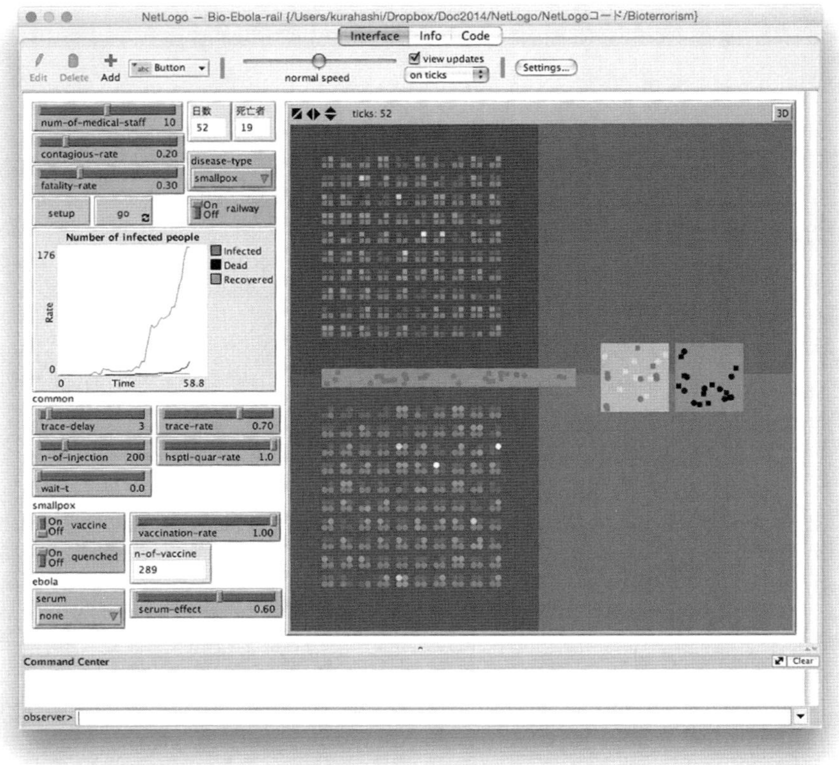

Fig. 1 Interface view of a health policy simulation model of infectious disease

4.3 Vaccination Strategies

The vaccination strategies we can select in the model are mass vaccination and trace vaccination. Each of them has advantages and disadvantages.

4.3.1 Mass Vaccination

As preemptive vaccination, the mass vaccination strategy adopts an indiscriminate approach. First all of the medical staff is vaccinated to prevent infection. When the first infected person is recognised, certain per cent of individuals in both towns will be vaccinated immediately. The vaccination rate and the upper limit number of vaccination per day are set on the model for the strategy.

4.3.2 Trace Vaccination

All of the medical staff is vaccinated as pre-emptive vaccination. Given a confirmed smallpox case, medical staff traces every contact of the infected person and vaccinates that group. In addition of the mass vaccination strategy, the trace rate and the delay days of contact tracing are able to be set according to the model for the trace vaccination strategy.

4.3.3 Trace Serum or Antiviral Medicine Dosing

All of the medical staff is given serum or antiviral medicine as TAP (Target antivirus prophylaxis). Given a confirmed Ebola hemorrhagic fever case, medical staff traces every contact of the infected person and provides the medicine to that group. In addition to the mass vaccination strategy, the trace rate and the delay days of contact tracing are set according to the model.

4.4 A Health Policy Game of Infectious Disease

A serious game approach has been introduced to various study and economic fields to analyse human behaviour and education including a medical field [9–11]. A health policy game was developed to decide a policy to prevent an epidemic based on the health policy simulation model. In this game, it is supposed that a new type of infectious disease occurs in which the disease is similar to smallpox and Ebola haemorrhagic fever. Vaccine and antiviral medicine for the disease are already developed and provided to the market in this model. Players as authorities of two countries decide the amount of both medicines stocks according to their restricted budget. They also need to decide the number of medical staff, blockade and restrictions on outings. The players should consider giving support medicine and staff to countries to prevent or control its epidemic for his/her own country, while taking account of economic cost and loss. Travel restrictions have a huge economic impact, while it is very effective in stopping an outbreak. Supporting to another country means decreasing its own preparations. Thus, this game has a complicated structure of trade-offs among cost, effect, cooperation and defence.

5 Experimental Results

5.1 Base Model of Smallpox

The process of infection in the base model is plotted in Fig. 2. The model employs non-intervention to the disease. A solid line, a dotted line and a line with marker indicate the number of infected, dead and people who have recovered respectively. When a player adopts non-intervention, it takes approximately 169 days until convergence of the outbreak and more than 350 people have died.

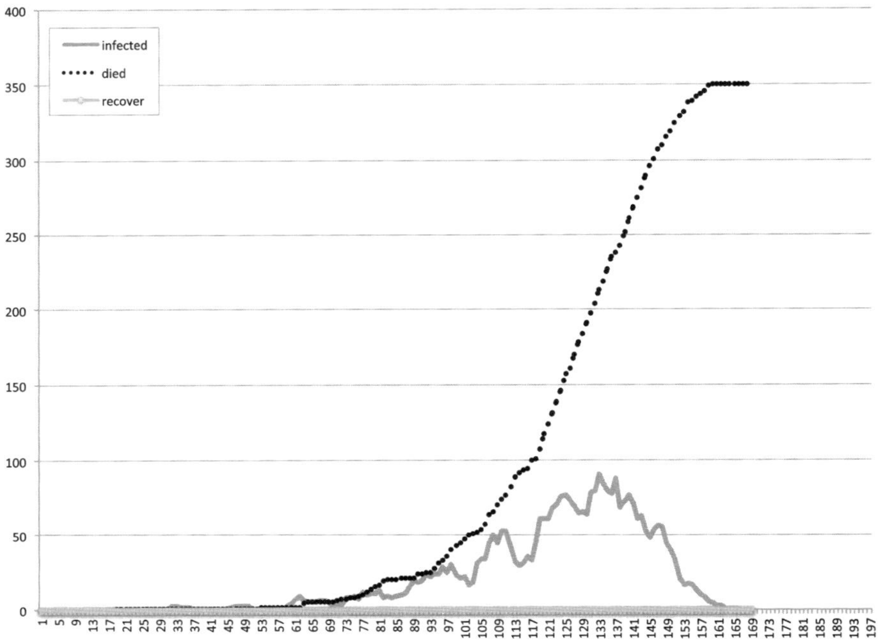

Fig. 2 The experimental result of the base model: non-intervention

5.2 Mass Vaccination Model of Smallpox

The process of infection is plotted with the mass vaccination strategy in which individuals are vaccinated randomly after three days when given a confirmed smallpox case. The policy succeeds to prevent the outbreak because the number of vaccination per day is 600 and three fourths of inhabitants are vaccinated per day (Fig. 3). On the other hand, the short ability of vaccination ends in failure because the number of vaccination per day is 400 and a half of inhabitants are vaccinated (Fig. 4). The ability of vaccination per day bifurcates the results.

5.3 Trace Vaccination Strategy

It was found that the ability of more than 50 vaccinations per day was able to control an epidemic in most cases. In the case of using a public transportation to commute, however, it makes a substantial difference. 400 vaccinations per day could not pre-

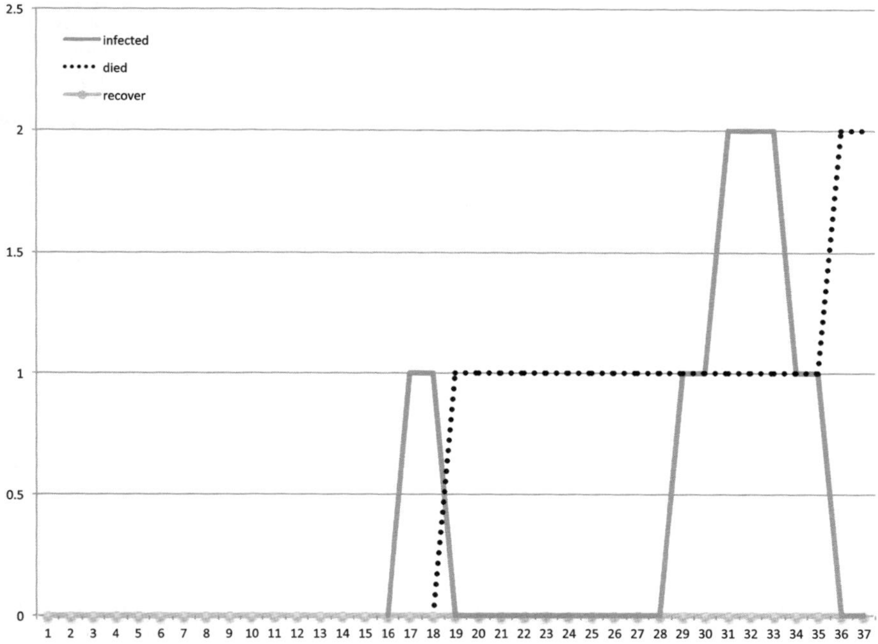

Fig. 3 The experimental result of Mass vaccination model: the number of vaccination per day is 400

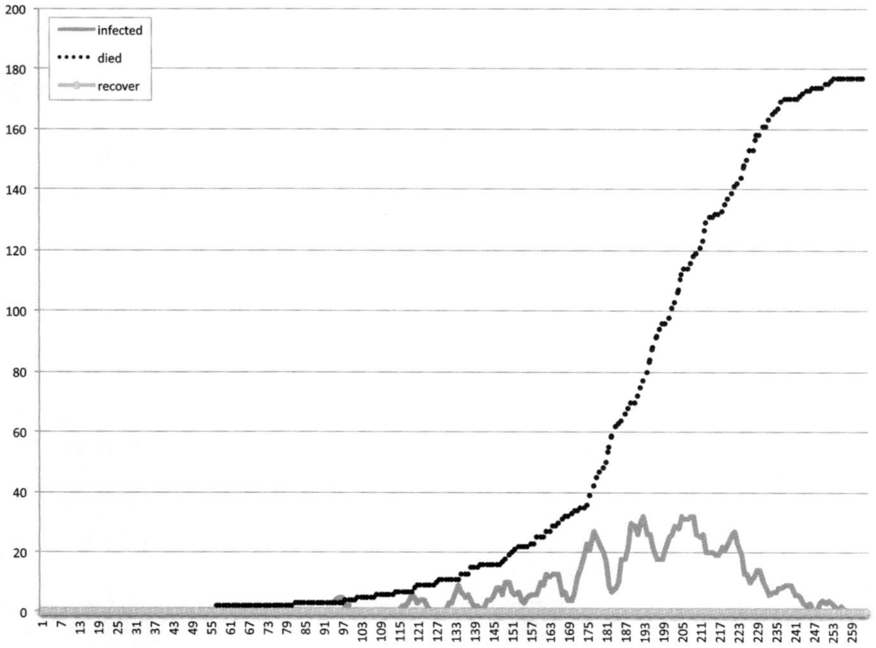

Fig. 4 The experimental result of Mass vaccination model: the number of vaccination per day is 400

vent the outbreak (Fig. 5), while 600 vaccinations per day succeeded to prevent it. Trace vaccination strategy is one of the most effective policies in a town where people commute by car, but a large number of vaccinations per day, at least a half of people, is required if most of people use public transportation systems like a railway and a bus.

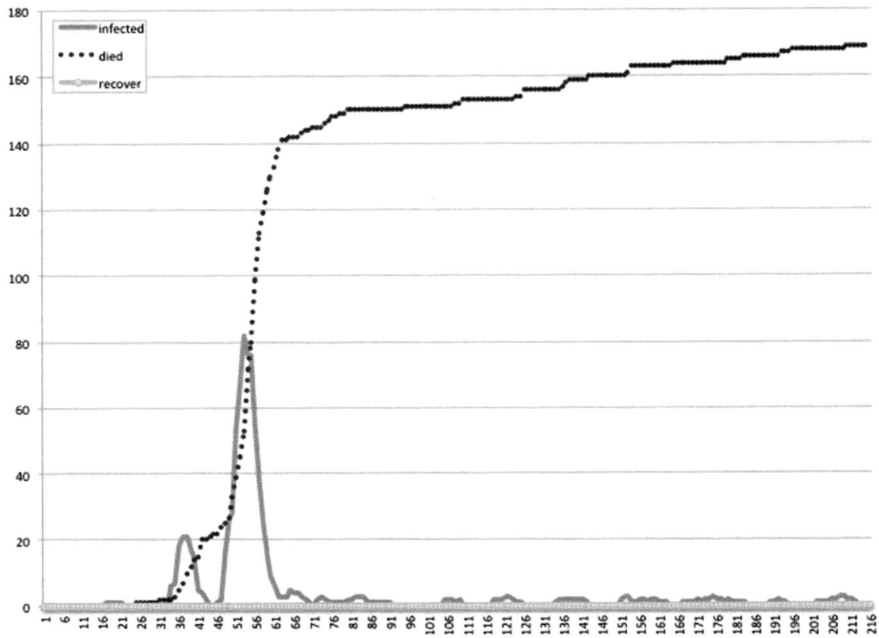

Fig. 5 The experimental result of Trace vaccination model: the number of vaccination per day is 400

6 Conclusion

This study proposes a simulation model of smallpox and Ebola haemorrhagic fever. It also evaluates health policies to prevent an epidemic. As health policies, vaccine stocks, antiviral medicine stocks, the number of vaccinations per day by medical staff, mass vaccination, and trace vaccination are implemented in the model. As a result of experiments, it has been found that vaccination availability per day and the number of medical staff are crucial factors to prevent the spread for the mass vaccination strategy. On the other hand, small quantities of vaccination for approximately 10 per cent of inhabitants are vaccinated per day are able to control an epidemic in most cases. In the case of using a public transportation to commute, however, even if half of inhabitants were vaccinated per day, it would not prevent the epidemic. Two thirds of vaccinations per day are required each day to prevent the epidemic.

In the future work, using a health policy game against a new type of infectious disease, a trade-off between policies will be evaluated as a serious game.

References

1. Burke, D.S., et al.: Individual based computational modeling of smallpox epidemic control strategies. Acad. Emerg. Med.**13**(1), 1142–1149 (2006)
2. Longini, I.M., et al.: Containing a large bioterrorist smallpox attack: a computer simulation approach. Int. J. Infect. Dis. **11**(2), 98–108 (2007)
3. Gilbert, N. Agent-based models. No. 153. Sage (2008)
4. Easley, D., Kleinberg, J.: Networks, crowds, and markets: Reasoning about a highly connected world. Cambridge University Press (2010)
5. Okabe, N.: Risk and benefit of immunisation : infectious disease prevention with immunization. Iryo to Shakai **21**(1), 33–40 (2011)
6. Epstein, J.M., et al.: Toward a containment strategy for smallpox bioterror: an individual-based computational approach (2002)
7. Epstein, J.M.: Generative social science: studies in agent-based computational modeling. Princeton University Press (2006)
8. Ohkusa, Y.: An Evaluation of counter measures for smallpox outbreak using an individual based model and taking into consideration the limitation of human resources of public health workers. Iryo to Shakai **16**(3), 275–284 (2007)
9. Lofgren, E.T., Fefferman, N.H.: The untapped potential of virtual game worlds to shed light on real world epidemics. Lancet infect. dis. 7(9), 625–629 (2007)
10. Kennedy-Clark, S., Thompson, K.: What do students learn when collaboratively using a computer game in the study of historical disease epidemics, and why?. games and culture, 1555412011431361 (2011)
11. Manfredi, P., Alberto d'Onofrio.: Modeling the interplay between human behavior and the spread of infectious diseases. Springer (2013)

Analyzing the Influence of Market Conditions on the Effectiveness of Smart Beta

Hiroshi Takahashi

Abstract Stock indices play a significant role in asset management business. Price indices are popular in practical business. However, recent empirical analyses suggest that a smart beta, which is proposed as a new stock index, could achieve a positive excess return. With these arguments in mind, this study analyzes the effectiveness of smart beta through agent-based modeling. As a result of intensive experiments in the market, I made the following finding that effectiveness of smart beta could be influenced by the extent of a diversity of investors' behavior. These results are significant from both practical and academic viewpoints.

Keywords Financial markets · Smart beta · Agent-based model · Asset management business

1 Introduction

The theories of financial markets have developed mainly in the area of finance and many prominent theories have been proposed [3, 9, 14]. Along with the advancement of these theories, many arguments regarding securities investment in practical business affairs have been actively discussed and various kinds of analyses focusing on stock indices have been conducted [6, 8, 14–16].

This research investigates the effectiveness of stock indices through agent-based modeling. Especially, this analysis focuses on smart beta, which has been proposed as a new benchmark for investments in place of price indices, which are currently employed in practical business affairs. Although smart beta is becoming popular in asset management business, the mechanism of the effectiveness of smart beta is not clear.

H. Takahashi (✉)
Graduate School of Business Administration, Keio University,
4-1-1 Hiyoshi, Yokohama 223-8572, Kohoku-ku, Japan
e-mail: htaka@kbs.keio.ac.jp

G. Jezic et al. (eds.), *Agent and Multi-Agent Systems: Technologies and Applications*, Smart Innovation, Systems and Technologies 38,
DOI 10.1007/978-3-319-19728-9_35

Considering these arguments, this analysis attempts to examine the effectiveness of smart beta through social simulation(agent-based modeling) [2, 18, 20, 21]. Among several types of smart beta, this study focus on fundamental index which is one of the most popular smart beta [1]. A fundamental index is calculated based on the value of a company's financial statements - profit and so on - instead of the market price that is commonly used in business affairs. Compared to price indexing, fundamental indexing is relatively unaffected by the decision bias of investors and could have better characteristics than traditional stock indexing. Additionally fundamental indexing could actually contribute to market efficiency. Therefore, the analysis of fundamental indexing is significant from both academic and practical points of view. The next section describes the model used in this analysis. Section 3 shows the results of the analysis. Section 4 summarizes this paper.

2 Model

A computer simulation of the financial market involving 1000 investors was used as the model for this research.[1] Shares and risk-free assets were the two types of assets used, along with the possible transaction methods. Several types of investors exist in the market, each undertaking transactions based on their own stock evaluations. This market was composed of three major stages: (1) generation of corporate earnings; (2) formation of investor forecasts; and (3) setting transaction prices. The market advances through repetition of these stages. The following sections describe negotiable transaction assets, modeling of investor behavior, setting of transaction prices, and the rule of natural selection.

2.1 Negotiable Assets in the Market

This market has risk-free and risk assets. There are risky assets in which all profits gained during each term are distributed to the shareholders. Corporate earnings (y_t) are expressed as $y_t = y_{t-1} \cdot (1 + \varepsilon_t)$ [13]. However, they are generated according to the process of $\varepsilon_t \sim N(0, \sigma_y^2)$ with share trading being undertaken after the public announcement of profits for the term . Each investor is given common asset holdings at the start of the term with no limit placed on debit and credit transactions (1000 in risk-free assets and 1000 in stocks). Investors adopt the buy-and-hold method for the relevant portfolio as a benchmark to conduct decision-making by using a one-term model.[2]

---

[1]I built a virtual financial market on a personal computer with i7 2600 K 3.4 GHz, RAM16 GB. The simulation background is financial theory [5, 10].

[2]'Buy-and-hold' is an investment method to hold shares for the medium to long term.

2.2 Modeling Investor Behavior

Each type of investor considered in this analysis is organized in Table 1.[3] Type 1–4 are active investors and type 5 is smart beta investor. This section describes the investors' behavior.

Smart Beta Investor This section describes the behavior of smart beta investor who try to follow a fundamental index. Smart beta investors automatically decide their investment ratio in stock based on the index's value. Smart beta investors decide their investment ratio in stock based on the value of the fundamental index.[4]

Table 1 List of investors type

No.	Investor type
1	Fundamentalist
2	Forecasting by latest price
3	Forecasting by trend(most recent 10 days)
4	Forecasting by past average(most recent 10 days)
5	Smart beta investor

Active Investors Active investors in this market evaluate transaction prices based on their own market forecasts, taking into consideration both risk and return rates when making decisions. Each investor determines the investment ratio (w_t^i) based on the maximum objective function ($f(w^i t)$), as shown below.[5]

$$f(w_t^i) = r_{t+1}^{int,i} \cdot w_t^i + r_f \cdot (1 - w_t^i) - \lambda(\sigma_{t-1}^{s,i})^2 \cdot (w_t^i)^2. \quad (1)$$

Here, $r_{t+1}^{int,i}$ and $\sigma_{t-1}^{s,i}$ expresses the expected rate of return and risk for stocks as estimated by each investor i. r_f indicates the risk-free rate. w_t^i represents the stock investment ratio of the investor i for term t. The investor decision-making model here is based on the Black/Litterman model that is used in securities investment [4]. The expected rate of return for shares is calculated as follows.

$$r_{t+1}^{int,i} = (r_{t+1}^{f,i} c^{-1} (\sigma_{t-1}^i)^{-2} + r_t^{im} (\sigma_{t-1}^i)^{-2}) / (1 \cdot c^{-1} (\sigma_{t-1}^i)^{-2} + 1 \cdot (\sigma_{t-1}^i)^{-2}). \quad (2)$$

Here, $r_{t+1}^{f,i}$ Cr_t^{im} expresses the expected rate of return, calculated from the short-term expected rate of return, plus risk and gross price ratio of stocks respectively. c is a

[3]This analysis covers the major types of investor behavior [16].

[4]When market prices coincide with fundamental value, both passive investors behave in the same way.

[5]The value of objective function $f(w_t^i)$ depends on the investment ratio (w_t^i). The investor decision-making model here is based on the Black/Litterman model that is used in securities investment [4, 11, 12].

coefficient that adjusts the dispersion level of the expected rate of return calculated from risk and gross current price ratio of stocks [4].

The short-term expected rate of return ($r_t^{f,i}$) is obtained where ($P_{t+1}^{f,i}, y_{t+1}^{f,i}$) is the equity price and profit forecast for term $t + 1$ is estimated by the investor, as shown below:

$$r_{t+1}^{f,i} = ((P_{t+1}^{f,i} + y_{t+1}^{f,i})/P_t - 1) \cdot (1 + \eta_t^i). \tag{3}$$

The short-term expected rate of return includes the error term ($\eta_t^i \sim N(0, \sigma_n^2)$) reflecting that even investors using the same forecast model vary slightly in their detailed outlook. The stock price ($P_{t+1}^{f,i}$), profit forecast ($y_{t+1}^{f,i}$), and risk estimation methods are described below.

The expected rate of return obtained from stock risk and so forth is calculated from stock risk (σ_{t-1}^i), benchmark equity stake (W_{t-1}), investors' degree of risk avoidance (λ), and risk-free rate (r_f), as shown below [4].

$$r_t^{im} = 2\lambda(\sigma_{t-1}^s)^2 W_{t-1} + r_f. \tag{4}$$

Stock Price Forecasting Method The fundamental value is estimated by using the dividend discount model, which is a well known model in the field of finance. Fundamentalists estimate the forecasted stock price and forecasted profit from the profit for the term (y_t) and the discount rate (δ) as $P_{t+1}^{f,i} = y_t/\delta, y_{t+1}^{f,i} = y_t$.

Forecasting based on trends involves forecasting the following term's stock prices and profit through extrapolation of the most recent stock value fluctuation trends. The following term's stock price and profit is estimated from the most recent trends of stock price fluctuation (a_{t-1}) from time point $t-1$ as $P_{t+1}^{f,i} = P_{t-1} \cdot (1 + a_{t-1})^2, y_{t+1}^{f,i} = y_t \cdot (1 + a_{t-1})$.

Forecasting based on past averages involves estimating the following term's stock prices and profit based on the most recent average stock value.

Risk Estimation Method In this analysis, each investor estimates risk from past price fluctuations. Specifically, stock risk is estimated as $\sigma_{t-1}^i = \sigma_{t-1}^h$ (common to each investor). Here, σ_{t-1}^h represents the stock volatility that is calculated from price fluctuation from the most recent 100 terms.

2.3 Determination of Transaction Prices

Transaction prices are determined as the price where stock supply and demand converge ($\sum_{i=1}^{M}(F_t^i w_t^i)/P_t = N$) [17]. In this case, the total asset (F_t^i) of investor i is calculated from transaction price (P_t) for term t, profit (y_t) and total assets from the term $t - 1$, stock investment ratio (w_{t-1}^i), and risk-free rate (r_f), as $F_t^i = F_{t-1}^i \cdot (w_{t-1}^i \cdot (P_t + y_t)/P_{t-1} + \left(1 - w_{t-1}^i\right) \cdot (1 + r_f))$.

2.4 Rules of Natural Selection in the Market

The rules of natural selection can be identified in this market. The driving force behind these rules is cumulative excess profit [7, 17]. The rules of natural selection go through the following two stages: (1) the identification of investors who alter their investment strategy, and (2) the actual alteration of investment strategy [19].

Each investor determines the existence of investment strategy alteration based on the most recent performance of each 5 term period after 25 terms have passed since the beginning of market transactions.[6] The higher the profit rate obtained most recently is, the lesser the possibility of strategy alteration becomes. The lower the profit, the higher the possibility becomes. Specifically, when an investor cannot obtain a positive excess profit for the benchmark portfolio, they are likely to alter their investment strategy with the probability below:

In actual market, investors do not always determine their investment strategies based on past investment performance. This study takes these aspects into consideration. In Sect. 3.2, the case where a certain percentage of investors changes its own investment strategy in a random manner (5 %) is analyzed. Those investors who altered their strategies make investments based on the new strategies after the next step.

$$p_i = \min(1, \max(-100 \cdot r^{cum}, 0)). \tag{5}$$

Here, however, r_i^{cum} is the cumulative excess profitability for the most recent benchmark of investor i. Measurement was conducted for 5 terms, and the cumulative excess profit was converted to a period of one term.

When it comes to determining new investment strategies, an investment strategy that has had a high cumulative excess profit for the most five recent terms (forecasting type) is more likely to be selected. Where the strategy of the investor i is z_i and the cumulative excess profit for the most recent five terms is r_i^{cum}, the probability p_i that z_i is selected as a new investment strategy is given as $p_i = e^{(a \cdot r_i^{cum})} / \sum_{j=1}^{M} e^{(a \cdot r_j^{cum})}$.[7] Those investors who altered their strategies make investments based on the new strategies after the next step.

3 Result

Firstly, this study analyzes the effectiveness of smart. Afterwards, the influence of market conditions on the effectiveness of smart beta is explored. Especially, this study focuses on the diversity of investors' strategies.

---

[6]In the actual market, evaluation tends to be conducted according to baseline profit and loss.

[7]Selection pressures on an investment strategy become higher as the value of the coefficient a increases.

3.1 Effectiveness of Smart Beta

This section analyzes the effectiveness of a smart beta strategy. Firstly, this section analyzes the case where there are no smart beta investors in the market; secondly, this study analyzes the effectiveness of a smart beta strategy.

Without Smart Beta Investors This section analyzes the case where there are no smart beta investors in the market Figs. 1 and 2 show the typical transition of market price and the number of investors with investors categorized into 4 types (Table 1 FType1-4). These results suggest that share prices closely correspond to fundamental value and the number of fundamentalists is increasing throughout the periods. These results coincide with traditional financial theory and suggest the validity of this model [9].

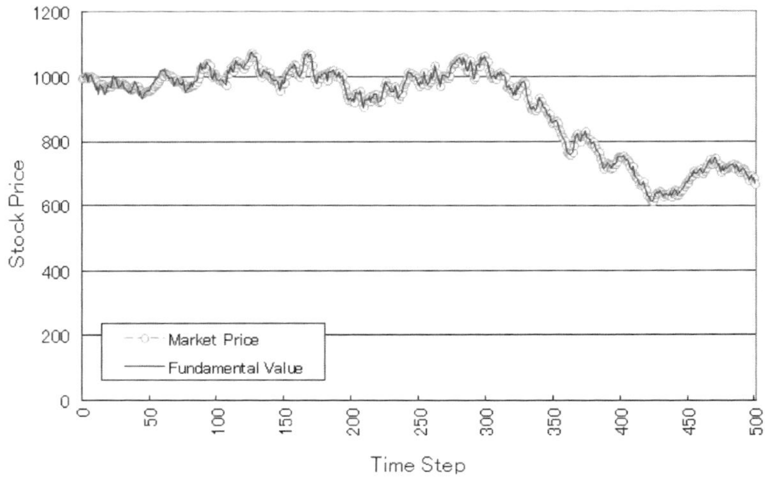

Fig. 1 Price transitions (without smart beta)

The Effectiveness of Smart Beta Figures 3 and 4 show the transition of market prices and the number of investors where there are the same number of 5 types of investors (Table 1 FType1-5), including smart beta investors. These results suggest that the market situation can be divided into 3 periods, as follows: (1) the period when the number of both fundamentalists and smart beta investors increase; (2) the period when the number of fundamentalists decreases; (3) the period when all investors employ smart beta strategies. These results suggest the effectiveness of smart beta strategies.

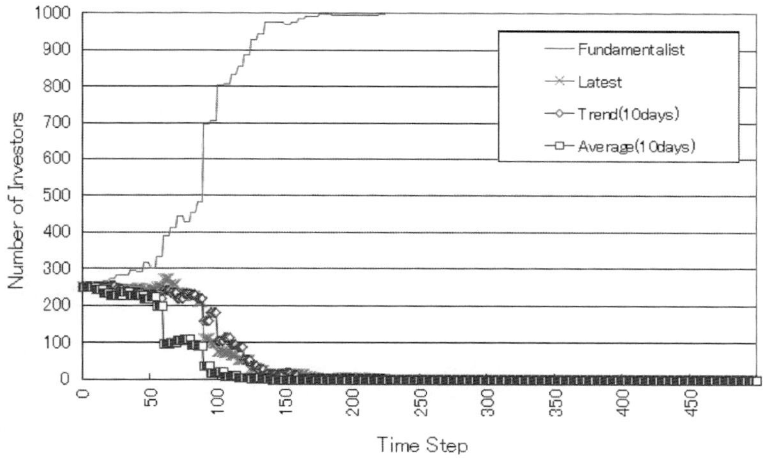

Fig. 2 Transition of the number of investors (without smart beta)

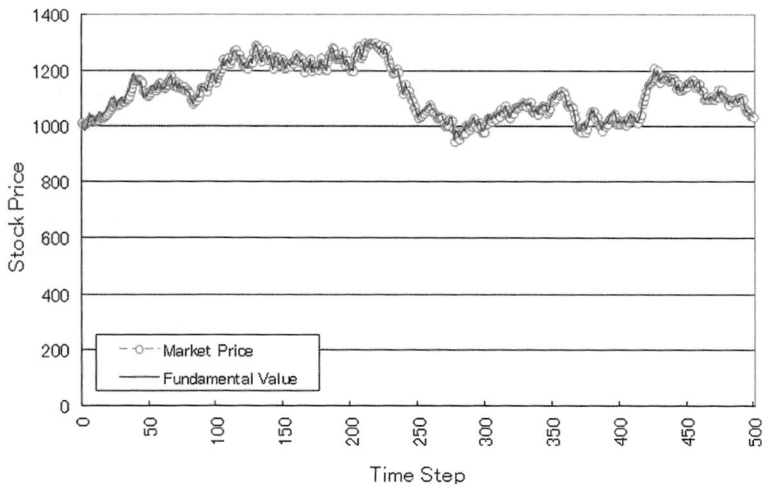

Fig. 3 Price transitions (with smart beta)

3.2 The Influence of Diversity of Investorsf Behavior

This section analyzes the influence of diversity of investorsf behavior on the effectiveness of smart beta strategies. This section analyzes the case where a certain percentage of investors changes its own investment strategy in a random manner (5 %). Figure 5 shows the transition of the number of investors where there are the same number of 5 types of investors (Table 1 FType1–5). From the figure, it is confirmed that the number of smart beta is smaller than previous results (Fig. 4). These results suggest that the effectiveness of smart beta could be influenced by diversity of investors' behavior.

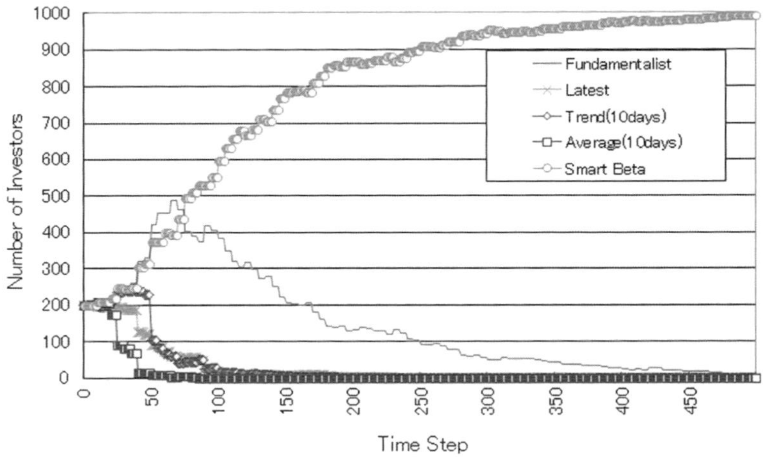

Fig. 4 Transition of the number of investors (with smart beta)

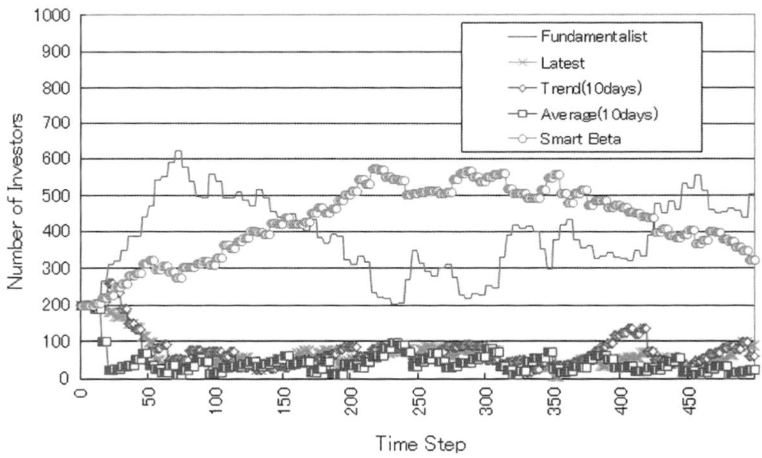

Fig. 5 Transition of the number of investors (Diversity of investorsf behavior become larger)

4 Conclusion

This research analyzes the effectiveness of smart beta strategies and made the following finding that effectiveness of smart beta could be influenced by the extent of a diversity of investors' behavior. These results indicate that the effectiveness of smart beta could be influenced by various kinds of factors. This research has attempted to tackle the problem that practical asset management business faces, and has produced results which may prove significant from both academic and practical points of view. These analyses also demonstrate the effectiveness of agent-based techniques for financial research. More detailed analyses relating to cost, market design and several other matters of interest are planned for the future.

A List of Main Parameters

This section lists the major parameters of the financial market designed for this paper. Explanations and values for each parameter are described.
M: Number of investors (1000)
N: Number of shares (1000)
F_t^i: Total asset value of investor i for term t ($F_0^i = 2000$: common)
W_t: Ratio of stock in benchmark for term t ($W_0 = 0.5$)
w_t^i: Stock investment rate of investor i for term t ($w_0^i = 0.5$: common)
y_t: Profits generated during term t ($y_0 = 0.5$)
σ_y: Standard deviation of profit fluctuation ($0.2/\sqrt{200}$)
δ: Discount rate for stock($0.1/200$)
λ: Degree of investor risk aversion (1.25)
σ_n: Standard deviation of dispersion from short-term expected rate of return on shares (0.01)
c: Adjustment coefficient (0.01)

References

1. Arnott, R., Hsu, J., Moore, P.: Fundamental indexation. Fin. Anal. J. **61**(2), 83–99 (2005)
2. Axelrod, R.: The Complexity of Cooperation—Agent-Based Model of Competition and Collaboration. Princeton University Press, Princeton (1997)
3. Black, F., Scholes, M.: Pricing of options and corporate liabilities. Bell J. Econ. Manage. Sci. **4**, 141–183 (1973)
4. Black, F., Litterman, R.: Global portfolio optimization. Fin. Anal. J. September–October, 28–43 (1992)
5. Brealey, R., Myers, S., Allen, F.: Principles of Corporate Finance, 8th edn. The McGraw-Hill (2006)
6. Fama, E.: Efficient capital markets: a review of theory and empirical work. J. Fin. **25**, 383–417 (1970)
7. Goldberg, D.: Genetic Algorithms in Search, Optimization, and Machine Learning. Addison-Wesley (1989)
8. Grossman, S.J., Stiglitz, J.E.: Information and competitive price systems. Am. Econ. Rev. **66**, 246–253 (1976)
9. Ingersoll, J.E.: Theory of Financial Decision Making. Rowman & Littlefield (1987)
10. Luenberger, D.G.: Investment Science. Oxford University Press (2000)
11. Martellini, L., Ziemann, V.: Extending Black-Litterman analysis beyond the mean-variance framework: an application to hedge fund style active allocation decisions. J. Portfolio Manage. **33**(4), 33–45 (2007)
12. Meucci, A.: Beyond Black-Litterman in practice. Risk **19**, 114–119 (2006)
13. O'Brien, P.: Analysts' forecasts as earnings expectations. J. Account. Econ. January, 53–83 (1988)
14. Sharpe, W.F.: Capital asset prices: a theory of market equilibrium under condition of risk. J. Fin. **19**, 425–442 (1964)
15. Shiller, R.J.: Irrational Exuberance. Princeton University Press, Princeton (2000)
16. Shleifer, A.: Inefficient Markets. Oxford University Press, Oxford (2000)
17. Takahashi, H., Terano, T.: Agent-based approach to investors' behavior and asset price fluctuation in financial markets. J. Artif. Soc. Soc. Simul. **6**, 3 (2003)

18. Takahashi, H., Takahashi, S., Terano, T.: Analyzing the influences of passive investment strategies on financial markets via agent-based modeling. In: Edmonds, B., Hernandes, C., Troitzsch, K (eds.) Social Simulation Technologies: Advances and New Discoveries (Representing the Best of the European Social Simulation Association conferences). Idea Group Inc. (2007)
19. Takahashi, H.: An analysis of the influence of fundamental values' estimation accuracy on financial markets. J. Prob. Stat. 17 (2010). doi:10.1155/2010/543065
20. Takahashi, H., Takahashi, S., Terano, T.: Analyzing the validity of passive investment strategies employing fundamental indices through agent-based simulation. In: The Proceeding of KES-AMSTA 2011 (LNAI6682), pp. 180–189, Springer (2011)
21. Takahashi, H.: An analysis of the influence of dispersion of valuations on financial markets through agent-based modeling. Int. J. Inf. Technol. Decis. Making **11**, 143–166 (2012)

Part VIII
IS: Learning Paradigms and Applications: Agent-Based Approach

Evaluating Organizational Structures for Supporting Business Processes Reengineering: An Agent Based Approach

Mahdi Abdelkafi and Lotfi Bouzguenda

Abstract This paper addresses an important issue in Workflow mining which is the evaluation of the discovered Organizational Structures (OS). By OS, we mean, the social structures defining the activity distribution among actors involved in the Workflow (namely federation, coalition, market or hierarchy). First, it recalls the events log model, proposed in our previous work, which extends existing models by considering the interactions among actors thanks to the FIPA-ACL performatives-based agent communication language. Then, this paper proposes a method allowing the evaluation of the discovered organizational structures in terms of flexibility, robustness and efficiency. Finally, we give an example to better illustrate the OS evaluation and we expose our DiscopFlow tool that validates our solution.

Keywords Workflow mining · Organizational structures · Evaluation · Agent approach · DiscopFlow

1 Introduction

A Workflow Management System (WfMS) is a software that defines, implements and executes business processes (BP or Workflow) [11]. The majority of existing WfMSs (Bonita [9], YAWL [11] and FlowMind [14]) focuses on design, configuration and enactment phases and ignores the diagnostic phase that is very important to improve the design of BP. Subsequently, the Workflow Mining (WM) research area has recently emerged to address these workflows related shortcomings. It aims mainly at analyzing the processes execution traces (or events log) in order to discover the key perspectives of the Workflow such as the

M. Abdelkafi (✉) · L. Bouzguenda
Miracl Laboratory, University of Sfax, Sfax, Tunisia
e-mail: mahdi.abdelkafi@hotmail.fr

L. Bouzguenda
e-mail: lotfi.bouzguenda@isimsf.rnu.tn

G. Jezic et al. (eds.), *Agent and Multi-Agent Systems: Technologies and Applications*, Smart Innovation, Systems and Technologies 38,
DOI 10.1007/978-3-319-19728-9_36

Organizational Perspective (OP), the Informational Perspective (IP) and the Process Perspective (PP). The scope of WM consists in monitoring, improving or proposing new workflows [12, 13]. Most existing WM systems (InWolvE [7], Workflow-Miner [4] and ProM [14]) focus only on the discovering of PP. Discovering and evaluating of OP were not considered. We believe that OP mining and evaluating are useful for improving the design of BP. Indeed, these processes require cooperation and coordination of several participants distributed and structured as federations or hierarchies for instance. Also, they require the use of sophisticated interaction protocols to support consistency of communications between participants. The only system that supports the OP mining is ProM but unfortunately limited to actors, roles and few kinds of social networks based on specific cooperation techniques such as work-handover, subcontracting, working together, reassignments and doing similar tasks [14]. In our work, the OP discovering is not limited to the previous elements but also includes further advanced organizational structures (OS) (such as federation, coalition, market or hierarchy), as well as other interaction protocols (IPs) (such as contract net, vote or negotiation) like these are deployed in real inter and intra organizational processes [6]. The rationales for existing propositions to deal only with PP mining are the use of an events log limited to the activities and to the actors performing those activities. No interactions among actors are ever tracked. In our previous work [1], we have proposed a solution to deal with OS and IPs mined from events log. More precisely, we have proposed an events log model to support this mining and a tool that implements our solution. The problem that is still asking for an answer; "*How to evaluate organizational structures discovered from events log*?" The goal of this paper is to propose a method based on performance indicator proprieties to deal with OS assessment. The remainder of this paper is organized as follows. Section 2 recalls our proposed events log model to accommodate organizational perspective mining and assessment. Section 3 shows how we evaluate the quality of discovered OS in terms of flexibility, efficiency and robustness. Section 4 presents the DiscopFlow tool that validates our approach. Section 5 concludes the paper with a summary of our contributions and some suggested future works.

2 Our Proposed Events Log Model

2.1 Motivations for Using Agent Approach

The agent approach can help the organizational structures and interaction protocols mining thanks to the following properties:

- Natural abstractions to deal with cooperation. A lot of sophisticated protocols like Contract-Net protocols, auctions and negotiation mechanisms are available and could be used to coordinate processes [3]. Agent technology also provides organizational concepts to abstract and structure a system as a computational community made of groups, roles and interactions.

- Social abilities of agents also facilitate the cooperation needed to enact complex Workflow and to provide an abstraction to high-level concepts like commitments, reputations and so on.

In our work, we focus on the social abilities and particularly on the two following multi-agent concepts:

- Performatives-based FIPA-ACL that defines clearly the semantic of messages and namely the agent's intentions (delegate, subcontract, negotiate…) [6];
- OS that model the behaviour of the actors' group i.e. they describe the macro-level dimension of the coordination among actors in terms of externally observable behaviour independently from the internal features of each participating component.

2.2 The Proposed Events Log Model

The proposed events log model is represented by the UML diagram (see Fig. 1). The Process is composed of one (or several) Process Instance(s) (or cases). Each Process Instance is composed of one (or several) EventLine(s). Each EventLine makes references to the following elements:

- An Activity which is described through the Act-Name, EventStream and TimeStamp attributes,
- An Actor which is described through the Act-ID and Act-Name attributes.
- A Performative which is described through the Perf-Name attribute.

The initiator Actor and the receiver actor are represented respectively by the Has_initiator_Actor and Has_ receiver_Actor relations ship.

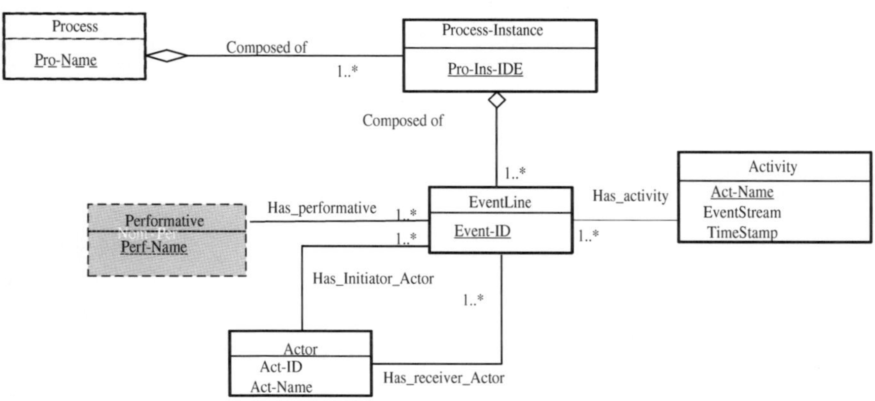

Fig. 1 The proposed events log model

3 How to Evaluate OS Discovered from Events Log?

3.1 Related Works

A very few methods have been proposed in the literature to evaluate an organization. The proposals of [2] are focused on the evaluation of the design quality of an organizational model based on the quality metrics such as the Control-Flow complexity, the coupling and the cohesion. For instance in [5], the proposed method consists in evaluating a multi-agent organization. It supposes an organization structured around roles and three types of relations between them: coordination, control and power. More precisely, the suggested method permits to evaluate an organization in terms of flexibility, robustness and efficiency thanks to the use of graph theory proprieties such as completeness, connectivity and so on. In [8], authors propose metrics to evaluate the distribution degree, the associability and the centrality of an organization. We think that the method used in [5] is convenient for OS assessment for the following reasons. Firstly, there is a strong matching between the concepts of a workflow organizational structure and a multi-agent organization (see Table 1). We define delegation relation-ship as hierarchical relation describing allocation policies of tasks. As for communication relationship, it describes knowledge handover between organization entities and monitoring relationship deals with the activity recovery functions and allows control and continuity for task's execution. The identification of the three relationships is based on the expressiveness FIPA-ACL Performatives. For instance, the delegate performative reflects the fact that there is a delegation relationship; the monitor performative expresses the meaning of supervising relationship. As for communication relationship, it is recognized by the other performatives such as inform, forward or ask.

Table 1 Corresponding concepts between multi-agent organization and Workflow Organizational structures

Concepts of a multi-agent organization	Concepts of a workflow organizational structure
Roles	Actors
Relation of coordination	Relation of communication (inform performative for instance)
Relation of power	Relation of delegation (delegate performative)
Relation of control	Relation of monitoring (monitor performative)

Secondly, the assessment can help the crisis managers to have an exact view of the reality (i.e., the detours or the weaknesses), to find fast and appropriate answers and to identify the matching process for the crisis resolution.

3.2 General Principles of the Adopted Method

In our work, we are interested by structural properties defined in [5] such as connectedness, chain or economy. These properties are considered as Organizational Performance Indicators (OPIs) and there are the bases of the flexibility, robustness and efficiency evaluation of an OS. By flexibility we mean the ability of an OS to adapt to the unexpected executions by the adding of new interaction links between performers. This growth avoids the blockage due to the inability of the organizational model to support at the design step all practices used. By robustness, we mean the capacity of an OS to stay unaffected with the change of the process model by the addicting of new activities. By efficiency, we mean the ability of an OS to reach the final goal with minimum interactions between resources. OPIs are measured according to the three main steps which are: (i) Creating oriented graph G (N, A) to model OS previously discovered (the interested reader can find more information about our previous contribution on the discovery of OS [1]), where N is the set of nodes which depict the performers and A is the set of arcs which represent the interaction links between them, (ii) from events log, extending G with addictive arcs which represent the remainder of interactions between performers that constitute the nodes of G. These interactions are based on FIPA-ACL performatives where each of one belongs to one organizational relation-ship K which can be delegation, monitoring or communication. According to this representation, we have three connected graphs Gk; each one is shaped by a set of Actors-K which represent the nodes of Gk and a set of A-k which are the links that relate Actors-K, and finally (iii) Inferring adjacency matrix M [n]*[n] to each Gk and figure out the values of A-k and Actors-K where |Actors-K| is equal to |n| and |A-k| is the number of cases of M which are equal to 1. Thereafter, to evaluate a workflow OS we use the formulas used in [5]:

`Completeness`: A complete graph G is a graph whose nodes are joined in twos. The Completeness of Gk is the joint degree of an organizational relation-ship K.

$$\text{Completeness}\,(G_k) = \frac{|A-k|}{|Actors-K| \quad (|Actors-K| - 1)}$$

`Connectedness`: A connected graph G is a graph in which each node has a way of degree 1 to another node of G. G (N, A) is connected if exists A ⊂ N x N where ∀ (x) ∈ N, ∃ (y) ∈ N with (x, y) ∈ A.

$$\text{Connectedness}\,(G_k) = \frac{|DISCON-k|}{|Actors-K| \quad (|Actors-K| - 1)}$$

The connectedness property shows the monitoring capacity of failure activities in order to find assistance to achieve work. The value of property DISCON is determined as follow: We check whether the graph G of K is connected, if it is connected then |DISCON-k| equals to 0 else |DISCON-k| > 0 and {CC} is the set of connected components of G with n = |CC| and i ∈ {1, …, n}. |CCi| is the number of nodes of the connected component CCi.

$$|\text{DISCON-k}| = \sum_{i=1}^{n} |CCi| * \sum_{CCj \in \{CC\} - CCi}^{n} |CCj|$$

`Economy`: the economy of a graph G is characterized by the fact that connectedness property is greater than completeness property. A highly degree of economy maximizes the circulation of information and promotes direct communication, there are no intermediary between two nodes and is not necessary to explore the whole network to convey the information.

$$\text{Economy} - \text{k}(G) = \frac{1 - (|A - k| - (|\text{Actors} - k| - 1))}{|\text{Actors} - k|(|\text{Actors} - k| - 1) - (|\text{Actors} - k| - 1)}$$

The evaluation of OS allows the enhancement of organizational performance in crisis field. The assessment stages are the following: to each instance of the crisis management process, we represent the matching inferred OS. Next, we evaluate and compare them in terms of flexibility, efficiency and robustness. For the evaluation of OS, we have to firstly get the values of corresponding OPIs to each organizational relation-ship K. After that, we give robustness, flexibility and efficiency values of G, using the following formula where x ∈ {robustness, flexibility, efficiency}.

$$x(G) = \alpha \,.\, (\beta . \sum_{p \in IP_0^x} \| p(G) \| + \gamma . \sum_{q \in IP_1^x} \| 1 - q(G) \|)$$

Where

$$\alpha = \frac{|IP^x|}{|IP_1^x|^2 + |IP_0^x|^2} \quad \beta = \frac{|IP_0^x|}{|IP^x|} \quad \gamma = \frac{|IP_1^x|}{|IP^x|}$$

Table 2 gives the set of OPIs and their optimal values {0, 1} to each X.

Table 2 Corresponding IP^x to each X

IP^x of robustness	Val	IP^x of flexibility	Val	IP^x of efficiency	Val
$\text{Overlap}_{\text{com-del}}$	1	$\text{Completeness}_{\text{del}}$	0	$\text{Connectedness}_{\text{del}}$	1
$\text{Chain}_{\text{mon-del}}$	1	$\text{Connectedness}_{\text{del}}$	0	$\text{Economy}_{\text{del}}$	1
$\text{Chain}_{\text{mon-com}}$	1	$\text{Chain}_{\text{mon-del}}$	1	$\text{Economy}_{\text{com}}$	1
$\text{InCover}_{\text{mon-com}}$	1	$\text{Completeness}_{\text{com}}$	1	$\text{Overlap}_{\text{com-del}}$	1
$\text{OutCover}_{\text{del-mon}}$	1	$\text{Connectedness}_{\text{com}}$	1	$\text{Overlap}_{\text{del-com}}$	1
$\text{OutCover}_{\text{del-com}}$	1	$\text{OutCover}_{\text{del-mon}}$	1	$\text{Univocity}_{\text{del}}$	1
$\text{Completeness}_{\text{com}}$	1			$\text{Unilaterality}_{\text{del}}$	1
$\text{Connectedness}_{\text{com}}$	1			$\text{Economy}_{\text{mon}}$	1
$\text{Univocity}_{\text{del}}$	0			$\text{Overlap}_{\text{mon-del}}$	1
$\text{Unilaterality}_{\text{com}}$	0			$\text{Overlap}_{\text{del-mon}}$	1
$\text{Univocity}_{\text{mon}}$	0				
$\text{Flatness}_{\text{mon}}$	0				

- IP^x Represents the set of the performance indicators OPIs which optimize the concerned X.
- IP^x_1 Represents the OPIs of IP^x which have a value equal to 1.
- IP^x_0 Represents the OPIs of IP^x which have a value equal to 0.
- p represents the property value of IP^x which belong to the set of IP^x_0
- q represents the property value of IP^x which belong to the set of IP^x_1

3.3 Crisis Scenario

To illustrate the discovered OS assessment, we have chosen the well known case study "forest fire". In this scenario, the authorities set up a crisis unit consisting of DFR representatives (the Department of Fire and Rescue), ambulance, police force, DDE (Directorate Department of Equipment) and hospitals. At the occurrence of forest fires after a strong wind episode, the following process is established. Police is involved to Release Secure Perimeter (RSP) to allow others to act. DDE then comes to Clear Trees (CT) that obstructs traffic. DFR Sends three Teams of Fire-fighter (STF : E1, E2 and E3): the first Evacuates some People (EP), while the second and third teams Deal Off (DO) the two identified houses. DFR informs the crisis unit of the presence large burned. The crisis unit launches a tender to hospitals to requisition one with available beds and burn specialists. The E2 team is declared "Failed" and the Crisis unit decides that team E1 has a new mission to help team E2 after evacuating the people of the village. Because of space limitation, we give a simplified extract of log file of the considered crisis scenario (see Table 3).

Table 3 A simplified extract of the log file

Case	Activity	Initiator	Receiver	Performative
C1	RSP	Jean	Mahdi	Delegate
C1	RSP	Mahdi	Jean	Execute
--	CT	Mahdi	Marie	Inform
--	CT	Jean	Marie	Delegate
--	CT	Marie	Jean	Execute
--	STF	Jean	Taher	Delegate
--	STF	Taher	Marie	Query
--	STF	Marie	Taher	Answer
--	DO	Taher	Marc	Delegate
--	DO	Marc	Jean	Inform
--	EP	Taher	Pauline	Delegate
--	EP	Pauline	Jean	Inform
--	DO	Marc	Taher	Failure
--	DO	Taher	Pauline	Delegate
--	DO	Pauline	Jean	Inform
--	SA	Jean	Sami	Delegate

We'll just consider the gray sequence of interaction which allows us to infer hierarchical structure (HS). According to the step (1) we represent HS with an oriented graph G (see Fig. 2a) where its nodes represent the actors and its arcs represent the delegation relation-ship.

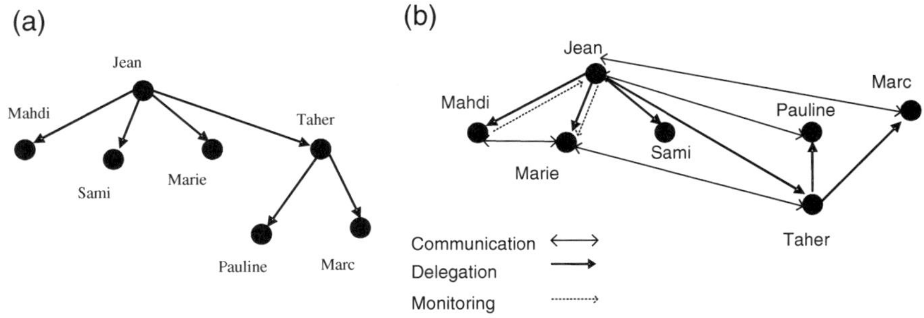

Fig. 2 Graphic representation of strict hierarchy structure and extended HS

3.4 Evaluation of HS

OS evaluation gives an idea about the efficiency, the robustness or the flexibility of an organization. As well as, helps to propose some organizational adjustments which are advisable to the situation imposed by the coming crisis. These adjustments allow important gains at time and resources. In what follows, we implement the other two stages. As explained in step 2, we extend the graph of Fig. 2a with the three organizational relation-ship K (see Fig. 2b) and then we measure OPIs of each K. All OPIs measures are based on A-k and Actors-K values. To get these values we infer the adjacency matrix Mk [n]*[n] of each K relation-ship and we count the number of cases to 1 of Mk in order to find the value of |A-k| and to give the value of |Actors-k| we do sums of nodes of the K graph.

To better illustrate the assessment process, we figure out the OPIs of the delegation relation-ship of Fig. 2b. The values of A-delegation and Actors-delegation are calculated from the adjacency matrix of Delegation relation-ship M_d [n]*[n] of the Table 4.

$|A - Delegation| = \sum_{i,j=1}^{n} Ci,j$ where Ci,j is the value of the case of column j and line i.

|*Actors-delegation*| = | n |, where n is the number of elements of Md [n]*[n].

Table 4 Adjacency matrix M of delegation relation-ship M [N]*[N]

	Taher	Mahdi	Sami	Jean	Marc	Paul	Marie
99Taher	0	0	0	0	1	1	0
Mahdi	0	0	0	0	0	0	0
Sami	0	0	0	0	0	0	0
Jean	1	1	1	0	0	0	1
Marc	0	0	0	0	0	0	0
Paul	0	0	0	0	0	0	0
Marie	0	0	0	0	0	0	0

Table 5 shows OPIs values of the OS of Fig. 2b. On the basis of these values we will estimate the flexibility, efficiency and robustness degree of the given structure.

Table 5 OPIs values of the extended hierarchical structure

OPIs	Relaxed hierarchy structure		
	Delegation	Monitoring	Communication
Completeness	0.14	0.15	0.16
Connectedness	1	0.4	0.4
Economy	1	1	1
Unilaterality	1	1	0.6
Univocity	1	1	0.66
Flatness	0.85	1	0.66

	Com-Mon	Com-Del	Mon-Del	Del-Mon	Mon-Com	Del-Com
Detour	0	0	0.5	1	0	0
Overlap	0	0	0.5	1	0	0
Incover	0.33	0.33	0.5	1	0.2	0.4
Outcover	1	1	1	1	0.25	0.25
Chain	1	1	0	0.5	0.37	0.62

According to the values of Table 5 and the formula given in the previous section, the efficiency value of the extended hierarchy structure is 0.25 and the robustness value is 0.73 and the flexibility value is 0.6. Bring to a close; the evaluation approach is very interesting since it allows the benchmarking of the organizational structures that each one corresponds to an instance of the same previous crisis process. Then, choose the matching process to better resolve the crisis.

4 Implementation

The current version of DiscopFlow aims at discovering and evaluating the organizational perspective (see Fig. 3). It is implemented with Eclipse platform. More precisely, the window 1 presents the general interface of the DiscopFlow. The window 2 shows the result of federation discovering while the window 3 exposes the result of OS assessment in term of efficiency (see Table 2). In fact, our datasets are synthetic because no existing system can generate a log file according to the proposed events log model. As we have mentioned in introduction, our log file integrates besides activities execution the interactions among actors. However, we have developed a LogGenerator which generates in automatic and random way the process instances. This study can help, on the one hand, the designers of BP to analyze the processes execution traces and as a consequence to improve the prescribed processes.

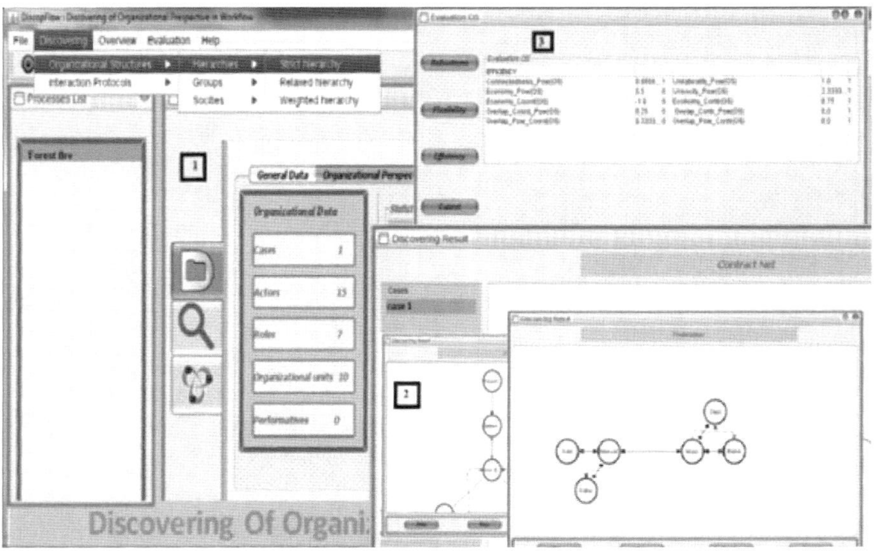

Fig. 3 The main interfaces of DiscopFlow

5 Discussions and Conclusion

This paper has (i) presented the proposed events log model which extends existing models by using the performatives based agent communication language FIPA-ACL, (ii) selected and adapted a method to deal with organizational structures evaluating and (iii) exposed a tool called DiscopFlow which supports the discovering and evaluating of organizational structures and (iv) a crisis scenario to better illustrate our contributions. Workflow mining has already led to several works and systems such as InWoLvE [7], WorkflowMiner [4] and ProM [14]. These works address Workflow mining by considering only the process perspective. For instance, [7] proposes the InWolve system which first creates a stochastic activity graph from events log and in the second step, it transforms the activity graph into a well defined process Workflow model. [4] proposes the WorkflowMiner system which aims at discovering Workflow patterns from events log by using a statistical technique. [14] proposes the ProM system, which only supports the discovery of process and organizational perspectives. Regarding organizational perspective, it provides three methods: default mining, mining based on the similarity of activities and mining based on the similarity of cases. On the one hand, this solution only supports the classical elements such as roles, organizational units and socials networks. On the other hand, it does not exploit the agent approach and as hence it does not support the organizational structures as proposed in our work. Our work differs from existing works for several reasons. First, we provide a solution for discovering organizational structures (like strict hierarchy in Fig. 3). Second, we evaluate the discovered organizational structures in terms of robustness, flexibility and efficiency. Third, even if is not presented in this paper, we can also discover interaction protocols taken individually or within process models. We believe that it is interesting to discover, in addition the activities and their coordination, the interactions among actors in order to build a real view on the crisis. As future works, we aim at exploiting the result of organizational structures evaluating in order to improve or propose a new Workflow (called processes reengineering).

References

1. Abdelkafi, M., Bouzguenda, L.: Discovering organizational perspective in workflow using agent approach: an illustrative case study. EOMAS in Conjunction with CAISE, Hammamet, Tunisia, pp. 84–98 (2010)
2. Ben Hmida, F., Lejouad Chaar, W., Dupas, R., Seguy, A.: Graph based evaluation of organization in multi-agent systems. In: 9th International Conference of Modeling, Optimization and Simulation, Bordeaux, France (2012)
3. Cernuzzi, L., Giret A.: Methodological aspects in the design of a multi-agent system. In: Proceedings of Workshop on Agent Oriented Information Systems. Austin, USA, pp. 21–28 (2000)
4. Gaaloul, W., Bhiri, K., Haller, S., Hauswirth, A.: Log-based transactional workflow mining. Distrib. Parallel Datab. **25**, 193–240 (2009)
5. Grossi, D., Dignum, F., Dignum, V., Dastani, M., Royakkers, L.: Structural aspects of the evaluation of agent organizations. Coordination, Organizations, Institutions, and Norms in Agent Systems II, Berlin, pp. 3–18 (2007)
6. Hanachi, C., Sibertin-blanc, C.: Protocol moderators as active middles-agent in multi-agent system. Int. J. Auton. Agents Multiagent Syst. **8**(2), 131–164 (2004)
7. Herbst, J.K.: Workflow mining with InWoLvE. Computers in Industry—Special Issue: Process/Workflow Mining, vol. 53, Issue 3, pp 245–264, Elsevier Science Publishers B. V. Amsterdam (2004)

8. Makni, L., Khlif, W., Haddar, N., Ben-Abdallah, N.Z..: Tool for evaluating the quality of business process models. ISSS/BPSC, pp. 230–242 (2010)
9. Miguel, J., Charoy, V., Bonita, F.: Workfow cooperative system. http://bonita.objectweb.org (2003)
10. Van der Aalst, W.M.P.: Process mining in CSCW systems. In: Proceedings of the 9th IEEE International Conference on Computer Supported Cooperative Work in Design (CSCWD), pp. 1–8. Coventry University/IEEE Computer Society Press, London, UK (2005)
11. Van der Aalst, W.M.P., Aldred, L., Dumas, M., Hofstede, A.H.M.: Design and implementation of the YAWL system. In: Advanced Information Systems Engineering CAiSE'04, Berlin, pp. 142–159 (2004)
12. Van der Aalst, W.M.P.: Process Mining: Discovery, Conformance and Enhancement of Business Processes, pp. 1–352. Springer Science & Business Media (2011)
13. Van der Aalst, W.M.P., Van Dongen, B.F., Günther, C.W., Rozinat, A., Verbeek, E., Weijters, T.: ProM: the process mining toolkit. In: Proceedings of the Business Process Management Demonstration Track (BPMDemos), Ulm, Germany, (2009)
14. Website. http://www.flowmind.org

Personal Assistance Agent in Programming Tutoring System

Boban Vesin, Mirjana Ivanović, Aleksandra Klašnja-Milićević and Zoran Budimac

Abstract E-learning systems must use different technologies to change the educational environment and perform the adaptation of educational material according to the needs of learners. Popular approach in designing and developing adaptive courses include employment of different kinds of personalized agents. In our previous research, we implemented a tutoring system named Protus (*PROgramming TUtoring System*) that is used for learning programming basics. This paper presents the architecture and methodology for implementation of personal assistance agent in programming tutoring system that dynamically tracks actions of the learner, determines his/her learning styles and adapts educational material and user interface to assigned individual. The role of the personalized agent will be to collect data on assigned learner, track his/her actions, learning styles and the results achieved in the tests. In a further process, the agent consults personal agents of other learners that have the same learning style. The engaged agent determines what actions and teaching materials brought the most benefit to these learners and in further learning process generates and displays the recommendations of the best ranked actions and materials to assigned learner. The main pedagogical objective of the personal assistance agent in Protus is to present learners the appropriate educational material, tailored to their learning style in order to efficiently and quickly learn the content.

Keywords Personal assistance agent · Protus · Programming course · Learning styles

B. Vesin (✉)
Higher School of Professional Business Studies, Novi Sad, Serbia
e-mail: vesinboban@yahoo.com

M. Ivanović · A. Klašnja-Milićević · Z. Budimac
Faculty of Science, Department of Mathematics and Informatics, Novi Sad, Serbia
e-mail: mira@dmi.uns.ac.rs

A. Klašnja-Milićević
e-mail: akm@dmi.uns.ac.rs

Z. Budimac
e-mail: zjb@dmi.uns.ac.rs

G. Jezic et al. (eds.), *Agent and Multi-Agent Systems: Technologies and Applications*, Smart Innovation, Systems and Technologies 38,
DOI 10.1007/978-3-319-19728-9_37

1 Introduction

The widespread use of computers and access to the Internet have created many opportunities for online education, such as improving distance-learning and classroom support [1]. Traditional content-delivery can be enhanced by tutoring systems with implemented intelligence for improving the effectiveness of a learning process. One among very popular approaches in designing and developing educational tools include employment of different kinds of agents [2]. These intelligent software agents have been recognized as a promising approach for the adaptation in e-learning environments [3] and can serve as a personal assistant to learner that will recommend him/her optimal actions and adequate educational material. Those assistance can bring benefits to learning efficiency, speed and quality [4].

Personal assistance agents (i.e. personal assistants) are computer programs that enhance the functionality provided by a software application implementing a two-way interaction with the user in which both, the assistant and the user, can execute tasks and initiate the interaction with each other [5]. This kind of software in the artificial intelligence community is also known as interface or embodied agents. Moreover, these agents in e-learning environments can detect when and how they could help learners and suggest an action or adequate education material. These agents could employ a variety of techniques to provide active assistance to a learner. The agent observes the user interaction with the application to learn about his habits and preferences with the purpose of adapting the interaction to his/her particular needs [5].

Protus is a tutoring system designed to provide learners with personalized courses from various domains. It is an interactive system that allows learners to use teaching material prepared for appropriate courses and also includes parts for testing acquired knowledge [6]. In previous work we demonstrated how programming tutoring systems can be enabled to provide adaptivity based on learning styles [7]. Learning styles can be defined as unique manners in which learners begin to concentrate on, process, absorb, and retain new and difficult information [8].

In this paper we will present a design of an extension module for Protus that will enable personal assistance agent to observe the actions the learner performs and suggest appropriate further actions based on the determined learning style of assigned learner. We will also explain the architecture of extension module, which will be constructed to act as a personal assistant within Protus.

The paper is structured as follows. In Sect. 2, we summarize the results of recent research on the effect of the use of assistance agents in educational systems. We show few examples of e-learning systems that use agents to enhance learning. Section 3 provides an overview of Protus and design and architecture of integrated module that will enable the use of personal assistance agent. In Sect. 4, conclusions and ideas for future work are given.

2 Related Work

The focus of computerized learning has shifted from content delivery towards personalized online learning with Intelligent tutoring systems [9]. Different techniques could be used to achieve personalization in e-learning systems. Interface agents enable learners to interact with tutoring systems in real time by replicating human communication, helping learners to build motivation and confidence [9].

Design issues and implications that relate to the use of software agents in tutoring systems are presented in some earlier publications [3, 4, 10]. The ideas of agents that assist the user while learning grew since then. This early work on agent-based intelligent assistance mainly focused on tutoring/learning environments and information retrieval systems. In these attempts, user modeling was employed to create intelligent adaptive user interfaces to assist learners in acquiring required educational resources using classical keyword based approaches [3].

Oscar CITS is an innovative Conversational Intelligent Tutoring System that can imitate a human tutor by directing a tutoring conversation and dynamically detecting and adapting to learning styles of individual learner during the conversation [1, 9]. Its pedagogical aim is to enhance the learning experience by providing the learner with tutoring material suited to their learning styles. Oscar's natural language interface is familiar and intuitive to learners, allowing the exploration of problems and helping to build confidence and motivation during programming courses [9].

Other systems with implemented assistance agents that takes into account learning styles are shown in [3, 10, 11]. Authors in [11] presented an intelligent argumentation assessment system based on machine learning techniques for computer supported cooperative learning. System is proposed in Moodle, an open source software e-learning platform, and it is used to establish the cooperative learning. Authors in [3] described a cognitive learner model as a composition of the learner's interest and behaviour models. Authors proposed intelligent assistance architecture to integrate these models with other environment models as well as inference, knowledge update and collaboration components. All these models and components collectively enable personal assistance agents to effectively work with the corresponding learners to achieve their goals in a collaborative environment. Authors in [10] described the design of an ontology-based speech interface for personal assistants applied in the context of cooperative projects.

A framework that allows attaching an interface agent to a conventional application without modifying the application in any way has been presented in [5]. This allows enhancing an existent application (for which there is no source code available) with an interface agent that will assist the learner.

INES (INtelligent Educational System) is an operative prototype of an e-learning platform, which combines essential capabilities related to e-learning activities and agents that communicate with learners in natural language [12]. The system not only models its learners, but is able to specify for each learner what to learn and how to do that, offering personalized formation without human intervention.

However, authors did not consider learners' learning style in order to provide adaptive feedback message for different types of learners.

The above-mentioned online environments emphasized either exchanging information or enhancing presentation. All of them attempt to mimic a human tutor, but as presented in [13], learners' experience of animated pedagogical agents is too important with respect to the goals to motivate and engage. Therefore, those attempts cannot be treated as a secondary issue.

Hence, we choose to implement non animated personal agents. Those agents will help learners during teaching process and recommend actions, but will not distract learners nor aggressively guide them out of their usual learning process (that they find more appropriate and effective).

3 Personal Assistance Agent in Protus

Protus is a tutoring system designed to provide learners with personalized courses from various domains. It is an interactive system that allows learners to use teaching material prepared for appropriate courses and also includes parts for testing acquired knowledge [14].

The ultimate goal of developing the Protus system has been increasing the learning opportunities, challenges and efficiency. Two important ways of increasing the quality of Protus services are to make it adaptive and if possible somehow intelligent.

Different techniques are needed to be implemented to adapt content delivery to individual learners according to their learning characteristics, preferences, styles, and goals. Protus provides two general categories of personalization based on learning styles identification and recommender systems [7] (Fig. 1).

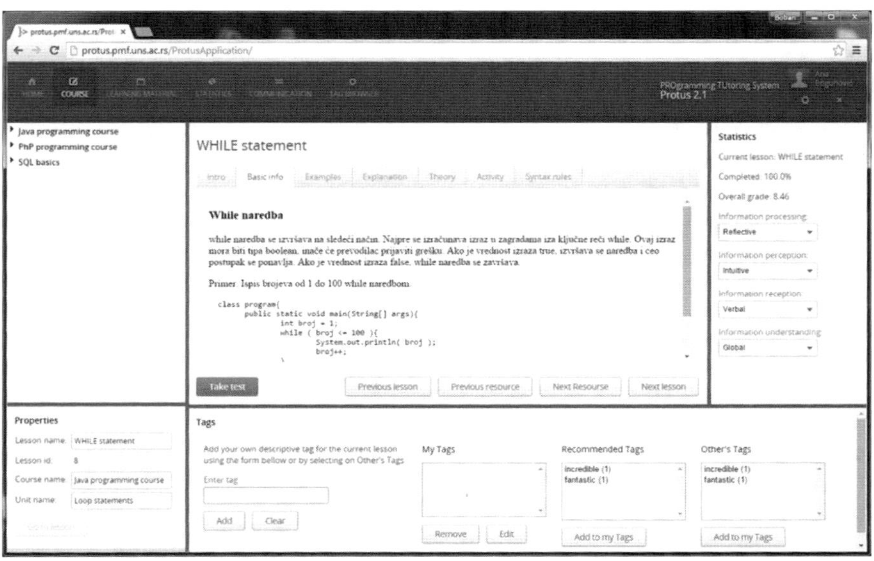

Fig. 1 Interface of Protus

The term learning styles refers to the concept that individuals differ in regard to what mode of instruction or study is most effective for them [8]. Proponents of learning-style assessment argue that optimal instructions require diagnosing individuals' learning style and tailoring instructions accordingly. While many learning style models exist in literature, in Protus system we implemented model proposed in [15] concerning the different cognitive styles of learning learners may have, which were described in [7]. Based on those cognitive learning styles, a graphical user interface was developed to enable the learner himself/herself to categorize his/her learning styles, set his/her learning goals and characteristics of work environment and the kind of course he/she wants to take. At run-time, the learner model is updated taking into account visited resources and test results.

3.1 Personal Assistance

A software application can be viewed as a set of tools. Each of these tools aims at helping its users to perform a specific task in the domain of the application [5]. It is very important for an assistance agent to know in every moment the task the learner is performing because it gives a context in which the user is moving through learning material. Taking this context into account, the agent may infer the user's intention and try to collaborate with the learner. Moreover, the context will allow the agent to improve teaching process and overall learner experience.

Our main idea is to design a personal assistance agent and assign it to each learner in the tutoring system. The intention is that personal assistant helps the learner during teaching process in several ways by:

- advising the learner, when needed, to take the appropriate actions,
- monitoring their progress,
- tracking cognitive learning styles of learner and
- creating a learner-adaptive environment.

The objective is to provide the learner with actions that are adapted and personalized according to its preferences and to offer educational material presented in appropriate form. To accomplish this task, every agent must build detailed overview of learner's characteristics to assist him/her in teaching process.

The personal assistance agent provides suggestions during the learner's interaction with a tutoring system. In addition, the agent exploits the personal learner model which maintains knowledge about the learner's current progress and learning styles.

Personal assistance agent, presented in this paper, can demonstrate or guide the learner through the tutoring sessions, suggesting relevant material to visit. It can also acquire knowledge from personal assistance agents of other learners and find similar learners regarding their learning styles. In effect, this agent serves as the communication medium for assigned learner and other learners that took the same course earlier.

Personal assistance agent in Protus is personalized to the learner. Each learner has his/her own personal assistance agent. The agent adjusts itself to the learners needs through time and learner's learning history. As the learner learns more, the agent gets more knowledge about learning habits of assigned learner and could make a better prediction of resources and actions that will correspond to learner's knowledge level and personal learning styles.

3.2 Personal Assistant Module Architecture

Architecture of Protus assistance agent module represents mixture of similar tutoring systems architectures with agent support [1, 3, 5, 11, 12]. Figure 2 shows an outline of our *Personal assistance agent module*. The observation of the learner's actions during courses is done by a personal assistant named *PersonalLearnerAssistant*. This Java class tracks visited resources and lessons, monitor test results for assigned learner and forms the personal learner model.

The personal assistance agent collects data on assigned learner, visited resources, current learning styles and the results achieved in the tests. In a further process, the agent consults agents of other learners that have the same learning style. The engaged agent determines what actions and teaching materials brought the most benefit to those learners and in further learning process generates and displays the recommendations of the best ranked actions and materials to assigned learner. Another role of personal assistance agent is to provide information to other personal assistance agents of other learners, what were the most effective resources for its assigned learner for particular learning material. In this way, every personal assistant collects data for the assigned learner and can retrieve similar data from other personal assistance agents. The main pedagogical objective of the personal assistance agent in Protus is to present learners the appropriate educational material, tailored to their learning style in order to more efficiently and more quickly learn the content.

This module aims at the construction of personal assistance agent network that will gather and exchange information about learners. In further learning process, all this personal agents will provide guidelines that will assist their own assigner learners with recommendation of lessons, resources or actions.

The core of the proposed model is a *Personal assistance agent module* that will be added to Protus. The role of this module is to assign specific agent to every learner. Module is composed of a *Learner performance model*, a *Learner interaction model*, an *Inference component* and a *Recommendation module* (Fig. 2). The *learner performance model* and the *learner interaction model* are designed to capture the assigned learner's results and behaviors in tutoring system environment during learning sessions. Based on these two models, *Personal assistance agent* uses the *Inference component* to recommend various strategies and educational material to assigned learner and facilitate collaboration with personal assistance agents of other learners.

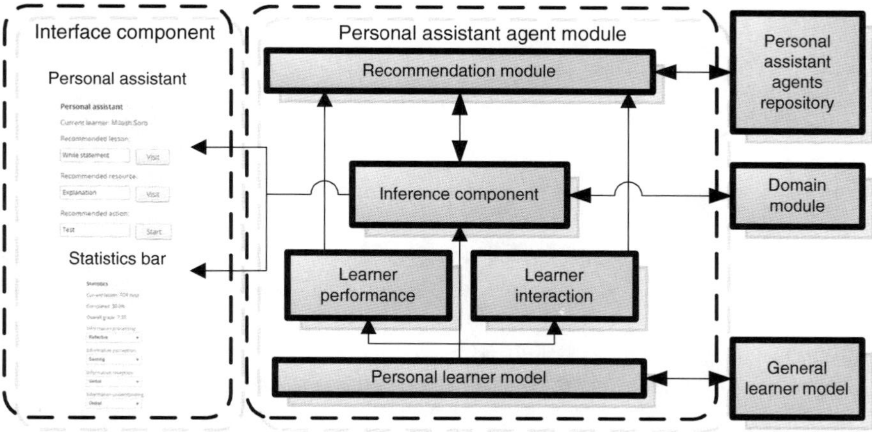

Fig. 2 Architecture of personal assistant module in Protus

The *Recommendation module* is used to determine current learning style of the assigned learner, and collect data from other agents assigned to learners with similar learning style and preferences. All personal assistance agents for all learners in Protus are stored in global *Personal assistance agent repository.*

The *Personal assistance agent module* assists learners in collecting the supplementary learning material, that provided best results in past to similar learners in Protus. This agent makes decisions via communication and gathering experiences from *Personal assistance agents* of other learners with similar learning habits. Learners are similar if they belong to same learning style categories within four domains of learning styles [7]. Briefly, the task of personal assistant is to consult other personal assistants (assigned to other learners), find similar learners and to obtain information about his/her learning progress. Therefore, it can be able to suggest to assigned learner specific tasks or learning resources that brought the most benefits to those similar learners.

The *Personal learner model* consists of two components, including the *Learner performance model* and *Learner interaction model.* Its role is to generate a specific segment of *General learner model* that belongs to assigned learner. In order to become aware of the learner's interaction with the Protus, a *Learner interaction model* tracks the visited resources and chosen options while *Learner performance model* tracks grades for specific lessons of assigned learner. Once the Personal assistant has detected an action of the learner, it should update personal learner model with data on actions or test results.

The main task of presented personal assistant module in Protus is to decide at every moment what educational content is the most appropriate to offer to assigned learner. To make these decisions, the personal assistant will take into account the information from other assistants, such as the defined learning paths, the learning styles of each learner, etc.

3.3 User Interface

The interface of *Personal assistance agent* in Protus is consistent and keeps a standard structure through different sessions. Therefore, the learner can get familiar with its form, position and suggestions. The presentation of the *Personal assistance agent* consists of two sections: the *Personal assistant* and the *Statistics*.

Personal assistant interface informs the learner, about the recommended educational material and recommended actions (Fig. 3). It provides links to recommended learning resources, lessons and possible actions that learner could take next. Furthermore, *Statistics* component displays current learning styles of the learner and his/her progress details (Fig. 4). Current learning styles categories of a learner are presented to him/her. Every student is given the opportunity to change the current learning styles, and thus changes the display of the lesson as presented in [7].

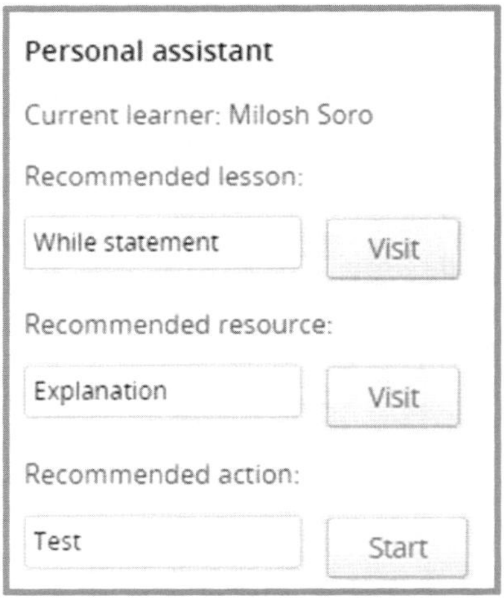

Fig. 3 *Personal assistant* interface in Protus

Personal agent in Protus is a customizable and non-intrusive. Customizable is because the learner can modify personalization options that are performed during courses (Fig. 5). All personalization options, implemented in Protus can be switched on or off regarding the preferences of learners. Non-intrusive is because the learner can disable the agent and continue course without its help.

After recommended educational material and actions has been provided by other assistance agents, personal assistance agent of current active learner provides timely feedback messages to him/her.

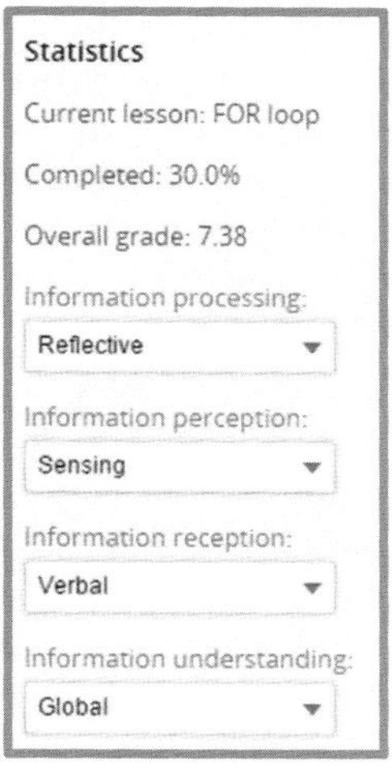

Fig. 4 *Statistics* options in Protus

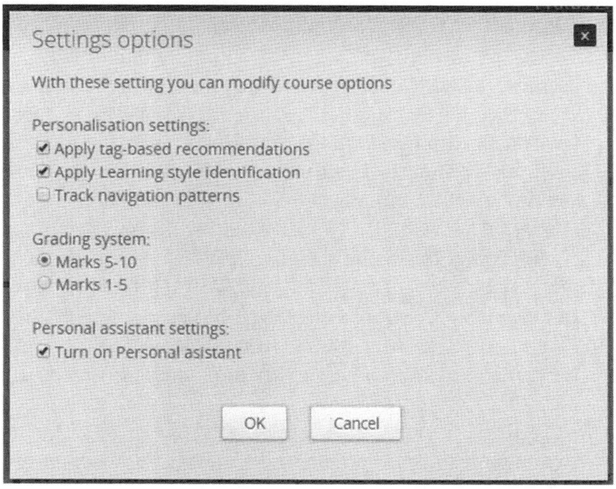

Fig. 5 *Setting options* in Protus

4 Conclusion

Software agents are radically changing the way people communicate with computers. As software agent technology becomes more commonly used, people will start to interact with computers in more natural ways. The computer will no longer be a silent participant in the conversation, but a more active supporter of a two-way dialogue.

In this paper we presented our design of an interface agent that recognizes the tasks a learner performs in a tutoring system. Personal assistance agent observes the actions of the learner in Protus and generates personal learning model for assigned learner.

The role of the Personal assistance agent is to collect data on assigned learner, track his/her actions, learning styles and the results achieved in the tests. In a further process, the agent consults agents of other learners with the same learning style, determines what actions and teaching materials brought the most benefit to these learners and in further learning process generates and displays the recommendations of the best ranked actions and materials to the assigned learner.

We also explain the architecture of an intelligent interface agent, which has been constructed to act as an assistant within the learning environment.

Protus aims to assist learners by implicitly modeling the learning style during sessions, personalizing the educational material and user interface to boost motivation of learner and improve the effectiveness of the learning process. Personal assistance agent in Protus serves as the communication medium for assigned learner and similar learners that already took the course. For the future work we plan to empirically evaluate its efficiency and impact to learning process.

References

1. Latham, A., Crockett, K., McLean, D., Edmonds, B.: A conversational intelligent tutoring system to automatically predict learning styles. Comput. Educ. **59**, 95–109 (2012)
2. Ivanović, M., Mitrović, D., Budimac, Z., Vesin, B., Jerinić, L.: Different roles of agents in personalized programming. New Horiz. Web Based Learn. **7697**, 161–170 (2014)
3. Wu, S., Ghenniwa, H., Zhang, Y., Shen, W.: Personal assistant agents for collaborative design environments. Comput. Ind. **57**, 732–739 (2006)
4. Farías, A., Arvanitis, T.N.: Building software agents for training systems: a case study on radiotherapy treatment planning (1997)
5. Armentano, M.G., Amandi, A.A.: A framework for attaching personal assistants to existing applications. Comput. Lang. Syst. Struct. **35**, 448–463 (2009)
6. Vesin, B., Ivanović, M., Klašnja-Milićević, A., Budimac, Z.: Protus 2.0: ontology-based semantic recommendation in programming tutoring system. Expert Syst. Appl. **39**, 12229–12246 (2012)
7. Klašnja-Milićević, A., Vesin, B., Ivanović, M., Budimac, Z.: E-Learning personalization based on hybrid recommendation strategy and learning style identification. Comput. Educ. **56**, 885–899 (2011)
8. Pashler, H., McDaniel, M., Rohrer, D., Bjork, R.: Learning styles concepts and evidence. Psychol. Sci. Public Interest Suppl. **9**, 105–119 (2009)
9. Latham, A., Crockett, K., McLean, D.: An adaptation algorithm for an intelligent natural language tutoring system. Comput. Educ. **71**, 97–110 (2014)
10. Paraiso, E.C., Barthès, J.P.A.: An intelligent speech interface for personal assistants in R&D projects. Expert Syst. Appl. **31**, 673–683 (2006)
11. Huang, C.J., Wang, Y.W., Huang, T.H., Chen, Y.C., Chen, H.M., Chang, S.C.: Performance evaluation of an online argumentation learning assistance agent. Comput. Educ. **57**, 1270–1280 (2011)
12. Mikic Fonte, F.A., Burguillo, J.C., Nistal, M.L.: An intelligent tutoring module controlled by BDI agents for an e-learning platform. Expert Syst. Appl. **39**, 7546–7554 (2012)
13. Gulz, A., Haake, M.: Design of animated pedagogical agents—a look at their look. Int. J. Hum Comput Stud. **64**, 322–339 (2006)

14. Vesin, B., Ivanović, M., Klašnja-Milićević, A., Budimac, Z.: Protus 2.0: ontology-based semantic recommendation in programming tutoring system. Expert Syst. Appl. **39**, 12229–12246 (2012)
15. Felder, R., Silverman, L.: Learning and teaching styles in engineering education. Eng. Educ. **78**, 674–681 (1988)

Agent-Based Approach for Game-Based Learning Applications: Case Study in Agent-Personalized Trend in Engineering Education

Dejan Rančić, Kristijan Kuk, Olivera Pronić-Rančić and Dragan Ranđelović

Abstract This article presents the effectiveness of agent-based approach in computer learning games. There are a lot of intelligent agents that are successfully used for educational purposes. However, some of them are very interesting because they can demonstrate behaviors that agents may possess; on the other hand they are used to display a set of requirements that must be met during their creation and design. Some of these requirements are similar to those used in the ITS, while the second type of behavior is quite different from them, and may even be unique. Based on this re-search, the effectiveness of games is illustrated in detail with regard to three current perspectives on agent-based approach to games: design game-based learning environment with agent, the process decisions of agent interaction and the reflection of agent recommendations and learning outcomes. Although, all perspectives are connect with the hope of better learning through games, it is criticized that the effectiveness cannot be simply answered by one of the three alone. The goal of the article is therefore to clarify the different views in case agent-personalized trend in engineering education.

Keywords Game-based learning · Pedagogical agent · Personalized agent system · Engineering education

D. Rančić (✉) · O. Pronić-Rančić
Faculty of Electronic Engineering, Niš, Serbia
e-mail: dejan.rancic@elfak.ni.ac.rs

O. Pronić-Rančić
e-mail: olivera.pronic@elfak.ni.ac.rs

K. Kuk · D. Ranđelović
Academy of Criminology and Police Studies, Belgrade, Serbia
e-mail: kristijan.kuk@kpa.edu.rs

D. Ranđelović
e-mail: dragan.randjelovic@kpa.edu.rs

G. Jezic et al. (eds.), *Agent and Multi-Agent Systems: Technologies and Applications*, Smart Innovation, Systems and Technologies 38,
DOI 10.1007/978-3-319-19728-9_38

1 Introduction

Educational research has shown that by discerning how one learns, one can become a more efficient and effective learner. There is no one single method of learning and the most effective approach depends upon the task, context and learner's personality. The learning will be more effective if learners can choose from a wide range of possible learning methods, if they know when to apply them and which approach is the best for them. Honey and Mumford [1] identify four distinct styles or preferences that people use while learning. Kolb et al. [2] suggest that in order to learn effectively one need to keep moving around the following cycle: *Experiencing* – doing some-thing; *Reviewing* – thinking about what has happened; *Concluding* – drawing some conclusions and *Planning* – deciding what to do in the future.

To become a more effective learner one should engage in each stage of the cycle, and that may imply resorting to activities and styles that one would not normally choose.

On the other hand, anthropologist Levi-Strauss predicates that brain uses a story oriented structure to store and recall life experiences [3]. Furthermore, Heo has asserted that even facts, ideas, and theories are learnt more effectively if these are linked as a narrative [4]. Storytelling provides a powerful model of effective communication because it links with a basic human need: to create emotional engagement and movement [5]. According to Hassenzahl and Tractinsky, "User experience is a strange phenomenon: readily adopted by the human – computer interaction (HCI) community – practitioners and researchers a like – and at the same time critiqued repeatedly for being vague, elusive, and ephemeral". However, they attempt to convey the experience to users by considering the following three aspects [6]: the experiential (dynamic, complex, unique, situated, and temporally bounded), beyond the instrumental (holistic, aesthetic, and hedonic) and emotion and affect (subjective, positive, antecedents and consequences).

Regardless of delivery medium, any training development process must identify key skills that promote organizational goals and build training around the tasks that constitute those skills. Be it games, virtual worlds, or social media, technophiles gravitate toward the latest cool trends—sometimes without considering whether and how best to leverage them in ways that support relevant learning. Effective guided discovery forms of e-learning, including simulations and games, engage learners both behaviorally and psychologically. Today, there is an impressive arsenal of instruction-al technologies that can be used, ranging from educational games played on mobile devices to virtual reality environments to online learning with animated pedagogic agents and with video and animation [7].

Interactive animated pedagogical agent provides a simple learning environment that allows users to gain knowledge, adapting to their own pace. These agents are trying to achieve a certain balance in relation to the factors of good human tutor and good aspects of computer intelligent tutoring systems. For example, it is human to encourage students as they learn well, as it is human that the student should not feel embarrassed if he needs repeated explanation of materials. This paper reports a case study that proposes the use of pedagogical agents in a Graphics Algorithm unit in the IT Fundamentals Course built under the agent-based approach in computer learning games. Its main purpose is to discuss how agent's recommender can be used to improve students' cognitive preferences and level of recommender with agent's content realized as a text message.

2 Related Work

A pedagogical agent is not a human teacher but is a computer-based animated interface character. Consequently, students might have different perceptions and expectations of a human teacher as compared to a pedagogical agent. Findings from research on the effect of animated pedagogical agents on learning achievement are not consistent. The study of Kim et al. [8] tested strategies incorporating change management, motivational, and volitional characteristics in order to facilitate positive attitudes toward engineering. In an introductory engineering course, the strategies were distributed via email to two groups: one received the strategies with an animated pedagogical agent (Agent-MVM) and the other received the strategies in a text-only format (Text-MVM). The independent variable was type of message with three levels. The first consisted of motivational and volitional messages (hereafter, the acronym MVM will be used for convenience) delivered by an animated pedagogical agent, the second was MVM delivered in a text-only format, and the third was a control group that received innocuous information in text. The effects of the strategies on attitudes were compared with the control group which received neither formats of the strategy message. Contrary to expectations, the results indicated that the attitudes of the Agent-MVM group were not significantly more positive than the Text-MVM group or the control group.

Other studies show that use of pedagogical agents has a positive impact on near and far knowledge transfer Atkinson (2002) conducted a study on helping undergraduate students solve word problems using a computer program [9]. 50 undergraduate students were randomly assigned to one of five conditions: voice plus agent, text plus agent, voice only, text only, or control. Students in the voice-plus-agent condition outperformed their counterparts in the control condition on both near and far transfers.

Paradoxically, in experiment by authors Hershey et al. [10] students detected higher degrees of social presence in both the text only and the fully animated social agent conditions than students in the voice only and the static image of the agent with voice conditions. Furthermore, students had more positive perceptions of the learning experience in the text only condition. The results support the careful design of social behaviors for animated pedagogical agents if they are to be of educational value, otherwise, the use of agent technology can actually detract from the learning experience.

3 Theoretical Background

Practical experience has shown that active participation of the learners is an important part of the learning process and that can be enabled by means of personalization. The learning process has to enable active work by learners, to allow learners to share knowledge in groups, and to enable teachers to cooperatively develop learning content. To accommodate differences between learners: different learning objectives, different prior knowledge and past experiences and different cognitive preferences to the learning experience, the personalization of the learning experience towards the individual requirements has to be supported. The personalized e-learning systems have to include variations in students' intellectual profiles, the influence of the ability theories, Gardner's theory of multiple intelligences, cognitive controls, and in turn, cognitive styles and the learning styles with the aim to provide the highest degree of education efficiency.

An interesting design of service-oriented reference architecture for personalized e-learning systems (SORAPES) and validation of the architecture is described in [11]. The SORAPES is designed by re-using web services and learning objects with layered architecture and highly-scalable personalized e-learning system. The work described in [12] present part of the open learning project, where business practitioners and university researchers aim to combine the most frequently used e-learning technologies with the benefits of customized systems to develop an innovative personalized learning content delivery system.

Systems like My Online Teacher (*MOT*) [13] assist in developing rich adaptive hypermedia but are technically complex and offer little pedagogical support. The Adaptive Course Construction Toolkit (*ACCT*) [14] provides pedagogy, activity, subject matter, personalization and learning re-source based support to the course developer in addressing some of the key barriers to the main-stream adoption of personalized e-learning.

A recommendation module of a programming tutoring system - *Protus*, which can automatically adapt to the interests and knowledge levels of learners, is described in [15]. Personalized approaches based on the technology of intelligent agents means that each student has his pedagogical agent who represents him in the system of agents. Readapting learning objects to different categories of learners constitutes a challenge for intelligent agents in their effort to provide a large scale of collaboration between different e-learning organizations. In order not only to have efficient access to learning objects, but also to offer to learners tutoring and mentoring help, collaborative and cooperative learning strategies, learning advancements, and social interactions, intelligent agents have been highly recommended by a number of researchers. Intelligent-agent strategies and learning objects were successfully joined together in Learning Management Systems (*LMS*). For the purpose of providing more flexibility in *LMS*, Santos et al. [16] introduced the concept of intelligent LMS (*iLMS*) through using an intelligent-agent strategy. van Rosmalen et al. [17] discussed the main tendencies of Instructional Management Systems (IMS).

In paper [18] authors described *InCA*, a modular agent-based architecture framework, which integrates a set of interactive features allowing personalized and adaptive curriculum generation. They present an e-learning framework, named Intelligent Cognitive Agents (*InCA*).

4 Relationship Between Computer Games and Learning

Upon An impact of visual design quality on the learning is relevant to the design of computer games-based learning. Graphics can support learning in a variety of ways, such as drawing attention to key elements, providing links to existing mental models and supporting the creation of new models, simplifying presentation to minimize mental effort, and supporting the transfer of knowledge. The field of visual design is complex and encompasses many areas. Vanderdonckt [19] describes a detailed taxonomy for understanding visual design, including physical techniques such as balance and symmetry.

4.1 Game-Based Learning Model in Engineering Education

Prior to creating a game, an outline for accomplishing educational objectives and tasks compliant to the *Bloom's Taxonomy* has to be drafted. Opposite to the linear sequence that makes Bloom's Taxonomy useful for curriculum planning, the approach based on the game is reflected in perceiving different directions, such as, for example how players acquire or utilize knowledge in the game. In the context of the environment based on the game, model known as *GBL* – Game Based Learning [20] is used. This model represents the experience integration process between game cycles and accomplishing learning results. The link between simulation and the real world labelled as "experience integration" shows link between events in the game and real events, and at the same time it merges experience acquired in the game and learning process. The first step in designing a game-based learning model in the field of computer engineering represents defining start-up values. Following steps of creating games and learning elements should be taken into consideration:

- To establish educational approach,
- Set the model task, to create the game context, respectively,
- Develop details between game and education,
- Insert basic educational back up,
- Create learning activity map as a link between game events, and
- Create learning concept map a link between objects in the game.

Deducing a game-based learning model in the field of computer and electrical engineering - *GBLm4CE* model [21] created out of learners' profile and teaching methodology. Deployment of start-up values with compliance to the given suggestions for creating *GBLm4CE* model shows the method and pathway of implementing computer games into curricula – in the field of IT Science. Game implemented in the learning curricula in the field of IT Science, created by *GBLm4CE* model application should rely on constructive and collaborative learning approach. Players shall learn to understand and combine various aspects. Abstract and multidimensional spaces in combination with various colors according to the aforementioned aspects could be easier to understand as per *GBLm4CE* model rather than image to be conceived before the eyes or that could be visible during the learning process. Hence, this model application enables easier approach to analyze and com-pare existing solutions, especially its ability of visualization and simulation to be translated to the game.

5 Pedagogical Agents

Pedagogical agents are on-screen characters who help guide the learning process during an e-learning episode. Agents can be represented visually as cartoon-like characters, as talking-head video, or as virtual reality avatars; they can be represented verbally through machine-simulated voice, human recorded voice, or printed text. An important primary question is whether adding on-screen agents can have any positive effects on learning. Even if computer scientists can develop extremely lifelike agents that are entertaining, is it worth the time and expense to incorporate them into e-learning courses? In order to answer this question, researchers began with an agent-based educational game called Design-A-Plant, described previously [22].

Moments in which is necessary to prevent the appearance of an interactive agent are those when the user is focused on solving tasks. Also, it should be done in situations when customers are introduced with global information, or direct access to the raw data is performed. Complex visual information provided by the agent, may interfere with clear presentation of basic information. In addition, when working with children there is a fine line between pedagogical agent and fun. Younger users may be too amazed by interacting with the image of the agent, instead of to be focused on the problem [23]. In the digital world, pedagogical agents most commonly used type of human teaching "one on one". This technique is a traditional form of learning and it is realized through three models [24]:

- Guided Learning - teacher has a didactic role in presenting the teaching material and training of students;
- Learning by discovery - it is a way of teaching where the student has complete control over the learning environment and control the progress of his knowledge;
- Guided detection - is a combination of the previous two ways. The teacher provides learning and progress, acting as a guide on the way to solving the given problem.

Pedagogical agents can be very useful for teaching forms such as guided learning and guided discovery. Learning by discovery is still an underdeveloped area in the domain of knowledge, since it is beyond the current capabilities of artificial intelligence. The pedagogical agent is increasingly popular in the process of training and teaching of younger age students.

In academic circles, many institutions slowly adopt pedagogical agent technology because of their expensive and complex implementations. The technology of intelligent intermediaries (agents), also known as "*Knowbots*" technology, shows an example of the use of pedagogical agents in academic institutions. However, most of the universities are still very far from the application of agents in laboratories. As this technology becomes more available, and numerous studies [25, 26] demonstrate strong positive beliefs about their application in the learning process, it indicates the willing-ness of many institutions to invest in them.

Numerous discussions in the field of feedback information provided by intelligent agents are based on the fact that the agent should not provide too much to them because it would thus burdening the students. In addition, Negroponte [27] suggests that in the process of waiting in the execution of the student's actions is better to give more simple instructions than not to have any instruction. To avoid giving too much instructions, a good knowledge and familiarity in dealing with pedagogical agents is necessary. To resolve this issue, part of the pedagogical task is to keep track of time and the number of shown information in the form of notice. With the principle of minimal assistance as defaults, the student should be allowed to choose the type of feedback depending on the amount of information, interaction and feedback when solving a given problem.

The basic terms that explain the principle of operations functioning are reached by selecting the Help option in modules. When a student starts learning with the use of the game or faces a difficulty during solving a task generated by the application, Help serves to accelerate finding the right solution [28]. This means that formulation of definitions and theorems within Help is the key moment in designing the entire application. Quality evaluation whether an operation is acquired or not is performed through visual indication of the number of successful and unsuccessful tasks (score) with the same operation, and comparison with preset criteria. In some educational application, through the Help window, realized as a text message, and generated by a pedagogical agent, student is given a 'piece of knowledge', in the form of a theoretical theorem or a definition [29], necessary for successful solving the problem given in the game-based application.

6 Case Study: Agent-Personalized Approach in an Educational Game

In starting from the point that two genres cover all levels of *Bloom's Taxonomy* we have created simple educative game with shooting effects within action genre. Since we have already decided about the game genre we have decided to integrate physics and smart enemies – (bots) that will be run by the computer as per artificial intelligence algorithm. In order to make the game more interesting for students, we have decided to place the game in Victorian Era, and the ground story is to be based on the clash of the gangs from that Era. The action of the game is located in the high street of imagined town from Victorian Era (Fig. 1).

The task of the game is to run a soldier who is fighting on two different fields under diverse conditions and situations. Two different scenarios were selected so that students could associate with two basic principles of *Z-buffer* algorithm performance red within Graphics Algorithm unit in the IT Fundamentals Course. The objective of the player is to recognize in the enemy who is running after the player and shooting at him recognize proper combination of enemy soldiers. Enemy soldier wears a badge implying which algorithm it is about. Also, the enemy outfit implies to the player if it is about real enemy or not, so the player has to decide whether to destroy the enemy or not. In that manner of the game, students should keep alive enemies that represent proper formation of the pixel colors recorded in Depth buffer of *Z-buffer* algorithm. In case that student as a player eliminates the enemy that should have stayed alive, the error information will be displayed together with the text message of wrongdoing. In this way the role of pedagogical agent that follows up the player giving him/her the instructions is generated.

Fig. 1 A screenshot of the educational game

Usually, the computer tutor needs to provide human educators' teaching strategies such as observing students' progress and giving appropriate feedback. The main reason for applying a pedagogical agent is to reduce the learner's feelings of isolation towards an instructor and to provide encouragement towards the content. Providing encouragement allows learners to feel as if the learning is more personalized like it would be with a human instructor or a one on one tutor. Because of pedagogical agents appearing in the same display with learning contents, the learner's attention to detail may be split between both objects (agent body and recommender content). Although the purpose of a pedagogical agent system in game-based learning contents is fostering a learner's motivation and eventually enhancing learning outcomes, sometimes the pedagogical agent is blamed for distracting learner's efficiency. From this rationale, we have designed three types of recommender contents, and tested them in various ways including measurement techniques of cognitive load to observe learners' help points by measuring agent recommender efficiency. Three types of pedagogical agent contents have been designed incorporating recommender. The recommender includes only theorem, help with text and figures, and introduction. The recommender views are shown in Fig. 2. Different types of content delivered by a pedagogical agent in recommender window are:

- **Type 1** - content from help option in formal style. In formal style important information (theoretical theorem or definition) are presented;
- **Type 2** - content from help option in conversational style. Use conversational style to present theorem or definition content based guide the personalization principle by Clark and Mayer [30];
- **Type 3** - content from interdiction. Interdiction has much important information about gaming and global goal in game.

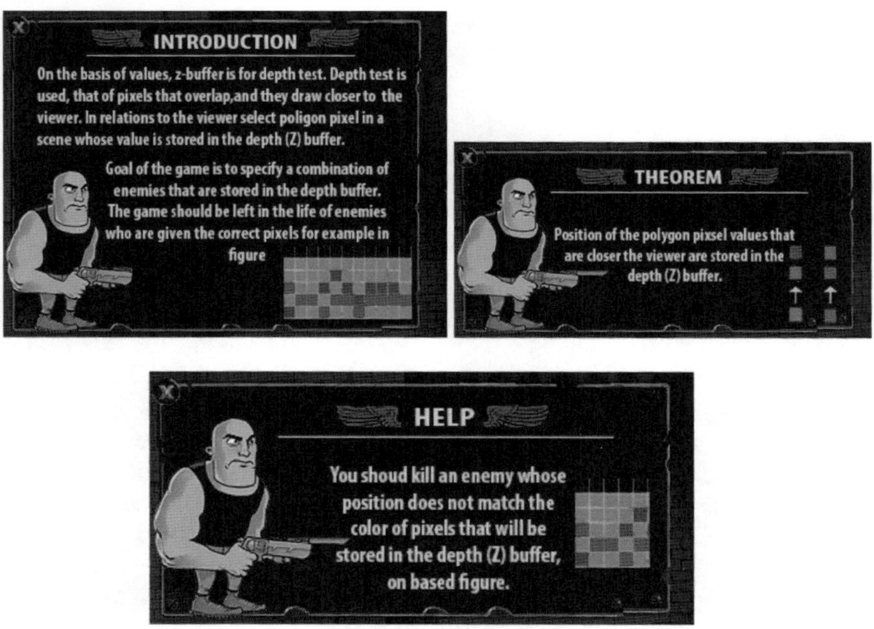

Fig. 2 Designed contents in recommender window

Among the measurement techniques of cognitive load such as physiological measurement, double task, and questionnaire survey, we have applied the questionnaire test of self-reporting style for allocation after the gaming period. The questionnaire was originally designed by authors. Questions covered the following basic concepts relevant to understanding agent interaction and recommender content:

1. *Is the help from agent was at the right time?*
2. *Is the content of assistance was appropriate to proceed successfully completing the game?*
3. *The content of help from agent is affected to better understand the whole task of the game?*
4. *The content of help from agent is affected to better understand only the current level task in the game?*
5. *The content of help from agent is affected to better understand the theoretical knowledge needed to successfully play the game?*

The cognitive load factor survey was implemented. It consists of 5 factors with twenty items, and 7 points Likert scale. As an effort of developing cognitive load measures, 2 sub factors of cognitive load measures were identified: (1) self-evaluation of learning, and (2) usability of learning material. The cognitive load sub factors and their reliability are given in Table 1.

The goal of this study was to examine how agent recommender with different content difficulties and levels of student knowledge affects learners' subjective judgment of cognitive load measures. Thirty-three college students are participated in experiments with different agent recommender and levels of student knowledge. The current knowledge level a student possesses after playing the education game is given with the following formula [31]:

$$P_i(X = \text{Mastered}) = \left(\mathbf{a} \times \frac{A_i}{N} - \mathbf{h} \times \frac{H_i}{N} - \mathbf{t} \times \frac{t_a \cdot (A_i - H_i) - t_h \cdot H_i}{T_{max}} \right) \times 100 \quad (1)$$

where values of coefficients a, h, and t are set 0.50, 0.30, 0.15. The number of correct answers of i-th student, A_i is present as right kill an enemy.

After that, the students are, according to their knowledge, divided into three levels (low-P(X) < 35 %, middle-35 < P(X) « 75 %, and high- P(X) > 75 %). Based on that, three groups of students were created. For first group G1, pedagogical agent, due to the low level of knowledge, is displayed by a recommender window content of help option, e.t. type 1 with formal presentation style theorem. Due to the partial knowledge that is not enough to understand the context (objective) of the game, as well as a lack of understanding of the concept of *Z-buffer* algorithm, for students of the second group G2, recommender window displays introduction which is normally shown at the beginning of the game. The agent is not sure whether students from this group misunderstanding the theory or the game, and it offers them both through the type 3. The group designated as G3 has the required knowledge and students from this group can respond to obstacles in the game in

Table 1 Elements of visual design quality

Factor	Meaning	Item reliability
Self-evaluation of learning (SEV)/the sense of accomplishment after studying	Self-evaluation is a personal perception of how successfully and/or efficiently a learner deals with a given problem to achieve desirable learning outcomes. The learner's subjective judgments are assumed to be an important factor for efficiency of learning. This factor is related to a learner's personal beliefs about his or her capabilities to produce the designated levels of performance. Learners, who measure highly on self-evaluation, tend to show low perceived task difficulty	0.770
Usability of learning material (USE)/ the effect of instructional design to learner's understanding	Usability measures how well the learning content is used towards the learning purpose. If a learner's perception of usability is high, it indicates that the learning content can facilitate learning or at least will not impede the learning process. When a learner is studying with a learning content with low usability, the learning content may hamper cognitive processes by increasing the unnecessary cognitive load. For this reason, this factor has a strong relationship with extraneous cognitive load	0.807

conversational style to. In all three cases, the display content recommender window by pedagogical agent have been fully complied with guidelines *EnALI* (Agent-Enhancing Learner Interaction) for the effective design of pedagogical agents that addresses social, conversational, and pedagogical issues [32].

To check the cognitive load differences of students, according to different types of contents, we have tested cognitive load and analyzed with MONOVA. The highest self-evaluation and usability factors are found in the G2 content with type 3 agent recommender. The technical statistics for each content are given in Table 2.

Significant interactions between agent recommender and levels of student knowledge were found in usability (USE, $F = 5.03$, $p = 0.017$) and self-evaluation (SEV, $F = 5.10$, $p = 0.016$) when an easy content was given in advance of a hard gameplay one. This result indicated that agent recommender interpretation is applied to the subjective scale of cognitive load. Usability and self-evaluation were very dependent on learners' performance.

Table 2 The technical statistics of cognitive load of each type agent recommender

Factor	G1		G2		G3	
	n = 10		n = 16		n = 7	
	Mean	SD	Mean	SD	Mean	SD
Self-evaluation	4.85	1.22	5.10	1.65	4.28	1.57
Usability	3.83	1.20	5.03	1.69	4.00	1.51

7 Conclusion

Pedagogical agents are autonomous software entities that provide support to the process of learning through interaction with students, teachers and other participants in the learning process, as well as cooperation with other similar agents. Similar to other approaches for personalization, also in the case of agents, the entire process is based on a model of student. A significant feature of pedagogical agents is that they have added a number of ontology in the field of personalized learning. The work of pedagogical agents depends not only on the existence of an ontological model of students, but also on the availability of other relevant ontology - domain ontologies, ontology of instructional approach, etc. which allows them to understand the demands of students and answer to them through different ways of personalization.

The learning task is an instructional goal to be learned. The main function of working memory is to integrate new information from the sensor memory and prior knowledge from the long-term memory. However, human can process only limited numbers of information. Because of this limited capacity the amount of learning content should be managed to avoid from being cognitive overload. In proposed approach, we have three types of recommender contents, and tested them in various ways including measuring techniques to observe learners' attention to detail. From the cognitive load analysis, the pedagogical agent helps the user to provide mental connection with recommender content by pedagogical agent. Even though the existence of an image of pedagogical agent's does cause significant differences in learning outcomes and cognitive load compared with recommender content using introduction, theorem and help. This implies that if the image conveyed more human-like feedback, it may contribute positively towards learners' overcoming the sense of isolation associated with an e-learning environment. Also, when considering the outperformance of learning content with pedagogical agents and recommender content, the pedagogical agent itself does not distract a student's attention, but rather can have a positive effect on creating a better e-learning environment.

The use of ontologies for the realization of personalized learning a few years is very actual direction of research in the area of the application of artificial intelligence in education (*AIED*), and our future work will be in that direction.

References

1. Honey, P., Mumford, A.: The Manual of Learning Styles. Peter Honey Associates (1986)
2. Kolb, D.A.: Experiential Learning. Prentice-Hall, Englewood Cliffs (1984)
3. Lévi-Strauss, C.: Myth and Meaning: Cracking the Code of Culture. Schocken Books, New York (1995)
4. Heo, H.: Storytelling and retelling as narrative inquiry in cyber learning environments. In Atkinson, R., McBeath, C., Jonas-Dwyer, D., Phillips, R. (eds.) Beyond the Comfort Zone: PROCEEDINGS of the 21st ASCILITE Conference (pp. 374–378). Perth, 5–8 Dec 2004
5. Sharda, N.: Movement oriented design: a new paradigm for multimedia design. Int. J. Lateral Comput. **1**(1), 7–14 (2005)
6. Hassenzahl, M., Tractinsky, N.: User experience—a research agenda. Behav. Inf. Technol. **25**, 91–97 (2006)
7. Clark, R., Mayer, R.: e-Learning and the Science of Instruction: Proven Guidelines for Consumers and Designers of Multimedia Learning, 3rd edn. Wiley (2011)
8. Kim, C., Keller, J.M., Baylor, A.L.: Effects of motivational and volitional messages on attitudes toward engineering: comparing text messages with animated messages delivered by a pedagogical agent. In: Kinshuk Sampson, D.G., Spector, J.M., Isaias, P. (eds.) Proceedings of the IADIS International Conference of Cognition and Exploratory Learning in Digital Age (CELDA), pp. 317–320. IADIS press, Algarve (2007)
9. Atkinson, R.K.: Optimizing learning from examples using animated pedagogical agents. J. Educ. Psychol. **94**, 416–427 (2002)
10. Hershey, K., Mishra, P., Altermatt, E.: All or nothing: levels of sociability of a pedagogical software agent and its impact on student perceptions and learning. J. Educ. Multimed. Hypermed. **14**, 113–127 (2005)
11. Palanivel, K., Kuppuswami, S.: Service-oriented reference architecture for personalized E-learning systems (SORAPES). Int. J. Comput. Appl. Found. Comput. Sci. **24**(5), 35–44 (2011)
12. Sciarrone, F., Vaste, G.: Personalized e-learning in moodle: the moodle LS system. J. e-Learn. Knowl. Soc. **7**(1), 49–58 (2011)
13. Cristea, A., De Mooij, A.: Adaptive course authoring: MOT, my online teacher. In: Proceedings of ICT-2003, Telecommunications+Education Workshop, Tahiti Island in Papetee—French Polynesia, IEEE LTTF IASTED (2003)
14. Henze, N.: Personalization services for e-learning in the semantic web. In: 2nd International Workshop on Adaptive Systems for Web-Based Education: Tools and Reusability (WASWBE'05) at AIED'05, Amsterdam, The Netherlands (2005)
15. Vesin, B., Klasnja-Milicevic, A., Ivanovic, M., Budimac, Z.: Applying recommender systems and adaptive hypermedia for E-learning personalization. Comput. Inform. **32**(3), 629–659 (2013)
16. Santos, O.C., Boticario, J.G., Barrera, C.: A machine learning multi-agent architecture to provide adaptation in a standard-based learning management system. (WSEAS). Trans. Inf. Sci. Appl. **1**(1), 468–473 (2004)
17. van Rosmalen, P., Brouns, F., Tattersall, C., Vogten, H., Bruggen, J., Sloep, P., et al.: Towards an open framework for adaptive, agent-supported e-learning. Int. J. Contin. Eng. Educ. Lifelong Learn. **15**(3–6), 261–275 (2005)
18. Razmerita, L., Nabeth, T., Angehrn, A., Roda, C.: InCA: an intelligent cognitive agent-based framework for adaptive and interactive learning, cognition and exploratory learning in digital age. In: Proceedings of the IADIS International Conference, Lisbon, Portugal, pp. 373–382 (2004)
19. Vanderdonckt, J.: Visual design methods in interactive applications. In: Albers, M., Mazur, B. (eds.) Content and Complexity: Information Design in Technical Communication. Lawrence Erlbaum Associates, Mahwah (2003)

20. Leo, L.C.: Bloom's taxonomy vs. game-based learning: toward a preliminary. Doctoral Student & Teaching Fellow (2003)
21. Kuk, K., Jovanovic, D., Joranovic, D., Spalevic, P., Caric, M., Panic, S.: Using a game-based learning model as a new teaching strategy for computer engineering. Turk. J. Elec. Eng. Comp. Sci **20**(2), 1312–1331 (2012)
22. Moreno, R., Mayer, R.E., Spires, H., Lester, J.: The case for social agency in computer-based teaching: do students learn more deeply when they interact with animated pedagogical agents? Cogn. Instr. **19**, 177–214 (2001)
23. Devedžić, V.: Semantic Web and Education. Springer, Berlin (2006)
24. Lester, J., Converse, S., Kahler, S., Barlow, T., Stone, B., Bhogal, R.: The persona effect: affective impact of animated pedagogical agents. In: Proceedings of the CHI '97 Conference, Atlanta, GA (1997)
25. Mărăcine, V., Scarlat, E.: Knowledge ecosystems' development in business and healthcare using knowbots. In: Proceedings of the 11th European Conference on Knowledge Management, Vila Nova de Famalicao, Portugal, pp. 663 – 675. Academic Publishing Limited, Reading, UK, 2–3 Sept 2010
26. Maracine, V., Iandoli, L., Scarlat, E., Nica, S.A.: Knowledge use and Sharing into a medical community of practice; the role of virtual agents (knowbots). Electron. J. Knowl. Manage. **10** (1), 64–81 (2012)
27. Negroponte, N.: Agents: from direct manipulation to delegation. In: Bradshaw, J.M. (ed.) Software Agents, pp. 57–66. MIT Press, Menlo Park (1997)
28. Kuk, K., Milentijević, I., Rančić, D., Spalević, P.: Designing intelligent agent in multilevel game-based modules for E-learning computer science course. Chapter in book: e-Learning Paradigms and Applications; Sub title: Agent-Based Approach. Springer-Verlag, Germany, Series: Studies in Computational Intelligence, vol. 528 (2014)
29. Kuk, K., Milentijević, I., Rančić, D., Spalević, P.: Pedagogical agent in multimedia interactive modules for learning—MIMLE. Expert Syst. Appl. **39**, 8051–8058 (2012)
30. Clark, R.C., Mayer, R.E.: E-Learning and the Science of Instruction: Proven Guidelines for Consumers and Designers of Multimedia Learning, 3rd edn. Wiley, San Francisco (2011)
31. Kuk, K., Spalević, P., Ilić, S., Carić, M., Trajčevski, Z.: A model for student knowledge diagnosis through game learning environment. Tech. Technol. Educ. Manage. **7**(1), 103–110 (2012)
32. Veletsianos, G., Miller, C., Doering, A.: EnALI: a research and design framework for virtual characters and pedagogical agents. J. Educ. Comput. Res. **41**(2), 171–194 (2009)

Author Index

G. Jezic et al. (eds.), *Agent and Multi-Agent Systems: Technologies and Applications*, Smart Innovation, Systems and Technologies 38,
DOI 10.1007/978-3-319-19728-9

Printed by Printforce, the Netherlands